# Organizing Enlightenment

# Organizing Enlightenment

*Information Overload and the Invention of the Modern Research University*

CHAD WELLMON

Johns Hopkins University Press
*Baltimore*

 Published 2015
Printed in the United States of America on acid-free paper
9 8 7 6 5 4 3 2 1

Johns Hopkins University Press
2715 North Charles Street
Baltimore, Maryland 21218-4363
www.press.jhu.edu

*Library of Congress Cataloging-in-Publication Data*

Wellmon, Chad, 1976–
Organizing Enlightenment : information overload and the invention of the modern research university / Chad Wellmon.
pages cm
Includes bibliographical references and index.
ISBN 978-1-4214-1615-1 (hardback) — ISBN 1-4214-1615-8 (hardcover) — ISBN 978-1-4214-1616-8 (electronic) — ISBN 1-4214-1616-6 (electronic)
1. Universities and colleges—Germany—History—18th century. 2. Universities and colleges—Curricula—Germany—History—18th century. 3. Education, Higher—Germany—History—18th century. 4. Research—Germany—History—18th century. 5. Enlightenment—Germany. 6. Germany—Intellectual life—19th century. 7. Education, Higher—Philosophy. I. Title.
LA727.W45 2015
378.43—dc23

2014018366

A catalog record for this book is available from the British Library.

*Special discounts are available for bulk purchases of this book. For more information, please contact Special Sales at 410-516-6936 or specialsales@press.jhu.edu.*

Johns Hopkins University Press uses environmentally friendly book materials, including recycled text paper that is composed of at least 30 percent post-consumer waste, whenever possible.

*to brooks*

# CONTENTS

# ACKNOWLEDGMENTS

Although my name is on the cover, this book is the product of four years of conversations and collaboration with a whole host of people and institutions. I would especially like to thank my friends and colleagues at the Institute for Advanced Studies in Culture and the University of Virginia: Ben Bennett, Tal Brewer, Emily O. Gravett, Jeff Grossman, Bill Hasselberger, James Davison Hunter, Volker Kaiser, Philip Lorish, Lorna Martens, James Mumford, Matthew Puffer, Chuck Mathewes, Allan Megill, Trenton Merricks, Bill McDonald, Jay Tolson, Joe Davis, Sophie Rosenfeld, Rita Felski, Brad Pasanek, Siva Vaidhyannathan, and Josh Yates. I would also like to thank the group of scholars and now friends organized by Tom Mole and Andrew Piper at McGill University as part of the Interacting with Print multigraph project. You all gave me a deep appreciation for the history of the book and print. Ned O'Gorman made this a better book. I would also like to thank my editor Matt McAdam for making me a better thinker and writer. This book would not have been possible were it not for the generosity of an ACLS Charles S. Ryskamp Fellowship. And, finally, I thank Brooks, who remains my best and most loving reader.

# Organizing Enlightenment

# Introduction

On Sunday, June 10, 2012, Helen Dragas, rector of the University of Virginia's Board of Visitors, announced that President Teresa Sullivan and the board had "mutually agreed" that Sullivan would resign. Citing a "rapidly changing" higher education environment, Dragas insisted that the university had to change, and fast. In the ensuing weeks, Dragas alluded to unspecified "philosophical differences" as the grounds for Sullivan's ouster and enumerated a litany of challenges facing U.Va., including the long-term decline in public funding, rising tuition costs, and—crucially—the changing role of technology.

I say "crucially" because, as a later Freedom of Information request was to reveal, these "philosophical differences" were actually anxieties about the potentially disruptive effects of digital technologies on the university.[1] In a series of e-mails, Dragas exchanged articles from the *New York Times* and the *Wall Street Journal* with fellow board members and well-placed alumni, all touting the revolutionary potential of MOOCs (massive open online courses).[2] New digital technologies posed an existential threat to the university. If the University of Virginia did not immediately adapt to the emerging online learning environment, then it would become irrelevant. Just as digital technologies had disrupted and remade the book, music, newspaper, and magazine businesses, so too would they remake the university.

But it was not just a cabal of University of Virginia trustees who insisted so breathlessly that the university had to reinvent itself in the image of new digital technologies. The president of Stanford University, John Hennessy, has also declared that "a tsunami is coming" that will wipe away any institution that does not adapt to the new digital reality,[3] and former Princeton University president William G. Bowen has predicted that such technologies will transform higher education by controlling costs and increasing productivity, all while preserving quality and other "core university" values.[4] A revolution is upon us: broad access to new digital technologies, argues media scholar Cathy Davidson, has "flattened" how knowledge is disseminated and has made it less a proprietary good of a "credentialed elite" and more the result of an increasingly democratic, collaborative endeavor.[5] Changes in digital technologies are not merely tinkering with the tools for distributing knowledge. The very authority structure of the university is at stake. Disciplinary distinctions, the way scholars are credentialized, how research-based knowledge is carried out—all these may become obsolete.[6]

These portentous predications belong to an ongoing debate about the raison d'être of the modern research university. Many have chronicled and diagnosed a "crisis" in higher education.[7] On the one side are the broad sociological critiques of the contemporary university. Exemplified most recently by Richard Arum and Josipa Roksa's *Academically Adrift*, they cast a pall on the university by focusing on particular problems: low graduation rates, skewed admission policies, indifferent faculty, disengaged students, and uncontrollable costs.[8] On the other side are the utopian voices warning that universities face existential threats and calling for intrepid entrepreneurs to offer bold, salvific solutions. Digital technologies can reinvent the university for the twenty-first century, say these voices. And in contradistinction to both these groups are those who defend a tradition of collegiate, residential learning and its celebration of humanist education over the endless accretion of research.[9] Finally, there are the democratic arguments. Following a tradition that extends from Thomas Jefferson's *Rockfish Gap Report* through William Harper Rainey's *The University and Democracy* to Martha Nussbaum's *Not for Profit*, these authors claim that the university should form democratic citizens.[10] The college model extends out to the formation of students for citizenship and civic responsibility by cultivating democratic virtues of sympathy, leadership, a work ethic, and respect for one's fellow citizens.

At their best, this array of arguments offers either incisive, sobering snapshots of contemporary universities or rousing exhortations about what universities could be, or could return to being. But few of them address a more basic question: What are we talking about when we talk about the modern research university? One of the few books that do address that question is Harvard Business School professor Clayton Christensen's *The Innovative University: Changing the DNA of Higher Education from the Inside Out*. His arguments are now regularly invoked by critics insisting that the university change now. Christensen applies his theory of "disruptive innovation"—innovations that threaten established providers by offering more affordable alternatives—to higher education. For the first time since the introduction of the printed textbook, he writes, "there is a new, much less expensive technology for educating students: on-line learning."[11] In order to survive, universities will have to come to grips with new technologies or risk obsolescence. They will simply work themselves out of a job.

But Christensen, unlike some who embrace his theories of disruption, recognizes that the university is not simply another business for which scale and efficiency are paramount. Higher education should not be conflated with other media businesses that distribute information. The research university is not simply a content delivery device; it is an institution unique in its capacity to produce and transmit a knowledge that is distinct and carries with it the stamp of authority. The research university has its own cultural logic and normative structure that allow it to generate and transmit a certain type of authoritative knowledge. It has, as Christensen puts it, a "DNA" that, in the United States at least, has organized it since the late nineteenth century. Since Johns Hopkins was founded in 1871, the American research university has been committed to extricating meaning from knowledge. Daniel Coit Gilman, the first president of Johns Hopkins University, wrote that "with their trained observers, their methods of accurate work, their habit of publication, and especially their traditional principles of cooperative study," research universities acquire, conserve, refine, and distribute knowledge.[12] They belong in the same historical lineage of technologies which extends from the invention of writing and the codex to the printing press and the modern scientific lab. By the late nineteenth century, the research university had become the consummate technology for organizing knowledge. It had also come to stand in for a whole way of configuring, managing, and cultivating the impulse to know.

Like other university presidents in late nineteenth-century America, Gilman adopted and adapted a German university ideal. They may have built their own structures—for example, the departmental organization of the university. But they clearly also inherited a particular ethos that began to take shape around 1800 in Germany at a cultural moment very similar to our own. Our contemporary anxieties about new technologies, the fate of the university, and what counts as authoritative knowledge echo similar cultural anxieties among late eighteenth-century German intellectuals about print technologies and epistemic authority which eventually gave rise to the modern research university. These two moments can illuminate each other. The anxieties and aspirations around 1800 can help us better understand our own situation and what is at stake in debates about the future of digital technologies, the university, and knowledge. And our contemporary situation can illuminate the history of the research university and help us better understand the norms, virtues, and purposes that have animated it for two centuries.

The ideal of the German research university was a response to a pervasive Enlightenment anxiety about *information overload*. This anxiety was particularly acute in late eighteenth-century Germany. Just as today we imagine ourselves to be engulfed by a flood of digital data, Germans of the late eighteenth century saw themselves as having been infested by a plague of books, circulating contagiously among the reading public. As in England and France, perceptions of information overload in Germany corresponded to a rapid increase in print titles in the last third of the eighteenth century, an increase of 125 percent from 1770 to 1800.[13] This growth corresponded to the broader proliferation of printed texts in Germany in the last third of the eighteenth century.[14]

In Germany, though, it was not just about the sheer numbers. The real issue concerned epistemological anxieties, and German intellectuals were unique in settling on the university as the solution. German-speaking intellectuals and writers were the first to connect in a systematic way the problem of overload with the institution that was the university.

Anticipating anxieties about information overload, technological change, and a crisis of the Enlightenment university in 1807, the German philosopher J. G. Fichte lambasted what he saw as the university's refusal to adapt to the new print environment. The first universities in Paris, Bologna, and Oxford, he wrote, had been an oral "*Ersatz*" for the general lack of texts.[15] More than two centuries after the invention of the printing press and the "overabundance of books" that followed it, however, the university's central pedagogical practice, the lecture, still consisted of professors reading the books of another, canonical scholar aloud, as if students could not read on their own. What was the purpose of the university in an age where print had reached a saturation point? If universities continued to present students "the entire world of books, which already lies printed before everyone's eyes," warned Fichte, they would soon become redundant.[16] Universities had not figured out how to respond to technological change, and if they could not distinguish themselves from printed books, they would fail.

For Fichte, the expansion of the book market had already altered the ways in which people could know and learn. If universities were to survive, they would have to change with the times and transform themselves. Fichte's contemporaries disagreed, however, about how universities should do so. Some argued that universities should be replaced by specialized schools devoted to training students in skills specific to particular professions. Others, like Fichte, argued against what he and his fellow German Idealists called the "utility message" of the Enlightenment—its singular focus on the technical and practical utility of all knowledge.[17] The survival of the university depended on the extent to which it could distinguish itself from the broader culture as the unique institution devoted to what Germans called *Wissenschaft*, or science as a practice.[18]

The underlying ethos of the research university was brought forth under the cloud of a crisis. What was the purpose of the university? How could it advance knowledge without being redundant, simply reproducing what print did more efficiently? Then as now, anxieties about the impact of new technologies on the future of the university were not simply demands for the university to better incorporate print technologies. They expressed confusions about the authority and legitimacy of knowledge as such. Did students need to pay a premium to listen to a lecture in person, or could they just buy a book?

The research university was an institutional response to structural changes in the media environment of the eighteenth-century German Enlightenment. The fomenting of new media with old proved in the end hugely fruitful. Overwhelmed by all that the modern print market had to offer, the original proponents of the research university sought to institutionalize practices and technologies for the generation of knowledge. Such a university had little in common with the American college of the eighteenth and nineteenth centuries, seeking as it did to inculcate a traditional set of Protestant virtues and moral character. Although it inherited some of its forms from medieval universities—the four faculties, for example, or the ritual that is the lecture—it did not share their organizing structure. Medieval universities were unitary corporations of students and masters bound together, in the broadest sense, by Christian values and an appeal to the authority of the Church, both legally and financially. Although particular universities resisted certain Christian theological doctrines, by and large they were grounded in a canon of Christian learning. When, in the late seventeenth and early eighteenth centuries, Enlightenment critics began to question the conflation of the Church and the university, they turned to another ethical resource: the state. The purpose of the university, they argued, was not contemplation of the divine or knowledge for knowledge's sake but rather service to the broader public and state. In the context of the history of the university, then, the modern research university's appeal to science, as a distinct research-oriented form of knowledge generation and dissemination, around 1800 was unprecedented.

Since its inception in Germany in the early nineteenth century and its reinvention in America later that same century, the research university has been the central institution of knowledge in the West. But now it finds itself confronted by the challenge of radical technological change; and because it does, its most ardent defenders need to be crystal clear about what they are defending. A particular institution? A commitment to knowledge in the abstract? Or merely their own place in a bureaucracy? The university's critics and defenders ought to focus not simply on fixing each university, one at a time, but rather on how universities overall can help us evaluate knowledge in an age of easily accessible information. The crisis of the university is part of a larger shift of epistemic authority in the modern digital age. The saturation of digital technologies, from *Wikipedia* to Google PageRank, is changing the ways by which humans create, store, distribute, and value knowledge in the twenty-first century. How today do we arrive at understanding? What constitutes authoritative or legitimate knowledge today? The future of the university will unfold in this context, as it did before.

In 2014 as in 1814, people find themselves compelled to decide which sources of knowledge to trust, and which not to, in environments of extraordinarily expanded production and unlimited access. It remains to be seen what practices, institutions, norms, and related technologies will ultimately emerge in our own digital age, but late eighteenth- and early nineteenth-century German intellectuals embraced the idea of the modern research university and its organizing ethic, disciplinarity. The "credentialized," disciplinary-based ordering of knowledge embodied in the research university was a new way of coping with a perceived proliferation of knowledge and the attendant crisis in epistemic authority.

The German Enlightenment is usually associated with a string of central concepts: *freedom*, *culture*, and the liberal *individual*. But, in its moment, Enlightenment also referred to an array of technologies—encyclopedias, dictionaries, taxonomies, philosophical systems—designed to manage the centrifugal experience of knowledge.

Yet the Enlightenment technologies designed to organize knowledge were not merely tools, material extensions of humans who controlled and determined their use. They were also value-laden metaphors for particular orders of knowledge and ways of managing the desire to advance and control knowledge. Scholars used "the book," "the encyclopedia," and eventually "the university" to stand in for normative conceptions of how humans should generate knowledge and manage their desire to know. "Encyclopedia," for example, referred not only to a printed reference book organized alphabetically but also to an array of practices, habits, norms, and virtues that were inseparable from the physical object.

But because those technologies were thought to have failed, buckling under the pressure exerted upon them, many intellectuals sought to reimagine the university as a better way of achieving the same end. This new university should be the institutional home of the disciplinary—or, as early nineteenth-century German figures put it, *wissenschaftlich*—arrangement of knowledge. The splintering of knowledge which characterized the Enlightenment was to be dealt with by the specialized work of distinct disciplines. A disciplinary-based order of knowledge prevented knowledge from becoming too abstract and unmanageable. It filtered and authorized the necessarily constrained and partial forms of modern, specialized knowledge; it legitimated limited knowledge by tying it to the endless, unceasing pursuit of "research." And it put the research university at the center of a modern media and knowledge environment. Somewhat autologically, what was produced, organized, and transmitted by the university was true knowledge.

This new order of knowledge, however, was a turn away from an Enlightenment order that valued utility and popularity. As a structure internal to the university, the new disciplinary order was more insular, self-referential, and increasingly distinct from the broader culture. Science gradually became a more distinct system, replete with its own practices, internal goods, norms, and virtues.[19]

*Organizing the Enlightenment* is not an account of individual disciplines—these types of studies already abound—but a conceptual and historical account of how and why this novel intellectual architecture came into existence in the first place.[20] In describing how disciplinarity emerged as the last technology of the Enlightenment, I sketch not only the university's ethical logic but more fundamentally the epistemological anxieties that precipitated its formation. I should point out at the start that my account of "disciplinarity" is more like a prehistory, because the word itself was rarely used in the eighteenth century. And its cognate "discipline," denoting a specialized science, was not used in Germanic countries with any regularity until the nineteenth century. But although the precise term was absent, the concept of disciplinarity was indeed present in a range of German concepts such as *Disziplin*, *Wissenschaft*, and *Enzyklopädistik*. The professionalization of the American research university in the late nineteenth century inherited the German invention of *Wissenschaft*, the key term I trace throughout this book. Examining its history will offer new insight into our current situation, as well as a better understanding of how and why the research university emerged. The story of the German research university, regularly upheld by American university presidents who cite Wilhelm von Humboldt, has given us not just the ideals of academic freedom and the unity of teaching and research; it has also lent us the logic of intellectual specialization that continues to form the contemporary university.[21] And while many critics have for two centuries decried intellectual specialization, I offer a defense of it.

Unlike medieval universities such as Paris or Oxford, and unlike the American college of the eighteenth and nineteenth centuries, from its inception in Germany the research university had as its primary purpose the pursuit of knowledge and the formation of those who engage in this enterprise. This is why the crisis of the contemporary university is not simply a moral one that pits liberal arts values against market values. It is rather a more fundamental ethical crisis concerning epistemic authority in the modern age. It is a crisis of the very means by which knowledge has been gleaned, distributed, and accorded worth from the nineteenth to the twenty-first centuries. And this raises several questions: Is the research university model worth defending? And what exactly would we be defending?[22]

The story of the research university is one not only of the generation of knowledge but also of the formation of a particular self, the development of a type of person. When I describe the solution to Enlightenment overload as "ethical," I mean that it was concerned with goods internal to the practice of science. The solution was a particular type of person, not a more comprehensive encyclopedia. Since its emergence in early nineteenth-century Germany, the research university has always been the bearer of practices with their own standards of excellence, ideals of conduct, and even virtues. And it was these upon which the university's epistemic authority was founded; it was these that enabled it to generate and transmit authoritative and legitimate knowledge. The following chapters recount how and why this ethic emerged in tandem with the research university, for which epistemology was always inextricable from ethics.

In 1798 Immanuel Kant described the university not as a place of Enlightenment or as an "exit from a self-incurred immaturity" but as a factory organized according to the division of intellectual labor for the purposes of producing both books and authors.[23] American research universities inherited this orientation toward the formation of a particular self from their nineteenth-century German counterparts. After material technologies—from encyclopedias to periodicals—failed to deliver on the promise of universal knowledge, a synoptic account of all knowledge, thinkers turned toward ethically formative technologies and practices that formed a particular type of person. Science would be about not writing books but creating a "disciplinary self," a particular identity or way of being in the world crafted and molded by distinct practices of mind and body.[24] Expanding on the work of Michel Foucault and what he termed "technologies of the self," Pierre Hadot and his studies of ancient philosophy, and Lorraine Daston and Peter Galison and their work on the "scientific self," I trace the emergence of a particular kind of self that was first conceptualized and crafted around 1800 in Germany amid anxieties about information overload and fears for the future of the university.[25] This disciplinary self was always aspirational, an ideal to be pursued. Normative and descriptive accounts were consistently conflated. But this conflation underlines how central the formation into the ethos of science always was for the research university.[26] The university was to be home to this self. And knowledge was to be embodied not in an ever more exhaustive encyclopedia but instead in the character of the student. The university's claim to winnow through all the dross of knowledge came to rely essentially on the cultivation of a particular persona.

The appeal to the "disciplinary self" was made more urgent in 1800 by invoking the specter of too much information, and this goes to the heart of claims about overload and their broader cultural functions. In 2006, Kevin Kelly, senior editor at *Wired*, predicted the advent of a universal library in which all the world's books would become a "single liquid fabric of interconnected words and ideas." He envisaged the digitization efforts of Google Books resulting in a searchable library that would connect every book ever written. Ideas would flow seamlessly. Others are less sanguine, as we have seen. Either way it is vital that we realize that the situation we face is not unprecedented. In both the optimism of Kelly's predictions and the pessimism of those who fear that Google is "making us stupid" we can hear echoes of late Enlightenment debates in Germany about the necessity of rescuing people from the glut of knowledge.[27]

But, as historians have noted, complaints about "too many books" echo across the centuries. Practically speaking, what is the difference between trying to read ten thousand books or ten million books?[28] Both are impossible tasks. What is most significant in these historical worries about excess, then, are the solutions inevitably offered to solve the purported problems.[29] Every shrill insistence brings with it a particular solution. The specific technologies we develop to manage information give us insight into not only how we organize, produce, and distribute knowledge but how we form ourselves. Historically, worries about "excess" have been fundamentally normative. They made particular claims not only about what was good or bad about print, for example, but about what constituted "true" knowledge. First, they presumed a normative level of information or, in the case of purported book plagues, some normative number of books. There are too many books; there is too much data. But compared to what? Second, such laments presumed the normative value of particular practices and technologies for dealing with all of this information. Every complaint about excess was followed by a proposal on how to fix the problem. To insist that there were too many books was to insist that there were too many books to be read or dealt with in a particular way and thus to assume the normative value of one form of reading over another. So I am concerned less with how empirically accurate such claims were—that is, the extent to which they corresponded to an actual, unique increase in printed material—than with the underlying cultural anxieties and attendant solutions. What do worries about, and the proffered solutions to, excess tell us about the historical modes of how knowledge was organized, produced, transmitted, and authorized? This book is, in part, a history of a cultural anxiety.

The sense of disconcertion at "information overload" holds sway only against a background assumption that we should know everything. If we do not assume that we should know everything, why would we mind not being able to appropriate all the information at our fingertips? This book is, then, preoccupied with the way the desire for universal knowledge has manifested itself over time. The eighteenth-century effort to achieve a unified account of knowledge by *capturing it in print* was very different from the attempt to unify knowledge institutionally in the form of the research university. When I refer to an Enlightenment information overload, then, I am referring less to a unique material condition—how many books were published in a particular year versus another year—and more to a complex of circumstances and motivations. Indeed, information overload refers to multiple projects, all of which were wrapped up with experiences of material conditions and technological change. "Overload," in my use, denotes experiences of excess.

Ann Blair distinguishes between information and knowledge and frames her interest in information management technologies in strictly functional terms. Information, she writes, is "distinct from data," which requires further processing to be meaningful, and from knowledge, which implies "an independent knower."[30] Information is "discrete and small-sized items that have been removed from their original contexts and made available as morsels ready to be articulated." Yet while these terms distinguish information and knowledge, they also obscure some nuances. First, "information" is a recent term and is thus very difficult to detach from contemporary uses. It was not introduced in its current use until 1948 in the context of Claude Shannon's mathematical work in the Bell Labs. Writing about an Enlightenment notion of information, then, will always run the risk of twentieth-century connotations overdetermining more historically specific concepts.

The more pressing concern is that sifting out knowledge from information is always normative; that is, it always entails historical and cultural assumptions about *what is worth knowing*. To identify X as information and thus not knowledge is to make a judgment about the value of X. Information is, after all, "mere" information. What amount of "original context," to use Blair's phrase, is sufficient to turn information into knowledge? How much "articulation" is required? But what then is knowledge—utility, wisdom, or something altogether different? As we shall see in the following chapters, the distinction between knowledge and information has its Enlightenment precedents in a range of distinctions: true and false learning, philosophical and historical knowledge, the aggregate and the whole. All of these distinctions were based on normative assumptions about what constituted true knowledge, as opposed to "mere" facts.

These historical debates concerning distinctions between knowledge and information revolved around knowledge as an honorific. To describe something as knowledge is to lend it a certain value. To claim that a belief constituted "true" knowledge was to claim that it met certain normative standards. Contemporary philosophers usually cite standards like justification, truth, warrant, coherence, or reliability. But during the late Enlightenment there were other standards governing the question of whether something was worth knowing. "True" knowledge ought to be nontrivial, worth the effort. Counting the blades of grass outside my window may be new knowledge (I currently don't know how many there are), but is it "worthwhile" knowledge? The demands for "true" knowledge in the late eighteenth century were tied to assumptions about the suspected triviality of scholarly practice that *merely* accumulated historical facts. The following chapters trace these debates about the relevant criteria for distinguishing trivial from nontrivial knowledge.

These debates concerned epistemic authority, that is, what counted as *authoritative* knowledge. What legitimates one form of knowledge over another? Which sources of knowledge are to be trusted? Which not? What practices and scholarly habits, techniques, and institutions render knowledge authoritative or worthy? Questions about distilling knowledge rely on assumptions about its value. Throughout this book, I write of a crisis in epistemic authority, a moment of uncertainty and possible change concerning the technologies and institutions that have traditionally generated, transmitted, and evaluated knowledge. Around 1800 in Germany, the proliferation of print and its associated technologies posed a challenge to the university's claim to be the dominant institution of knowledge. Today, digital technologies from *Wikipedia* to blogs and social media pose a similar challenge to the authority that the research university has enjoyed and defended for almost two centuries. But what most of the debates about these changes in media miss is that the research university is not just another content delivery device; it was and continues to be a bestower of epistemic authority. The university does not just transmit knowledge. It legitimates and authorizes knowledge.

At the center of both of these moments of crisis were changes not only in technology but also in the very notion of technology. Throughout this book, I use "technology" to refer not only to physical tools but also to different forms of print media, institutions (like the university), and practices of the self and how they shape each other. My use of the term "technology," then, refers to particular artifacts, as well as the complex interaction of humans with their various tools and the interaction of technologies and media with each other. I track how these relationships change over time and how the boundaries between them are never absolute or fixed. In this sense, the forms of technological change that I describe are not, as Neil Postman puts it, "additive or subtractive," but rather "ecological."[31] The research university did not supplant print media, books, periodicals, encyclopedias, but its emergence did generate changes in how people interacted with print media, as well as in the very conception of print. I use the term "media ecology" to describe these interactions among various technologies and human agents and how they always shape each other in irreducibly complex ways. Technology refers to this complex environment of interactions, replete with its own norms, practices, and emergent properties.[32]

The last technology of the Enlightenment was the modern research university and its organizing concept—disciplinarity or specialized science. The following chapters take us back to a German context immediately preceding the founding of the University of Berlin in 1810 and trace the prehistory of disciplinarity by outlining a much broader shift in the orchestration and classification of knowledge over the course of the eighteenth century. The way knowledge resolved into disciplines became possible under the aegis of the research university. This reorganization was an attempt to come to terms with the fragmenting character of modern knowledge and to manage its proliferation in and as print.

My account of the emergence of the research university begins with the history of a particular concept: "science" [*Wissenschaft*]. Until the last third of the eighteenth century, the German words for "knowledge" [*Wissen*] and "science" were the same. They both denoted an individual-dependent form of knowledge as a property or a state of an individual mind. Science, as one early Enlightenment lexicon defined it, was a "particular insight or knowledge." It was idiomatic to say that a person had a "science" of some object or state of affairs. Only in the last decade of the eighteenth century did "science" begin to be used as a general concept to denote a body of shared knowledge and research practices designed to generate it.

The gradual shift in the meaning of "science" to a body of knowledge had two important consequences with respect to the emergence of the research university. On the one hand, as a body of knowledge, sciences addressed an abstract "general public." In principle, anyone could have access to it. On the other hand, as internally coherent epistemic practices with their own norms, particular sciences were distinct from other sciences, as well as from the "general public." They stood opposed, as Kant wrote, "to common knowledge." Once the unity of knowledge was grounded not in mental faculties common to all but in objectified systems of knowledge, general access could not be presumed. Instead, it had to be cultivated through institutionalized habits, practices, and disciplines. Particular scientific cultures emerged that distinguished between expert and the layman. Disciplinarity gradually arose in this context as a system for managing distinct sciences and the people who labored within those sciences.

But the conceptual twists and turns of "science" only make sense against the backdrop of eighteenth-century print technologies and the norms that guided people's interactions with them. Over the course of much of the eighteenth century, as the various forms of print proliferated—journals, periodicals, encyclopedias, lexica—what was referred to as "aids to erudition" came to substitute in for knowledge itself. Scholars cultivated a broad cultural confidence in the capacity of print technologies to advance the sciences and bring about a universal, timeless knowledge. They attempted, as Adrian Johns puts it, "to invest" print with a capacity to transcend time and space.[33] They imagined an "empire of erudition" that was homogenous, complete, and easily accessible to all scholars, that is, to those who knew how to interact with print. In this virtual world, knowledge was imagined as an interconnected body of learning embodied in printed texts, what Novalis called a "chain of learning," and the impulse to understand led unavoidably to print.[34]

At the beginning of the eighteenth century, the Enlightenment gentleman scholar [*der Gelehrte*] exemplified this culture of knowledge. It was this man of letters, and not the expert, who reigned over an "empire of erudition." Christian Jöcher's *Universal Literature Lexicon* exemplified an early Enlightenment genre that reported "all the scholar's work" in one annual volume. This and related scholarly print aids were then consumed by a public that itself had not yet been sharply differentiated from the scholarly world.

Over the course of the eighteenth century, however, this empire of erudition gradually fractured into a world of specialized disciplines and concomitant experts under two parallel pressures. The first was an unrelenting attack on the very concept of erudition. Beginning in the early eighteenth century and continuing through the 1770s, many intellectuals, themselves members of the empire of erudition, sought to detach learning from the erudite, who had enjoyed the privileges of a distinct social class. Humankind stood in need of enlightenment; knowledge, it followed, ought to be liberated, made available to the whole of society.

The second, related pressure was ongoing changes in a print market trying desperately to cope with its radical expansion. Around midcentury, for example, new journals emerged catering to specialists. Now scholars could reimagine themselves as part of more limited communities consisting not of gentlemen scholars, all-rounders, or polymaths. Such specialization threw wide open the question of the presumed homogeneity and authority of the "empire of erudition." The polymath was replaced by the botanist, the philosopher, and the theologian, the specialized scientist.

But just as some scholars were realizing the popular potential of the bibliographic order of knowledge, others had become increasingly anxious that knowledge had been reduced to its print technologies. Print had been fetishized. Intellectuals began to worry that the proliferation of print objects had outstripped people's capacities to interact with them. Readers, writers, and publishers had been overwhelmed with floods of journals, barraged by books, and they could no longer digest the material at their disposal. They did not know what to do with all the learning that had fallen into their lap, so to speak. The "plague of books" that swept German readers referred not only to the increasing circulation of texts but also to the increased chatter about these books and their print cousins. And this ultimately led to a trenchant critique of print as a distinct Enlightenment culture.

Worries about a glut of texts were not simply irrational fears, however. They were, in part, cultural responses to an actual increase in the production of printed texts over the last three decades of the eighteenth century. The catalogue of the Leipzig book fair, the center of Germany's book trade, gave a sense of this rapid growth. The number of titles listed per year went from 755 in 1740 to 1,144 in 1770 and then to 2,569 in 1800—an exponential increase of more than 240 percent in just sixty years.[35]

One of the most influential accounts of this cultural situation was given by Immanuel Kant. The writer of "What Is Enlightenment?" not only diagnosed the malaise of modern print culture but also presented a clear solution that paved the way for a reimagination of the university. Sharply critical of what he considered the fetishization of humanist print technologies, he feared that books were beginning to think, as he put it in 1784, for humans.[36] To counter this cultural trend, he proposed a critical philosophy that would better manage the excess of books by focusing on the moral integrity and formation of the person. The proliferation of print posed a threat not only to the imperative to educate oneself but also to the ethical integrity of the subject. And describing the problem in those terms meant that he created a void that no encyclopedia or lexicon could fill. In a world awash with words, only a reimagined institution that could sustain norms and practices that would form particular types of persons could fill that void.

Increased anxieties about epistemic authority coincided with the near implosion of the German Enlightenment university, which had come under increasing pressure to offer more practical training and to justify its very existence in the age of the proliferation of print. By 1800, enrollments had dropped 50 percent from midcentury peaks, and the calls for the abolition of these so-called scholastic guilds of privilege and social and economic irrelevance had become a mantra. It is certainly no coincidence that contemporary fears, however alarmist, about the impending collapse of the university or demands that universities focus on training students for the job market are taking place at a similar moment of epistemic confusion and media surplus. In Germany between 1795 and 1810, Prussian scholars and civil bureaucrats addressed these anxieties and fears in a wide-ranging debate on the future of the university. For most commentators, the crisis of the university was directly related to the proliferation of print. How could the university not only accommodate new technologies but flourish in this new environment?

In this sense, the crisis of the university was another manifestation of the broader ethical crisis brought on by the proliferation of print and the attendant fragmentation of knowledge. At the University of Jena in 1803, the German philosopher F. W. J. Schelling opened his lectures on the methods of university study with a description of a young man who, upon arriving at the university, is thoroughly disoriented. Confronted with the dizzying array of subjects on offer, he finds himself at sea with no compass to guide him through the fragmented world of knowledge. Thus, either he dedicates himself to one particular field and forsakes any attempt at a knowledge of the whole, or he wanders among the various sciences and, as a result, becomes knowledgeable in none at all. For Schelling, information overload produced distracted, ethically unreflective people. It not only constrained the advancement of knowledge but also threatened the integrity of the human mind. The only sufficient response was a fundamentally different conception of the university's purpose. It had to be reconceived as not merely a more efficient or technologically capable institution but as the source and embodiment of a distinct way of life, namely, science. Only science as a practice with its own goods and ends—not better textbooks or more exhaustive encyclopedias—could address the effects of information overload. The task of the university was to form better people of knowledge who could navigate the oceans of print.

What figures like Schelling, Kant, and Fichte realized, however, was that the fundamental shifts in the media environment, combined with the atrophy of the theological basis of the medieval university, had forced the university to reimagine its justification. What normative resources could orient the university in a modern age in which the traditional authorities of knowledge—the church, the state, and humanist erudition—were dissipating?

In Prussia, these underlying questions provoked a range of responses, from bureaucrats and philosophers alike. They culminated in Wilhelm von Humboldt's plan for a new university devoted to science itself. While this plan was in part concerned with the relationship of the university and the state, it was also about the relationship of the university to the new media environment. For Humboldt and others like Fichte and Schleiermacher, the task was not simply the conservative defense of a particular institution that had come under attack by a growing bourgeois public emboldened by print. It was more an attempt to discern what now constituted authoritative knowledge. The new university would have to embody a new order of knowledge which was self-organizing, internally coherent, and distinct. It would have to distinguish itself from the market and more public forms of information distribution. In this modern research university, the bibliographic order of knowledge would give way to a disciplinary order of knowledge with its distinctions between general reader and specialized scientist [*Wissenschaftler*]. This disciplinary order organized knowledge by forming those who produced it.

Humboldt envisioned an intellectual architecture for science as a way of life. Enlightenment solutions to information overload, such as lexica, encyclopedias, and books, had failed, he contended, because they had focused too strictly on the objective task of knowledge—its advancement through technologies—having neglected at their peril the formation of the student himself. The research university would meld the two by harmonizing the technologies themselves with the persons who interacted with them. Echoing his contemporaries, he embedded academic professionalization—the imperative to publish, division of intellectual labor according to specialization, a focus on details—in a set of ideals. Implicit in this was the claim that these institutional practices together constituted a distinct way of life. Specialization gave the student an orientation, a source of meaning, a ground of authority. By tying the logic of science to the institution of the university, science became a viable form of life replete with its own set of virtues, practices, and ends. And above all science stood for a devotion to something that exceeded the self.

Over the course of the nineteenth century, one discipline in particular came to embody the logic and practice of specialized science: classical philology. For generations of German scholars in every field, philology—not physics, chemistry, or biology—was the consummate discipline, exemplifying the virtues of modern science: industriousness, attention to detail, a devotion to method, precision, exactitude, a commitment and facility to open discussion, and a critical disposition.[37] And it pioneered the site where these virtues were inculcated and ultimately crafted the modern disciplinary self—the *seminar*. It cultivated a set of virtues, those of specialized science or what we today call disciplinarity.

The ideal of the research university did not solve the problem of information overload, but its romantic and idealist advocates reimagined both the university and the very parameters and criteria of what counted as real, authoritative knowledge. They imagined a different way of conceiving of the problem of overload and epistemic authority. The research university emerged out of a particular moment of cultural anxiety about media surplus and a perception of a crisis in the authority of knowledge.

My focus on the history of cultural anxiety and the ethics of the modern research university is what distinguishes my story from two recent books to which I am deeply indebted. In his magisterial book on the history of the research university, William Clark ascribes the "shamelessly" political and economic roots of the modern university to universities in Halle and, especially, Göttingen, which translated Enlightenment notions of fame into institutional prestige, cash and credit, and the bureaucratic regime of a university oriented toward the state. Whereas Clark shines a detailed and exhaustive light on the rational structures of the modern research university, I focus on what he calls romanticism's substitution of a "cultural criterion" for the economic one.[38] Insofar as the research university combated the phenomenon of information overload, its advent cannot simply be reduced to the ineluctable march of a modern bureaucratic rationality.

Similarly, Ann Blair has recently given a detailed account of another age of information overload in which fifteenth- and sixteenth-century humanist scholars developed a range of print-based strategies—including cut-and-paste and note-taking techniques—to manage a daunting surfeit of information. I continue Blair's story through the eighteenth century and the Enlightenment to the point at which intellectuals and scholars began to criticize these forms of learning as merely technical solutions to a problem that had become ethical in nature. Improved print technologies or more efficient managerial skills could not solve the problem of fragmentation, both social and epistemic, that the proliferation of print had come to represent. My focus is on these cultural conditions and their ramifications for the emergence of the research university.

I draw on recent media theory, especially attempts to reconceptualize the relationship between humans and their technologies as inextricably bound to one another and to emphasize human interactions with (and not just use of) technologies;[39] I rely on the exhaustive historical work of historians of the university[40] and historians of the book;[41] I extend the work of recent historians of science;[42] and, finally, I engage ethical theory, both normative and historical.[43] Part history and part theory of media, part institutional and book history, part history of science, and part ethical theory, this book tries to combine these different approaches in pursuit of a better understanding of the relationship among technological change, anxieties about epistemic authority, and the birth of new structures for organizing and cultivating the desire to know.

I consider the place of the disciplinary arrangement of the university and knowledge in our own digital age. The crisis of the contemporary university concerns its place in the generation and dissemination of knowledge. The modern variegated university was a historical institution that emerged to meet specific needs. Is this historical arrangement, with all its productive efficiencies and specialization, the most apposite for our current situation? More importantly, if the university's monopoly on knowledge has already ended, as critics suggest, then what distinguishes it from other sources of knowledge in an age of Google and *Wikipedia*? What is the purpose of the university in an age in which academic expertise has been eroded by the democratization of the tools for distributing knowledge?

My contention is that we can only answer these questions if we first recognize that the research university was from its beginnings in Berlin an institution designed to sustain a particular practice and its virtues, habits, and purposes. It was never merely a content delivery system. It was a source of epistemic authority in an age of media surplus and cultural anxiety about what counted as real knowledge. It was an institution devoted to science. This was its ethos. Only when we have acknowledged this reality can we move beyond the idealizations and recriminations that hamper current debates about the university's future. Only then can we begin to consider the future of the university and the future of knowledge.

CHAPTER ONE

# Science as Culture

MOST GERMAN UNIVERSITIES of the eighteenth century were renowned not for their scholarship or commitment to science but rather for the bad behavior of their students.[1] Shortly after attending Fichte's lectures at the newly established University of Berlin in 1810, Friedrich W. Zimietzki, a young student who had yet to publish anything, offered a summary condemnation of German university life and a vision for a different institution. He summoned his "brothers" to forswear their dueling and boozing and embrace a new university culture, what he called science. "The purpose of the university," he wrote in his anonymously published book, "is the preservation and advancement of science among the citizens of the state and humanity in general." A university should be a "scientific community," not a band of dueling boys. Fichte's young student exhorted his "academic brothers" to embrace "the habits" and "morals" [*Sittlichkeit*] of science and to participate in a common life that would transform themselves and the institution of the university.[2]

For Zimietzki, science was not just a collection of facts and ideas. It was, as he put it, an "outlook on life." It was a particular way of life with its own "customs" and virtues that bound a community together and sustained an institution. For the "brothers" of the university, scientific knowledge was a "common good." It was something beyond themselves, and its pursuit required rigor, intellectual freedom, and, above all, a "love" that drove them toward the "good and true" of science.[3] Science was a distinct culture with its own authority and virtues. And it was embodied in the university.

Zimietzki's description of science as an "outlook on life" and a distinct culture with its own "morals" was representative of a shift not only in the history of the university but in the very concept of science as well. The story of how science came to be a cultural, institution-dependent form of knowledge is inextricable from the story of the research university. In order to understand how historically unique this appeal to science around 1800 was, an appeal exemplified by Zimietzki's exhortation to his fellow students, we need a better understanding of how the term had previously been used.

To invoke the authority of science today is to appeal to an institutional form of knowledge, one tied to books, periodicals, practices, and the modern research university itself. Scientific knowledge, of course, is also produced in a range of government- and corporate-sponsored research labs, independent research institutions, and a host of other places, and it is disseminated and publicized through modern media from newspapers and magazines to Web sites and blogs. The authority and legitimacy of scientific knowledge over the past two centuries, however, have been tied to the institutional authority of the modern research university and the ways in which it legitimates knowledge as scientific by awarding advanced degrees, facilitating the peer review process, or actually producing and preserving knowledge through research. Modern scientific knowledge is an institutional knowledge. It does not typically refer to a type of knowledge that a particular individual possesses. According to its more modern use, science enjoys its authoritative status precisely because it can be communicated, transmitted, objectified, and ultimately institutionalized.

But this was not always the case. The term "science" has not always referred to a supra-individual form of knowledge, and only within the past two hundred years has it become intertwined with the university. In order to understand how science became, as Zimietzki described it, the underlying culture of the university, we first need to consider some of the various conceptions of "science" which were available to late eighteenth-century German scholars and writers as they sought to claim it for their own purposes. In attempting to recount such a long and complex story, I focus on only one aspect of "science," namely, its broadest definition. What kind of knowledge has a particularly *scientific* knowledge designated? What have been the most basic conceptions of science to which intellectuals and scholars have appealed? Following Gyorgy Markus and drawing directly on his work, I want to highlight in particular a persistent ambivalence with respect to "science," a "constant vacillation between a subjectivist and objectified understanding," between what I term an individual-dependent form of knowledge and an institutional-dependent form of knowledge.[4] Throughout the Western philosophical and cultural tradition, "science" has wavered between a knowledge tied to an individual person and one embodied in material forms, such as writing, print, and institutions.

## From *Episteme* to *Disciplina*

Most eighteenth-century German lexica traced the etymology of the German word "science" [*Wissenschaft*] to the Latin *scientia* and the Greek *episteme*.[5] These two classical antecedents referred primarily to an individual-dependent form of knowledge, whereas the German term, by the beginning of the nineteenth century at least, referred to an institutional-dependent form of knowledge of the sort that Zimietzki so vividly invoked. These semantic shifts are significant because they were the background against which the debates central to our story—debates about what constituted real knowledge and the future of the university—were carried out.

In contrast to Zimietzki's more modern, institutionally dependent use of science, Aristotle defined scientific knowledge, *episteme*, primarily in terms of the individual. In contrast to mere opinion [*doxa*] or particular facts [*historia*], *episteme* involved logical necessity and truth. It referred to a form of knowledge that an individual achieved in a particular manner. "A man knows a thing scientifically," wrote Aristotle, "when he possesses a conviction arrived at in a certain way, and when the first principles on which that conviction rests are known to him with certainty."[6] To know something in this sense is to know its cause, "to know in a deep sense what it is and how it has come to be."[7] Such knowledge referred less to a possession that one could acquire and more to the way in which one knew something.

As a person acquired scientific knowledge, he did not simply know more; instead, as Aristotle put it, the "state of [his] soul" changed. Science was an acquired aptitude [*hexis*] for demonstrative or syllogistic reasoning; it referred to a capacity to deduce the true causal order of the world from principles. In this sense, *episteme* referred not to particular principles but rather to the mental state of the subject who knows something scientifically. "Science" designated a quality or characteristic of the knowing subject, namely, a capacity for a necessary and sure form of knowledge.

Aristotle's more individual-oriented notion of science as a mental aptitude did not, however, discount the idea that scientific knowledge needed to be organized into an objective body of knowledge. If knowledge were to be improved, he contended, it required collaboration over generations, because "individually we contribute little or nothing to the truth."[8] Individuals profit from the insights and errors of their predecessors, whose knowledge is transmitted and communicated in material forms, such as writing. In order for knowledge to advance, then, it has to be made communicable.[9]

And for Aristotle, the primary form of communication, both temporally and conceptually, was oral speech as practiced in dialogue and person-to-person interactions. Anticipating theories of signification and language which would become central to Western philosophical traditions, Aristotle defined writing as "the signs of spoken words," that is, as signs for other signs.[10] Writing was one step further removed than were the signs of oral speech from the "affections or impressions of the soul." It was, therefore, reasoned Aristotle, subordinate to oral speech. This did not mean that he simply dismissed writing, but he did criticize "technographers," people who taught and practiced the art of writing, for what he saw as their failure to conceive of writing as anything other than a mere storage device that fixed ideas in material form. Technographers did not understand the rational ends of rhetoric and the ways in which it could potentially facilitate complex patterns of thought and speech.[11] Writing, for Aristotle, was not an inherently and irredeemably deficient technology. It was just a new and untested one.

Plato had similarly considered writing a mechanism for memory, but he contrasted it even more sharply than Aristotle with oral communication and insisted that it was not an adequate replacement for the dynamism of oral interaction. Because they were fixed and had material form, written forms of communication could be broadcast widely, well beyond the small circle of interlocutors in which the intentions of a speaker could, or so Plato thought, be better monitored.[12] Plato's *Phaedrus* served as a source text for Western anxieties about technological changes in the means of communication and the ways in which a new technology is presumed to supersede an older, more established technology that is oftentimes regarded as more trustworthy and suitable to rational communication. For Plato, "writing parodies live presence; it is inhuman, lacks interiority, destroys authentic dialogue, is impersonal, and cannot acknowledge the individuality of its interlocutors; and it is promiscuous in its distribution."[13] Whereas in a Socratic dialogue one interlocutor engages another in love and thus with attention to the other's soul, and both are required to respond directly to questions, writing is indiscriminate and, as Socrates put it, "silent"; it disregards the particularity of its audience.

The classical notion of science as an individual-dependent form of knowledge was in large part a function of the emphasis that Aristotle and Plato placed on scientific knowledge's formative potential. *Episteme* was "edificatory" and could transform a person and cultivate excellence of the soul.[14] The pursuit of scientific knowledge was a matter of harmonizing the individual soul with the order of the cosmos. To acquire the habits of mind characteristic of science involved the acquisition of higher truths about the universe.

Ancients like Plato and Aristotle worried that to focus on the distribution or transmission of knowledge in a material form was to treat knowledge as an inert object that could be exchanged and cultivated like any other thing. And this concealed what they considered a basic truth: true knowledge is ultimately grounded in the act of thinking. Focusing on particular technologies risked concealing the human element of this act. It made the material technologies of transmission, and not the human soul, the agents of knowledge. Fearful that communicative technologies would become detached from individuals, Plato and Aristotle emphasized the interpersonal transmission of knowledge over the collection and aggregation of knowledge in material forms. Scientific knowledge, in this ancient sense, always implied a person who knows. It was a more personal conception of knowledge.

These ancient arguments about communicative forms became mainstays of the Western philosophical and cultural tradition, and they proved especially important to late eighteenth-century German critiques of early modern and Enlightenment forms of scholarship and their reliance on print technologies. These later eighteenth-century anxieties about print, however, were exacerbated by differences in scale. Those who worried about the ascendance of print over the course of the eighteenth century were primarily concerned about its unprecedented capacity to indiscriminately distribute information and ideas through mechanical means and thus the possibility that print technologies would outstrip human norms and practices. In the face of these epistemological anxieties, late eighteenth-century German intellectuals sought to recover and combine more modern, material notions of science with these more ancient notions that emphasized habits and the formation of individuals.

## The Middle Ages, Science, and Discipline

One consequence of the ancients' emphasis on the subjective or individual-dependent aspect of science was a long-standing hierarchy in which the authority of real, scientific knowledge was tied to an individual's capacity to ascend to ever-higher forms of purified knowledge. Knowledge that assumed material form was secondary to knowledge that could be traced back to the soul, a knowledge not subject to the constraints of the body.

This predominant hierarchy was in part responsible for the fact that the ancients never developed a robust concept of scientific knowledge that was broadly communicable beyond person-to-person exchanges. It was also why the dialogue remained the exemplary communicative form for most ancient commentators.

With the collapse of the ancient urban centers, the disintegration of the Roman Empire, and the dissolution of established networks of communication and distribution, much of antiquity's cultural and intellectual heritage and texts were lost, and with them ancient conceptions of *episteme* and *scientia*. With the gradual rise of Christianity in the early Middle Ages, a different, more text-focused conception of scientific knowledge emerged, one that was tied to material practices and institutions designed to guarantee the continuity and transmission of a relatively new cultural heritage. Early Christian monastic learning, for example, was primarily oriented not toward person-to-person interactions on the ancient dialectical model or theoretical discussions of *scientia* but toward interactions with sacred and canonical texts. Practices of glossing, interpretation, manuscript production, and silent reading were central to the transmission of knowledge in early and medieval Christianity.[15] In antiquity the reading of written texts focused on the ideal of the orator and his masterful oral delivery. Students were trained to read so that they could become proficient orators. In the early Middle Ages, in contrast, Christian-based practices of silent reading and reading aloud lent written texts, especially sacred scripture, a distinct value. Reading was not just a preparation for speaking; it was an opportunity to encounter the divine. Over the course of the early Middle Ages, basic attitudes toward "the nature of the written word" shifted.[16] The collapse of the Roman Empire and with it the schools of antiquity meant, as the sixth-century Roman statesman and monk Cassiodorus explained to his fellow monks, that books had to serve "in place of a teacher. You have masters of a bygone generation to teach you not so much by their tongues as by your eyes."[17] For centuries, these interactions with texts facilitated the kinds of insight and intellectual growth that Plato and Aristotle had assumed were limited to person-to-person exchanges through oral speech. The written word was not just a pale comparison of the spoken word. In the absence of living authorities, it was the medium that transmitted authoritative knowledge over time.

As Aristotle's logic texts, such as the *Prior* and *Posterior Analytics*, gradually made their way into Western Europe via Islamic scholars in the twelfth century, however, ancient conceptions of an individual-dependent notion of scientific knowledge returned. One consequence of the rediscovery of Aristotle was a reconceptualization, especially among scholastic scholars, of science as a mental state or *habitus*. Echoing Aristotle's use of *episteme*, Thomas Aquinas, in thirteenth-century Paris, used *scientia* to refer to a particular disposition in which the soul's powers are developed toward the perfect knowledge of the sovereign truth, that is, to the intellectual apprehension of God.[18] Throughout the Middle Ages and well into the early modern period, scholars used *scientia* to refer to a mental disposition toward rigorous thought tied to the demonstrative method.[19] Scholars debated whether there was one shared *habitus* that united all sciences or each science had its own *habitus*, but they all conceived of science as primarily a mental disposition and not a material form of knowledge or an intellectual heritage.[20]

Like the ancient meaning of the term, then, *scientia* connoted how an individual person was formed, and thus was often used, as Gyorgy Markus points out, in relation with terms such as "discipline" [*disciplina*] and "doctrine" [*doctrina*], which referred both to the process of instruction and to its results.[21] Both "discipline" and "doctrine" had a long history in Latin Christendom, which served as the basis for their adoption in high scholastic and humanist traditions.[22] The twelfth-century English monk Andrew of St. Victor, for example, understood discipline as a form of knowledge attained from another through instruction. Whoever allows himself to be properly trained, he wrote, is "wise, prudent, and disciplined."[23] "Discipline" was also used in these contexts to refer to the type of teaching and learning in the liberal arts and to describe a pedagogical relationship between teacher and pupil.[24] Discipline expressed the spiritual link between the master and the pupil. *Disciplina in disciplo, doctrina in magistro* was the motto: doctrine equals authority.[25]

The explicit connection between discipline and doctrine held over the following centuries. In 1692, Etienne Chauvin, French scholar and rector of the University of Paris, defined discipline as a "conception accepted from a master such that disciples follow the master's example through his teaching."[26] Discipline was inextricable from the subjective formation of individual students. The doctrine that was to be disseminated and the discipline that was to be exercised were inextricable and complementary concepts.[27]

There is, then, an important resemblance between medieval and early modern concepts of *disciplina*, classical notions of *episteme* and *scientia*, and the disciplinary order of knowledge which eventually organized the modern university.[28] Despite these continuities, however, neither classical nor medieval notions of science and discipline emphasized the production of new knowledge as later notions of disciplinarity eventually would. This is largely because ancient, medieval, and early modern cultures had no conception of research and thus did not tie science to a belief in the progressive, historical development of knowledge over time.[29] The unification of a subjective pedagogical context with the production of new knowledge—that is, the formation of subjects whose primary purpose it was to produce new knowledge—was an innovation of the research university. These earlier notions of discipline did not and were never expected to provide a particular form of life organized around a particular science; instead, they prepared students for further education in the higher faculties and thus life as a lawyer, doctor, or priest. Later notions of disciplinarity, in contrast, were never limited to their pedagogical function. They entailed a complex of institutional and social practices and an underlying ethos aimed at producing a particular type of person who could produce a particular kind of knowledge.[30]

Although concepts such as *scientia*, *disciplina*, and *doctrina* focused on the formation of the individual soul toward the apprehension of the divine, unlike their ancient predecessors their authority was closely tied to texts, especially in the authority of sacred scriptures.[31] Even the medieval university's tradition of the oral disputation relied on central, authoritative texts. Especially in its High Scholastic form, scientific knowledge, *scientia*, was grounded in the authority of *auctores*—past authorities.[32] The authority of scripture and the related texts that constituted the canon of the medieval education system was bound up with the person of their production, be they the church fathers or Greeks like Aristotle. The oral disputation and its related textual practices were extensions of one another; both assumed the existence of a truth that only became clear in a collective encounter of traditions and the meeting of arguments. The primary goal was not a better understanding of the "process of thought" but rather the engagement with a tradition of established, authoritative knowledge that had been fixed and "textually objectified."[33]

In the fifteenth and sixteenth centuries, humanist scholars further developed the textual practices and cultural assumptions about textual authority of medieval scholars as they attempted to harness the productive power of new print technologies. Their work was characterized, as historian Ann Blair writes, by "a newly invigorated info-lust that sought to gather and manage as much information as possible."[34] This "info-lust" was not just a vague attitude but a new "cultural conception of scholarly method," defined by the "stock-piling" of textual information intended to create a "treasury of material."[35] The early modern masterpieces of learning—such as Conrad Gesner's *Bibliotheca Universalis* (1545) or Johann H. Alsted's *Encyclopaedia septem tomis distincta* (1630)—sought primarily to compile and accumulate dispersed information and were less concerned with forming particular habits of the intellect. As sixteenth-century scholar and bibliographer Conrad Gesner put it, scholars left the "free selection and judgment"—the determination of good and bad—to readers.[36] These projects were technical and designed to store and organize the very material of learning and gave less attention than the ancients and medieval scholars to the formation of the individual person. The historical breadth and detail of these texts displayed a different conception of what constituted science and authoritative knowledge, one that was bound to texts and could be transmitted through books of all sorts from generation to generation.

Many of the authors of these Latin encyclopedias and compendia were motivated by a desire to protect ancient learning and what they considered to be its integrity and authority. They "hoped to safeguard the material they collected against a repetition of the traumatic loss of ancient learning of which they were keenly aware."[37] This loss had rendered Greek and Roman antiquity inaccessible until its gradual recovery in the fifteenth and sixteenth centuries. Humanist scholars saw encyclopedias and related reference works that collected ancient learning as guarantees that knowledge could be quickly reassembled should all books be lost again.[38] Hence, they emphasized that real, authoritative knowledge was something that could be stored and preserved in texts for posterity. In the broadest terms, then, the ancient notion of *scientific* knowledge as a state of mind or habit was gradually superseded by a notion of knowledge and "science" that was textually objectified in manuscripts and books. Medieval and humanist scholars' deep appreciation for the book in manuscript or printed form challenged the ancient hierarchy according to which the written word, etched or otherwise objectified, was always subordinate to oral speech.

## The New Science and Books

Not all early modern thinkers, however, were enthused by the long-term shift toward more text-based notions of knowledge. And some even wondered whether scholars had begun to fetishize the materials of knowledge. One such skeptic of book learning was the sixteenth-century English natural philosopher Francis Bacon, who formulated what he termed the "new science." Bacon's attempt to define a "new science" is important for our story, because it reanimated the ancient ambiguity with respect to "science." Bacon also presaged the more pointed conceptual conflict that emerged over the course of the eighteenth century in Germany between science as a material collection of knowledge and science as an intellectual habit. This conflict proved crucial for the emergence of the research university, which its romantic and idealist champions celebrated as the ultimate harmonization of the subjective and objective aspects of science. It was the romantic research university, they suggested, that would attempt to resolve these tensions and unify the persons and things of science.

For Bacon, the "aids" for the "regeneration and renewal of the sciences," which included not only books but also scientific instruments of all sorts, were not merely repositories of an authoritative tradition of knowledge. They were also tools for its advancement.[39] The technologies of the "new science" should improve the "force" of the individual mind, which, for Bacon, was the ultimate technology for producing new knowledge and mastering nature. For Bacon, the human mind had to be distinguished from such technological "aids" and what he termed the idols of learning, foremost among them books. Whoever has not only considered "the immense variety of books" but examined them, he wrote, cannot avoid their endless "repetitions."[40] In the name of his "new science," Bacon exhorted his readers to turn away from bibliographic details and seek natural historical facts.

The difference between Bacon's conception of knowledge and that of medieval and humanist traditions was evident in how each conceived of the unity of knowledge. Whereas for the latter such unity was grounded in an authoritative textual tradition, for Bacon it was grounded in the mind's three faculties: memory, reason, and imagination. All forms of knowledge could be traced back to mental capacities and thus to the fundamental unity of the mind. Bacon visualized this unity in his famous tree of knowledge, which stood in for a detailed classification of all learning. The various branches, limbs, and boughs of the tree presented a classification scheme of the various divisions and subdivisions on the basis of the mental faculties: history, which corresponds to memory; poetry, to imagination; and philosophy, to reason.[41] Science, for Bacon, was a method for honing this mental unity to produce new knowledge.

Even though he emphasized science as an individual mental capacity, Bacon consistently acknowledged how necessary it was that scientific knowledge be given material form. "Present and future generations would be better off," he wrote, if the philosopher, natural or rational, made his thoughts known to them.[42] Past achievements of the sciences had to be stored and transmitted not only to preserve them against possible calamity but to facilitate their future growth. Any advancement in knowledge required a taking stock, a review of the sciences. In his efforts to distinguish himself from what he considered humanism's obsession with accumulating bibliographic details, Bacon had to confront a few basic questions: What becomes of the newly created truths and natural historical facts once they have been discerned and propounded? How are they to be transmitted and shared beyond the individual? And what is the status of these transmissions once they are given material form and become subject to the conditions of space and time? Are they simply parodies of true knowledge?

Despite Bacon's apparent dismissal of bookish knowledge, however, the natural historian's passage into the "antechambers of nature" was possible "not only from the shock of the new worlds laid open by explorers and observers, but also from the shock of the old books recovered and explicated by scholars working patiently in their studies."[43] Erudite books, as well as their cultivation, which he sometimes denigrated, were critical to Bacon's larger project. Humanist methods—in particular, care with textual detail, selecting and recombination, commonplacing, cutting and pasting—were skills in clear and elegant transmission and thus essential for the production and distribution of new knowledge. Eighteenth-century scholars would adopt and adapt these methods and technologies and blur the line between knowledge and technologies. Thus, even as Bacon warned that books could be a scholarly "defect" that hindered true learning, he acknowledged that material forms of knowledge not only were important but needed to be cultivated. He was particularly concerned that in their diligence to collect information and protect it from loss, scholars failed to add anything new or distinguish between good and bad books. "In order for learning to advance," wrote Bacon, "a different kind of thinker is required," one who could account for the history of knowledge and make judicious use of it.[44]

In order for the "new science" to advance, knowledge had to be made communicable. But this recognition involved a conceptual bind that would become especially acute over the course of the eighteenth century as worries about the proliferation of print became more common. Bacon's imperative to produce new knowledge required that it be captured in material forms. As it was, however, knowledge and science came to refer less to individual aptitudes and more to the material forms in which they were produced and transmitted—books, encyclopedias, journals, sciences, systems, and, by the end of the century, disciplines.

## The Collapse of Latin Learning and the Pedant

As these print technologies proliferated, scholars began to worry that they were exceeding the rational control of humans, that they had somehow gained their own authority independent of humans. Eighteenth-century German writers and intellectuals struggled to reconcile the need to store and transmit knowledge in material forms with the increasing sense that these texts were overwhelming the human agents who authored and edited them. They worried that the authority of knowledge had been usurped by technologies of knowledge. In Germany, these epistemological anxieties peaked in the final decades of the eighteenth century with dystopian fears of book plagues and floods, reading addictions, and a widespread attempt to reconceive of science as a cultural solution.

Before these epistemological and cultural anxieties spread, however, there was a shift in the kinds of reference and scholarly books that were actually available. Over the course of the eighteenth century, there was a marked decline in the Latinate learning and scholarship that Bacon had both mocked and praised. After over two hundred years of regular printing, the publication of Latin works of learning and reference books came to an abrupt end in the first decades of the eighteenth century.[45] This collapse in Latin reference books was part of the more general decline in the publication of Latin texts over the course of the seventeenth century, as Latin scholarship was gradually replaced by vernacular scholarship written and published in German.[46] It also corresponded to the broader, if more gradual, decline in the use of Latin in schools and universities.[47] In 1690, the German philosopher and jurist Christian Thomasius delivered the first German-language university lecture in Leipzig, and by 1730, Prussian secondary schools had ceased using Latin as the language of instruction.

The forms and genres of Latin humanist scholarship did not simply disappear, however. Eighteenth-century German scholars reinvented early modern projects of universal learning and reimagined their purpose between 1700 and 1800 in a range of print genres that adopted and adapted traditions of humanist learning.[48] Printed encyclopedias, lexica, and bibliographies proliferated not only in number but also in kind, but they all focused on collecting textual information and were committed to what German scholars of the eighteenth century termed *Vollständigkeit*—completeness or comprehensiveness. Their goal was not just the gradual accretion of a "treasury of material" over time, as it was for humanist scholars, but a complete accounting of knowledge in printed form—its organization into a seamless whole that omitted nothing. The primary methods of handling such a mass of material were bibliographic, that is, organizing and creating bibliographic accounts of print.

These later-day humanist scholars were committed to a particular notion of what constituted the unity of knowledge—how it should be organized and transmitted. Their pursuit of a complete knowledge was sanctioned by a confidence in a natural or divine order. These projects, like their early modern predecessors, assumed an external guarantee. Nature and everything in it were assumed to be ordered and imbued with a divinely ordained meaning. The task of learning and scholarship, therefore, was less the production of new knowledge and more the recovery of reason—the divine, rational order of the universe—from the material stuff of sensual experience.[49] The goal of scholarship was the restoration of a pre-lapsarian epistemic condition and a defense against another calamitous loss of texts.

Almost as soon as these vernacular reinventions of humanist learning technologies appeared, however, they were beset by criticisms. One of the earliest and most pointed sources of criticism was German scholar Johann Burkhard Mencke's widely read *On the Charlatanry of the Learned* (1715). Mencke, a professor of history in Leipzig, mercilessly mocked his fellow scholars as vainglorious charlatans. He chided "erudite compilers" of the sixteenth and seventeenth centuries such as Johann Heinrich Alsted, Giulio Bordoni, Athanasius Kircher, Conrad Gesner, and Julius Scaliger for producing reference works that accumulated masses of information and still "scratched only the surface" of knowledge.[50] He scoffed at the long, puffed-up titles and honorifics that scholars collected—*Clarrisimus*, *Magnificus*, *Consultissimus*, *Excellentissimus*—simply to sell their books and impress fellow members of the scholarly guild. And he dismissed the ridiculously long and bloated titles of their books, such as "The Ampitheatre of the Only True Eternal Wisdom, Christian-Cabalistic, Divine-Magical, and Yet Physical-Chemical, or The Catholic Three Times Three in One, Arranged by Heinrich Cunrath." Mencke also denigrated their purported pansophism as little more than superficial knowledge and derided their worthless atlases of historical facts and compendia of isolated "kernels."[51] "What worthless paper!" he exclaimed. And yet, Mencke's book was a model of erudition, the very detailed and textual scholarship that he mocked. It began with a discussion of the etymology of *charlatanería* and throughout displayed a vast knowledge of the history of learning. Mencke's book exemplified an increasing ambivalence about humanist scholarly practices and the limitations of the scholarly guild in the eighteenth century.

Mencke's critique was but a prelude to broader cultural anxieties about the reduction of thought to printed objects which would become increasingly common over the course of the eighteenth century. As the Enlightenment vestiges of Latin learning grew in number and complexity, many later eighteenth-century scholars began to doubt widespread epistemological and ethical assumptions about what counted as real knowledge. Critics complained that the excerpting, compiling, note taking, and summarizing that characterized earlier Latinate scholarship had created generations of scholars who ignored original texts, had nothing new to say, were ignorant of textual and historical contexts, and thus arbitrarily aggregated textual tidbits. Scholarship had been reduced to little more than printed curiosity cabinets, and knowledge itself consisted of a body of bibliographic facts.

Over the course of the eighteenth century, Latin forms of scholarship and their Enlightenment successors became narrative counterpoints for critics seeking to establish new theories and practices of knowledge by dismissing what had come before them. In the first third of the eighteenth century, for example, the so-called moral periodicals—fashioned after their English predecessors such as the *Spectator* and addressed to broad audiences and the edification of moral character—tied the perceived failings of humanist-inspired forms of scholarship to the figure of the pedant, or what in German was referred to as the *Polyhistor*, or "humanist polymath."[52] In 1726, one writer worried that a "swarm of pedants" had "obscure[d]" "true" learning.[53] To battle this trend, readers needed to learn to distinguish "true scholars" from the various "species" of pedants, who, fortunately, were easily identified.[54] They obsessed over books and philological detail, tended toward wild speculation and abstract problems, labored under an affected seriousness, and dressed slovenly. They suffered from a general alienation from the world. As another writer put it, the pedant "deals only in books, . . . with actual thinking he has little to do."[55]

Underlying the satirical portraits and scathing rebukes, however, was an increasingly coherent critique of humanist scholars as so focused on questions of style, eloquence, and accumulation that they were uninterested and thus had no facility in addressing what critics increasingly considered the ethical and epistemological failures of an entire culture of learning. Humanists misapprehended the actual purpose of returning to the ancients and were therefore unable to engage these texts fully. They were unable to perceive contradictions, grasp contexts, and understand basic philosophical assumptions. They were plagued by problems of anachronism and the intransigence of authoritative readings.[56] Like their early modern predecessors, these scholars were so busy guarding against future calamity that they never stopped to consider the ends of knowledge: what was the purpose of all this accumulating? These criticisms assumed that safeguarding and transmitting knowledge was an insufficient end. The pedant initially signified the humanist scholar, a member of the Republic of Letters, who was interested in the broader world and the free circulation of knowledge, but the pedant gradually came to stand in for a truncated, more scholastic humanist, whose interests did not extend beyond particular texts and ever more particular interests. The pedant stood for the erudite scholar more generally.

By the middle of the eighteenth century, the pedant had become a satirical persona. But whereas the initial critiques focused on the failures of a particular person, the newer critiques targeted humanist scholars as a social class, the class of erudites. Until the late eighteenth century, the figure of the pedant served a double function as both an epistemological and ethical critique of an entire class's refusal to engage broader society. As Damis, the young scholar in Lessing's drama *The Young Erudite*, exemplified, the pedant saw himself not as a member of a local or even national community, but as a member of an imagined "Republic of Letters. What does Saxony, Germany, Europe mean to us scholars? A scholar like me is for the entire world. He is a cosmopolitan."[57] To be a scholar was to be an unmoored cosmopolitan, who was blind to the demands of a particular place, time, and people.

As Enlightenment ideas spread, critics increasingly contrasted the "pedant" to the broader Enlightenment imperative that sciences be pursued not for their own sake but for the happiness of fellow citizens.[58] Scholars should not only display erudition but, as one writer put it, conduct themselves as "virtuous humans and good citizens."[59] The proper end of erudition was the state and the welfare of its citizens. The value of knowledge was its social utility. Midcentury satires and popular critiques extolled values of rationality, morality, and usefulness and praised forms of scholarship and scientific knowledge that explicitly integrated the scholar into productive social roles. The idiosyncrasies of the scholar—the learned fool with his peculiar habits—had become a liability. Such criticisms were meant to level distinctions between scholars and the broader public and to integrate erudites into a more generally educated social structure.

## Science as System

In the last decades of the eighteenth century, however, the imperative to distribute knowledge more broadly and make it accessible began to conflict with another Enlightenment imperative to advance knowledge by producing more of it. As knowledge expanded and became more fragmented, this dual imperative made it increasingly difficult to transmit knowledge in a broad and accessible manner. As knowledge grew, it became too disparate and unwieldy. Enlightenment efforts to sustain a common base of knowledge, to bind not only scholar to scholar but scholar to citizen, were failing.

In order to manage the dual imperatives of the Enlightenment and the sense that knowledge itself was splintering, some scholars, such as the Berlin theologian Johann August Eberhard, suggested that a comprehensive term like "knowledge" [*Wissen*] was no longer capable of describing the various ways in which people produced, transmitted, and stored knowledge. Even the most common eighteenth-century synonyms for knowledge—"enlightenment," "erudition," and "science"—had to be more sharply distinguished from one another. "Enlightenment," as Eberhard put it, was knowledge that was useful to a broad public, whereas "erudition" [*Gelehrsamkeit*] was knowledge that was of interest only to a small group of scholars. Science [*Wissenschaft*], in contrast to both, was characterized not by its audience but rather by how it was produced. It was, wrote Eberhard, knowledge produced "according to a technical method, through which the highest level of certainty and thoroughness is demanded."[60] By identifying science with a technical method, Eberhard tied the rational necessity historically associated with ancient notions of science [*episteme*] to the productive potential of technical or artistic skill [*techne*]. Scientific knowledge, he suggested, could only be produced according to a technical method that could be replicated. On this account, science referred not to the cognitions or capacities of an individual person, or even to particular bodies of thought, but rather to a type of collective knowledge produced according to a rational method or the rules of an art. Science was the sum of practices, concepts, and methods designed to generate a particular kind of knowledge with its own authority and norms. It was not simply a particular intellectual disposition. Eberhard's was one of many attempts in the last two decades of the eighteenth century to manage the perceived fragmentation of knowledge by articulating a concept of science which combined more ancient notions of an individual-dependent form of knowledge with more institutional-dependent forms of knowledge; it pointed to the ways that scientific knowledge could be sustained over time not only through individuals but in books, methods, and practices that ultimately could be organized in institutions.

None of these attempts were more influential than Immanuel Kant's. Kant developed his concept of science out of his broader critique of modern analytic methods that he claimed merely accumulated the basic elements of knowledge—be it Baconian facts, Linnaean specimens, Lockean ideas, or bibliographic data.[61] For Kant, all of these methods lacked any sense of the ends of knowledge and the potential of reason to set such ends. If they did have a purpose, it was one given externally by some authority other than reason itself. On Kant's account, the problem of eighteenth-century overload and proliferation lay not in too much stuff but in a failure to recognize the ethical dimensions of knowledge. "The final end of the sciences," he wrote, was not to advance them infinitely but rather "to find the destiny of the human."[62]

Without theology or an appeal to a given natural order, Kant recognized that his more ethically oriented conception of science would require a secure ground. And, like his forbearers Bacon and Descartes, he sought a self-enclosed, rational order distinct from a natural order. What "makes common knowledge science," he wrote, is systematic unity or the "unity of manifold cognitions under one idea."[63] "System" or "science" denoted a unity not only of form but of purpose, and it was this common purpose—the idea of unity—through which manifold parts were related to the whole. Kant's solution to the eighteenth-century conundrum of maintaining a unity of knowledge amid its proliferation was to tie science to a systematic, self-enclosed and self-sufficient determination of ends.

Whereas Bacon, Diderot, and d'Alembert had considered the unity of knowledge to lie in the unity of the mental faculties, Kant placed it in the internal coherence of a rational system. For Kant, the system or the unity of a science was an ideal manifestation of the unity of reason, independent of individual minds.[64] A system or a science is not a subjective capacity of the mind; it is a supra-individual product of reason. One must think of a science as if it existed independent of the individual as a distinct, self-regulating entity.[65] Not tethered to individual minds, sciences exceed the capacity even of their supposed "founder" to give a comprehensive account of their development. Sciences were not bound to an individual knower.[66]

## Science as Ethic

Looking back on the semantic shifts that "science" had undergone over the course of the eighteenth century, the anonymous author of one lexicon article from 1811 summarized the changes thus: "science," he wrote, had been used initially to describe a subjective "condition in which one knows something" and to refer to the "clear and distinct ideas" of an individual. But it had also been used "objectively" to describe "general truths that were grounded in each other" and not in the mental processes of an individual.[67] Whereas the first usage had become "antiquated" by the end of the century, he concluded, the second had become standard. "Science" had gone from referring to a "particular insight" or mental capacity to an internal relationship among ideas themselves.[68] This shift represented in important ways a departure from ancient notions of *episteme* and *scientia* as an individual-dependent form of knowledge. It also gave way to various attempts to articulate a concept of science as a type of knowledge that could be communicated beyond person-to-person exchanges.

In an attempt to categorize these changes in the very notion of "science," many late eighteenth-century encyclopedists and lexicographers distinguished between the "subjective" and "objective" aspects of science. By "subjective" they did not mean mere opinion; rather, they used the term to identify ideas that were tied to an individual knower. Likewise, by "objective" they did not mean an actual correspondence between a proposition and a certain state of affairs in the world; rather, they used the term to draw attention away from the knowing subject and toward knowledge as something that existed independent of a knower and assumed some material or ideational form in the world.[69] As an objective form of knowledge, science was extrinsic to any one person and its transmission necessarily exceeded person-to-person exchanges. In fact, in order for knowledge to count as scientific and therefore be authoritative, it had to assume some more public form. To use the historical terminology, the objective aspect of science had come to predominate such that the strictly subjective use of the term largely disappeared by the first quarter of the nineteenth century.[70]

These lexical clues about the history of a concept help identify a time frame in which scholars first used "science" in its more contemporary sense, namely, as a realm of material works and ideas which is "publically accessible and detached from those who initially produced them."[71] It was also around 1800 that scholars increasingly used science in the collective singular to describe not only the objective internal relationship of ideas and concepts but a distinct ethic. They began to use the term in the sense that Fichte's young student used it in *The Academic Life in the Spirit of Science*, to refer to an internally coherent culture of knowledge replete with its own customs, habits, and norms. Science represented the organization of "general truths" through the material communicative forms and structures of knowledge—the habits, practices, and, ultimately, institutions. It was a distinct cultural sphere into which people could be inculcated. Whereas the ancient scholar was to be integrated into a harmonious relationship with the cosmos, the "scientist," which was first used in this same period, was to be integrated into a community of fellow scientists who organized themselves around the material forms of science and embodied the particular habits, virtues, and practices that ultimately, if belatedly, came to correspond to them.

These semantic changes represent a historical shift from science as an individual-dependent form of knowledge to a more institutional-dependent form of knowledge over the course of the eighteenth century. Unlike an individual-dependent form of knowledge, an institutional-dependent form of knowledge is not exhausted by psychological, subjective attitudes (justified true beliefs) and capacities of a given individual (even a hypothesized individual), although it may well include such individual knowledge. Institutional knowledge extends out, so to speak, to encompass a community of practitioners, texts, equipment, habits, practices, sensibilities, and organizing institutions. This shift to science as an institutional-dependent form of knowledge reached its apogee, as we shall see in the following chapters, in the modern research university, which organized the material of knowledge—books, encyclopedias, periodicals, and lexica—and formed the very persons of knowledge.

Late eighteenth-century scholars and critics described the emergence of an institutional-dependent form of knowledge, science as a distinct cultural sphere, in explicitly ethical terms. In much the same language as Zimietzki, they extolled science as a distinct source of meaning and value. The sciences were embraced as central elements of culture which, as Kant put it, "reduce the tyranny of sensible tendencies, and prepare humans for a sovereignty in which reason alone shall have power."[72] Science as culture combined the subjective and objective aspects of science—the ancient opposition between knowledge as an aptitude and knowledge as tied to material forms in space and time—to form particular types of people. Science was both an individual capacity that had to be learned as though it were an art and the sum total of all those products, both material and ideational, that the practitioners of science produced. This new conception of science was in some ways a recovery of ancient views of scientific knowledge as intimately related to the formation of the soul, but it combined these ancient notions with more modern, material aspects of science.

By casting science in these explicitly cultural and ethical terms, figures like Kant reframed the anxieties about information overload and the fragmentation of knowledge. The problem of excess lay not in the proliferation of things—too many books—but in the incapacity to recognize human moral agency. It was an ethical failure to recognize the capacity to think for oneself. The ascendance of "science" as an institutional-dependent form of knowledge and its invocation as a solution to epistemic anxieties entailed a cultural critique, which went something like this. Over the course of the eighteenth century, German readers and scholars had alienated themselves from their true moral nature, their capacity to set ends for themselves as autonomous, rational beings, and subordinated themselves to technologies of their own making, be they encyclopedias or wrongly conceived sciences. Eighteenth-century scholars in particular had fetishized the tools and instruments of knowledge. They had done so out of an admirable, if misguided, perceived duty to advance knowledge, but they mischanneled their desire to know into an unbounded confidence in the capacity of technologies to facilitate the production of ever-new knowledge and a deep misapprehension of what constituted authoritative knowledge. They were motivated to accumulate and collect knowledge by a long-standing fear that it would disappear, a fear born out of the loss of ancient manuscripts in the Middle Ages. Scholars assumed that better-organized, comprehensive encyclopedias or lexica would both expand and preserve knowledge. This also meant that they began to confuse the material forms of knowledge, objective technologies, with knowledge itself.

This narrative echoes through the work of well-known figures such as Johann Gottfried Herder, Novalis, Goethe, and other late eighteenth-century writers who derided the tradition of Latin learning from early modern reference works to eighteenth-century lexica as "merely" encyclopedic. Overwhelmed with footnotes, citations, and commentaries, scholars had become dependent on their technologies. Why bother to become enlightened, asked

Kant, if you have a book that can think for you?[73] Laments about too many books or too much to know were the misguided worries of scholars who simply reproduced and recirculated already-existent knowledge.

These late eighteenth-century critics argued that fetishizing instruments of knowledge was not the solution to information overload. Taking the ancient duty to know seriously in an age of technological change required more than new tools. It required a certain disposition toward the ends of knowledge. The use of scholarly aids and technologies had to be directed by clear accounts of the purpose of knowledge and norms for how people should interact with technologies of knowledge. True knowledge was not simply a function of unbounded expansion; it also required a capacity to judge what was worth knowing and what was not.

In this sense, these late eighteenth-century criticisms and anxieties were not simply a function of confident assumptions about technological or epistemic progress; rather, they were the products of deep shifts in normative assumptions about the value of different forms of knowledge and learning. By 1800 figures such as Kant, Fichte, the Schlegel brothers, Friedrich Schleiermacher, Novalis, and Goethe cast these distinctions in a series of value-laden metaphors and distinctions: whole versus aggregate, science versus erudition, a unified body of knowledge versus a concatenation of distinct parts, and university versus vocational school. At stake in these oppositions was a new ethics of knowledge—what constituted real knowledge and who should define, produce, and disseminate it. Seemingly arcane debates about scholarly technologies were actually a "search for criteria by which to distinguish true learning from the appearance of learning."[74] Every knowledge technology of the eighteenth century, from the book to the encyclopedia to the university and the ethos of science, stood in for a particular ordering of the desire to control and master knowledge, for a particular organization of the practices, habits, and visions of what counted as real knowledge.

When late eighteenth-century figures like Kant suggested that knowledge projects were as much about ethical questions as epistemological ones, they did not simply mean that they concerned rules for moral behavior. They were attempting to fuse ethics and epistemology and thus link the coherence and integrity of knowledge to particular types of people and communities. They sought to ground the authority of knowledge in the goods and virtues of a particular form of life and practice. In this sense, late eighteenth-century German intellectuals and writers saw in information overload not only an epistemic threat but an ethical one as well. Excess and overload confused the authority of knowledge. It confused what counted as real knowledge and thus threatened the capacity of humans to appeal to authoritative knowledge.

The first ethical threat could be considered in relation to the classical imperative that Kant adapted in his famous "What Is Enlightenment?" essay: "Dare to Know!" Over the course of the eighteenth century, the perception that knowledge was proliferating at an uncontrollable pace became common, and many scholars and commentators considered it a threat to the integrity of knowledge, that is, the coherence of knowledge about the world. Proliferation produced fragmentation, which, worried eighteenth-century critics, threatened people's ability to abide by long-standing imperatives to know the world. From Plato and Aristotle to the German idealists and romantics, scholars had almost universally assumed a basic ethical imperative to seek knowledge of oneself, the world, and others. Ignorance was not an ethically justifiable option. Over the course of the eighteenth century, however, there was an increasing fear that the centrifugal pressures of the proliferation of knowledge had begun to undermine the ability of any one person to know with any coherence—that is, with any sense of the legitimacy of any particular way of organizing the practices, skills, and technologies of knowledge. As books, periodicals, and encyclopedias proliferated, there was simply too much to know and little hope of giving a unified authoritative account of it. It was becoming impossible to integrate disparate kinds of knowledge. In 1787, the jurist Reinhard F. Terlinder worried that as the "discoveries" and knowledge of humans increased, the "corresponding truths" were becoming "incalculable."[75] In "modern times," in which these sciences grew "to such an extensiveness," this meant not only that there was so much material to manage, but that the truths that distinct sciences and kinds of knowledge embodied had proliferated as well, so much so that no human lived "long enough" or had "enough energy" to familiarize himself with "the country of learning, with its extent, borders and districts."[76] There was no way to gain a view of how all types of knowledge were related.

The truth claims, assumptions, and evidentiary grounds of various sciences or types of knowledge seemed increasingly incompatible. The human knowledge project was becoming, worried many, an incoherent and even futile enterprise. Knowledge, which for so many eighteenth-century figures was supposed to be a unified and coherent whole, had fragmented into distinct and often competing claims and truths.

These anxieties about the coherence, unity, and harmony of knowledge were ultimately about epistemic authority. There was a persistent fear among late eighteenth-century German scholars that Enlightenment knowledge technologies, especially those related to print, produced forms of knowledge that were arbitrary and lacked any underlying order. The desire for a new concept of science was precipitated by an incipient skepticism concerning previous ways of ordering knowledge and the desire to control and master it. The proliferation of print and its indiscriminant transmission of knowledge exceeded the capacities of previous ways of organizing knowledge. Despite these criticisms, early modern and most early eighteenth-century projects of learning were not, of course, simply arbitrary. However degenerate and incomplete human knowledge was, it was ultimately thought to be guaranteed by some external order, be it divine or natural. But as cosmological confidences—be it in the form of a Leibnizian rationality or a Linnaean taxonomy—eroded, figures like Kant began to fear that knowledge risked becoming arbitrary, lacking a rational and comprehensible order. If reason did not directly correspond to a cosmological or natural order, then it needed principles of organization internal to itself, principles that were immanent and not merely signs of an externally guaranteed order. On this account, something was arbitrary—be it nature or an encyclopedia—if it could not be accounted for on its own terms.[77]

The second ethical threat that information overload was thought to pose was to the integrity of the person. The fragmentation of knowledge about the world was the mirror image of the ethical fragmentation of the person who struggled to know the world. There was a clear ethical value to a coherent account of the world, in the harmony of knowledge, and the perceived fragmentation of knowledge threatened it. With so much to know, people—and especially those who read and exposed themselves to the surfeit of print—were becoming distracted and confused. And this lack of intellectual coherence inevitably led to an ethical confusion about what one ought to do and who one ought to be. Epistemological confusion produced ethical confusion.

## Science as Culture

In light of these anxieties, scholars and writers longed for a different source of epistemic authority, and many of them turned to the emerging concept of science as a distinct and unifying culture with its own practices, virtues, and internal goods. This promise of unity and coherence was what prompted Friedrich Zimietzki, after hearing Fichte's invocation of science at the University of Berlin in 1811, to exhort his "comrades" to embrace science not as an individual capacity or aggregation of facts but as a practice. Echoing his mentor and teacher, he called his "brothers" in science to recognize that the practices of scholarship could not be separated from the ethics of scholarship; epistemological questions could not be detached from the type of person who was formed and the type of knowledge produced. What was needed, he suggested, was a new ethics of knowledge and an institution that could sustain it and form ethically coherent persons oriented toward a unifying end and characterized by a distinct set of virtues. It was in this context that the idea of a modern research university emerged as a possible solution to the perceived fragmentation of knowledge in a new media age. A "scientific community" that embodied an authoritative knowledge, the university came to stand in for a new ethics of knowledge, a new way of organizing and cultivating the desire to know.

This late eighteenth-century conception of science as a distinct practice and culture differed in important respects from ancient notions of *episteme*, medieval notions of *scientia*, and Enlightenment notions of science. Whereas "science" had traditionally connoted an individual capacity or simply a category of knowledge, by 1800 figures such as Fichte and Kant used "science" in the singular [*Wissenschaft*] to refer to a coherent cultural system with its own logic and norms common to all sciences. In the German-speaking tradition, scholars had almost always used "science" to refer to just one of many "sciences." There was no one science, only various sciences [*Wissenschaften*].

In the following chapters, I consider the material, conceptual, and cultural conditions from which this new notion of science emerged, as well as some of the consequences of its emergence. As science developed into a distinct cultural and social system, scholars and writers began, as Markus puts it, to associate it with increased demands for entry and accessibility. These demands compensated, in part, for the theoretical openness of access afforded by print by protecting those on the inside of the system, but they also lent science greater coherence and authority. For figures such as Bacon or Diderot and d'Alembert, different types of knowledge were ultimately accessible to everyone because they considered the unity and authority of knowledge to be grounded in the unity of the mental faculties, that is, in the mind of an individual knower. Thus, every person shared the basis for a unified knowledge and could "participate in the range" of knowledge.[78] If, however, the unity and authority of knowledge were grounded in a system of science irreducible to individual minds, general access could not be presumed. In fact, entire habits and cultures had to be articulated to facilitate such access. The emergence of science as a distinct, self-regulating sphere entailed, then, a divide between those who had access to these objectified forms of knowledge and those who did not. The habits, cultures, and practices of these distinct realms and forms of knowledge became the key elements of *Wissenschaft* as a distinct culture. This development was crucial for the emergence of science as an institution-dependent form of knowledge, because science needed an institution to sustain its practices and distinctions.

As science became more self-regulating and distinct, it also became more internally focused. It did not point back to some divine or natural order or even to nature itself but rather to its own operations and history. It became a practice defined by its own internal goods that could be endlessly pursued. Science so conceived was a good in itself, or, in Wilhelm von Humboldt's famous phrase, it was a "never fully solved problem," an endless process of inquiry, what Humboldt called research [*Forschung*].[79] Such a pursuit was also endlessly motivating. It offered a distinct account of intellectual motivation. As a dynamic and interminable activity, science developed and expanded in new and surprising ways. The prospect of discovery provided a distinct ethical motivation for the scholar as researcher, whose worth and value were a function not of his individual apprehension of some metaphysical, God-like view of the relationship of all knowledge, but rather of his participation in the endless development of knowledge over time, a development embodied and transmitted in a community of fellow scientists. Communication with and participation in this historical community was the source of meaning, motivation, and identity for the individual scholar.

As already intimated by Kant's account, science became science only when it came to entail objective forms as well as subjective technologies that formed particular types of people, namely, when it became an ethics. For German intellectuals around 1800, the question soon became how to institutionalize science so as to guarantee its sustainability and continuity not only as a realm of material objects but as a coherent set of practices, habits, virtues, and values that defined the "scientific community." The research university was imagined as a technology that could reconnect an objectified world of knowledge, which so many feared had begun to exceed rational control, with the subjective development of individual persons. If classical conceptions of science sought to harmonize the individual with the cosmos, the research university sought to harmonize the individual with science as a distinct social sphere, a community dedicated to science. It sought to form scientists.

CHAPTER TWO

# The Fractured Empire of Erudition

IN THE PREVIOUS CHAPTER I sketched a partial semantic history of a key Enlightenment term, "science." Until the last third of the eighteenth century, the German words for "knowledge" [*Wissen*] and "science" [*Wissenschaft*] were basically synonymous. Echoing their roots in ancient and medieval concepts such as *episteme* and *scientia*, they both denoted an individual capacity to know. Only in the last two decades of the eighteenth century did German intellectuals and writers begin to use "science" as a more general concept to refer to a shared type of knowledge that entailed its own customs, habits, and practices supported by the university.

The story of how science in this institutional sense came to be so closely aligned with the university, however, cannot be told simply as the history of an idea leading from Aristotle to Kant. The appeals of figures such as Zimietzki to science as an institutional practice or distinct culture also arose out of more local, historical concerns in Germany around 1800 about the authority of knowledge in a media environment saturated with print. Zimietzki addressed his exhortations to join the "scientific community" to a group of students who lived in a time in which knowledge and science were nearly synonymous with print. Over the course of the eighteenth century, most German intellectuals had come to think of print as constituting a world of its own. To know meant to know how to interact with an ever-expanding realm of print, what writers and scholars referred to as a "world of books" in which the "language of books" dominated and a "knowledge of books" was vital. Whereas Zimietzki held up the university and its "scientific community" as the consummate order of knowledge, most eighteenth-century writers considered print and its various forms—from encyclopedias and lexica to periodicals and newspapers—to stand in for a certain kind of desire for the control and mastery of knowledge. The technology of print made knowledge manageable, accessible, and available.

The frontispiece to Johann Burkhard Mencke and Christian Jöcher's

*Compendious Lexicon of the Learned*, first published in 1715 and then republished and expanded through several editions under the sole editorship of Jöcher, offered a striking visual image of this eighteenth-century organization of knowledge. Extending from the bottom of the page to the top right side is a bookcase filled with books of all shapes and sizes. Its immensity is highlighted by a marked contrast with two much smaller men standing opposite it on the left-hand side of the page. One of the two men, cast in bright light, stands with an outstretched arm gesturing toward the towering shelf of books, whose spines are visible in a single glance. A vaulted ceiling extends across the top of the page and imbues the entire image with a sense of light-filled openness in which everything is illuminated and knowable. This image is a visualization of what the lexicon of the learned promised: an accessible, complete, and clear view of the entire scholarly world. It is an image of what eighteenth-century scholars referred to as the "empire of erudition"—a unified, homogenous realm of knowledge, theoretically accessible to all fellow scholars in print.

And yet, the sheer immensity of the bookshelf, how it dwarfs the two men opposite it, is ominous. With the unlimited vertical horizon, the bookshelf could expand infinitely. With every newly published book, it could continue to grow upward, but the two men would be left to gesture in vain toward a proliferating mass of print. Furthermore, the shelf may well be visible, but there is no ladder, no way to actually reach the books. Would its shadow not ultimately obscure those two men gazing skyward? And what was to prevent the bookshelf, teetering under the weight of ever-more print, from tipping over and crushing them?

This image tells one story of an eighteenth-century ecology of knowledge. Before the emergence of the research university and the ascendance of its disciplinary order, the one that Zimietzki summoned his fellow students to enter into, another environment was dominant: the empire of erudition and its bibliographic order. The story of the empire of erudition and the world of print that it organized, the way it sought to present an accessible and immediately visible display of knowledge to its erudite citizens, was the necessary prelude to the story of the research university. The long history of the concept of science cannot fully explain how or why the modern research university emerged as the last Enlightenment technology or why it was heralded by its romantic and idealist champions as the solution to information overload, nor does it explain how or why science underwent such a radical shift in meaning. This is because the rise of the research university was more particularly a story about the fracturing of the empire of erudition and the overgrowth of its bibliographic order. It is a story about the interactions of scholars, publishers, editors, and readers with print technologies and the normative assumptions about print and what counted as real knowledge which guided them.

At the beginning of the eighteenth century, the erudite scholar [*der Gelehrte*] and not the scientist [*Wissenschaftler*] embodied true knowledge and learning.[1] The erudite reigned over an empire of erudition in which "erudition" referred not only to a category of knowledge but also to a class of people. Over the course of the eighteenth century, however, the erudite was gradually replaced, on the one hand, by the botanist, philosopher, and theologian, the specialized scientist with his specialized journals and books. On the other hand, he was replaced by the generally educated person [*der gebildete Mensch*], the morally serious citizen who participated in public debate about social goods.[2] The empire of erudition faced an internal crisis on two fronts. While some scholars began to fragment into distinct circles of specialized discourse, others began to criticize the empire's pedantry and traditional notions of knowledge in the name of Enlightenment. Mencke and Jöcher's towering bookshelf of erudition was replaced by many smaller bookshelves, each representing its own specialized world that could be more efficiently managed as distinct scientific disciplines.

*Der Gelehrte* or "erudite" was originally the German translation of the Latin *literatus*, Renaissance humanism's term for a person devoted to the study of the humanities [*studia humanitatis*]. Erudites of the sixteenth, seventeenth, and eighteenth centuries constituted a distinct social class or a traditional estate, the learned estate or the *Gelehrtenstand*, which included university professors, as well as university-trained professionals such as lawyers, clergy, doctors, and secondary school teachers. Membership was based not on birth but primarily on knowledge of Latin and, usually, a university education, and it entailed a right to engage the work of fellow erudities as social equals.[3] The social distinction and honor accorded erudites were based not on personal professional achievement but rather on the broader social recognition of their "exclusive possession of the common intellectual culture upon which all professional learning was based."[4] This "common" intellectual heritage was the culture of a late European humanism that valued oral skill, mastery of Latin, knowledge of the Latin canon and scripture, and, above all, a commitment to that diffused, loosely organized set of contents among individual citizens of what since the fifteenth century had been referred to as the *Respublica literaria*, the Republic of Letters.[5] These skills, virtues, and types of knowledge were initially cultivated and transmitted in Latin secondary schools and then in university arts and philosophy faculties, both of which introduced students to scholarly social networks and the objective material forms of learning which were the basis of this common cultural heritage.[6] Erudition stood in for what counted as real, authoritative knowledge: a thorough, detailed familiarity with an established core of humanist learning and a rejection of intellectual specialization.

Over the course of the eighteenth century, most German erudites, scholars, and thinkers had some affiliation with a university or secondary school. These institutional ties were especially important in Germany, which lacked a metropolitan center like London and Paris which could more easily sustain a high concentration of unaffiliated writers and intellectuals. "Most of our great erudites," observed the German biblical scholar J. D. Michaelis, "actually live in universities, and the rest either actually taught as professors at universities, or they had the privilege in their youth to devote themselves to academic life."[7] In addition to his editorial work, Christian Jöcher, for example, was a professor of history at the University of Leipzig. The university with which most eighteenth-century erudites were associated, however, was a very different institution than the research university that would emerge in the early nineteenth century. With the exception of more modern universities like Göttingen and to a lesser extent Halle, eighteenth-century German universities were extensions of the class of erudites, which was committed primarily to maintaining its own juridical status and the legitimacy of erudition as established knowledge. The university itself, however, was not the source of epistemic and social authority; rather, it was the erudites' possession of a unique cultural heritage of which they were the guardians.

In the first half of the eighteenth century, the unity and authority of the empire of erudition and its cultural legacy became inextricably tied to print technologies—lexica, encyclopedias, books, and periodicals. Print, and not universities or the epistolary network of the Republic of Letters, stood in for the unity and authority of knowledge. As one scholar put it, the "fashionable aids to erudition" multiplied so dramatically over the first half of the eighteenth century that knowledge in general had become inconceivable without them.[8] By "recording" knowledge, they had also generated unprecedented "growth" in the sciences, so much in fact, worried the scholar, that they might one day stand in for knowledge itself.[9] As sciences "grew" and the aids to erudition continued to multiply, scholars began to worry that their print technologies, which they had assumed would unify them, would ultimately overwhelm them—too many journals, too many lexica, too many books. The bookshelf of Mencke and Jöcher's frontispiece, they feared, would tip over and subsume scholars in a mass of print and undermine the authority of erudition. Underlying these ominous intimations was a fear that print technologies might one day constitute a realm of their own, independent of the human agents who had created them.

## Erudition and Its Technologies

Early in the eighteenth century, Johann Georg Walch, a professor of theology and rhetoric at the University of Jena, defined erudition as the unity of sciences and scholars.[10] This unity was maintained, he contended, not through an epistolary and social network of scholars, as in the Republic of Letters or in the institution of the university, but rather through the individual scholar's capacity for managing the "means" of erudition, which, for eighteenth-century erudites like Walch, increasingly meant printed matter and the technologies devised to interact with it. The network of scholars depended on the network of material forms of communication. Throughout the eighteenth century "empire of erudition" referred to this network of humans and their print technologies.

Mencke and Jöcher's *Compendious Lexicon of the Learned* exemplified an early Enlightenment genre of print which cultivated the empire's unity by collecting and summarizing the lives and works of its members in a single printed volume.[11] Lexica like Mencke and Jöcher's did not focus on particular topics or sciences; instead, they sought to construct a timeless community of scholars bound together by a commitment to maintaining the unity of knowledge as erudition. These texts consisted of a few biographical details about a particular scholar followed by a list of his (and occasionally her) published works. The entry on Leibniz in Jöcher's lexicon of 1733, for example, read as follows: "a famous polymath, who excelled in almost all sciences, especially mathematics. . . . He wrote: . . . ."[12] Jöcher curated a printed collection of erudites, whom he detached from any particular historical context or tradition.

The title page of Jöcher's lexicon from 1726 promised an expansive but unified account of knowledge:

> Compendios Lexicon of the Learned / In which the learned of all classes as princes and statesmen, who are well-versed in literature, theologians, ministers, lawyers, politicians, doctors, / philologists, philosophers, historians, linguists, mathematicians, scholastics, / orators and poets, male as well as female, who / have lived primarily in Europe from the beginning of the world until today, and have made themselves known to the learned world through texts or otherwise, numbering more than 20,000 . . . according to their birth, death, most noble texts, lives and odd stories from the most trustworthy scribes, which are diligently observed, / briefly and clearly are described in alphabetical order.[13]

The world of erudition may have been peopled by a disparate and historically diverse array of learned men and women, but it stood in for a unity of knowledge which printed texts made visible and navigable. But Jöcher made even broader claims. Knowledge, he wrote, existed only to the extent that scholars had externalized it and thus "made themselves known to the learned world through texts." For scholars like Jöcher, true knowledge was embodied in a material form that could be added to over time. Texts like Jöcher's consolidated the loosely organized information and social knowledge of the Republic of Letters into a printed form that would be circulated more broadly. These texts were a virtual ordering, albeit in print, of a community, the Republic, that was already virtual—a community that was distant, in constant circulation, and existed primarily in the minds of its members. As a printed document theoretically visible to all, Jöcher's lexicon intensified a sense of belonging among erudites by reflecting on their community and making them aware of each other and their work. These types of texts, however, also flattened the history of knowledge out into a single, seemingly timeless world of print. According to this logic, Plato and Leibniz were contemporaries brought together by the printed page. Leafing through the *Lexicon of the Learned*, a reader could encounter ancient and modern thinkers as members of the same empire of erudition participating in a timeless conversation.

Vernacular texts like Jöcher's attempted to render the world of the erudite into a more homogenous and surveyable printed form—a lexical order of persons and their work. They generally eschewed the more conceptually complex memory systems of early modern encyclopedic works, like Alsted's Lullian-inspired encyclopedia, and embraced what they saw as the potential of print as itself a technology of memory. A more complete account of learning in print would alleviate the burden to remember.

Just as central to the cultivation of this unified empire of erudition were the scholarly periodicals that first appeared in the late seventeenth century and flourished between 1680 and 1800.[14] The *Journal des Şcavans* (1665) was the prototype for the first scholarly journal in England, *Philosophical Transactions* (1665), and in Germany, *Acta Philosophica* (1665–1670). These early scholarly journals were unique not because they were the first texts to contain critical discussions of other books, but because they established a new kind of stand-alone printed text devoted to such discussions.[15] This allowed for the development of a particular genre of writing and communication about the print of the empire. And unlike Jöcher's lexicon of persons, these early periodicals printed news about contemporary debates, ideas, and events.

These scholarly journals had a twofold purpose. First, they aimed to consolidate and make available, as one editor put it, the necessary "materials" to construct the "edifice of the history of erudition."[16] Just as Jöcher had done, the editors of these periodicals assumed that by collecting and organizing raw material for other scholars, they contributed to the construction of a unified structure of knowledge. Second, these new journals filtered out properly erudite materials from the increasingly diverse and plentiful forms of print. They helped scholars decide what to read, as one editor put it, "among the great mass that appear[s] almost daily."[17] Journals included detailed bibliographic descriptions that located a book, an article, or a periodical according to a combination of data: author, date, place of publication, format, and length. They communicated to their readers that a synoptic view of print, and thus erudition, was still possible.

The frequency with which these journals, or even lexica like Jöcher's, were published also offered a more efficient transmission of information than had the Republic of Letters' epistolary network.[18] They distributed information about newly published books, announcements about deceased scholars, summaries of discoveries in the natural sciences, and reports on scholars' publication plans. They collated and distributed in print news about the empire of erudition. Their very titles—which in German tended to include *Nachrichten*, *Anzeigen*, *Auszüge*, *Berichte*, *Zeitungen*, or *Journal* in some combination with *gelehrte* or *Gelehrsamkeit*—were indicative of these functions. They attempted to sustain the unity of the empire amid the perceived increase in print, which, as one editor complained, had become a "kind of epidemic disease."[19] Unlike lexica like Jöcher's, then, these periodicals tied print directly to a contemporary world in which new knowledge was happening every day.

By directly addressing and commenting on the empire of erudition, journals also cultivated a consciousness among erudites that they were participating in a common epistemic project. Every volume of the *Göttingen Journal of Learned Things*, for example, reported on scholars' work that it organized according to a scholar's city of residence (Leiden or Paris, for example). Its editors also published an annual supplemental index that gave an overview of the "news [of the] empire of erudition" in a single volume.[20]

In addition to his lexicon, Jöcher also edited from 1719 to 1739 the journal *Deutsche Acta Eruditorum*, whose subtitle—"or the history of the learned, who apprehend the current state of literature in Europe"—attested to the purported univocity of erudition. Its primary purpose was to distribute information about "literature," which throughout the eighteenth century referred to everything that had been printed and not, as it would much later in the century, to a specific type of writing. *Deutsche Acta Eruditorum* was an amalgamation of previously distinct textual genres: book catalogues, trade fair catalogues, correspondences among scholars, and review periodicals. Together, Jöcher's journal and lexicon assembled the elements of what he termed "book history," an account of all knowledge available in print.[21]

In both his periodical and lexicon, Jöcher insisted that his ultimate goal was to provide a "complete" account of the world of learning.[22] This aspiration for "completeness" was one of the distinguishing features of the empire of erudition and defined the empire's conception of the unity of knowledge. "Completeness" for most eighteenth-century scholars meant material completeness. According to Jöcher, his lexicon or journal would have been "complete" when it included in print all available knowledge on a given topic. In practice, however, such claims to completeness were inevitably reduced to the bibliographic details of printed material: author names, publication dates and locations, and titles. A "complete" scholarly work would have afforded a bibliographic account of the empire of erudition. It would have given an inventory, as Jöcher put it, of the empire's "books." Such a bibliographic order stood in for the desire to control and organize knowledge.

But Jöcher repeatedly acknowledged his failure in practice to achieve even this bibliographic completeness. In a later edition of his lexicon published in

1750, he admitted that none of the previous editions had been "complete."[23] Rather than abandoning his pursuit, however, he promised that each new edition would be "more complete" and contain more and more extensive entries.[24] In a final capitulation to the futility of his pursuit, however, he changed the title of his journal from *Deutsche Acta Eruditorum* to *Reliable News on the Current State, Transformation, and Growth of the Sciences.* Because the journal's volumes had become "too numerous and confused," he admitted, he could no longer report on a homogenous, unified empire of erudition. The new title acknowledged that the sciences, and thus knowledge, were not static. As print expanded, so too did knowledge. They were constantly developing and expanding. If scholarly periodicals were going to remain relevant, they could not simply display established erudition and aim only for completeness. They would have to chart the development of the sciences and acknowledge that their attempts at completeness were provisional.

Jöcher's experience was typical of early eighteenth-century attempts to organize knowledge in print. Theoretical claims to material completeness were continually thwarted by the failure of actual projects to achieve anything like comprehensive accounts of everything in print. Lexica were endlessly updated, encyclopedias were never finished, and periodicals proliferated. Gradually, however, scholars began to question not only the feasibility but also the ethics of a conception of knowledge premised on accumulation and material completeness. They wondered what kinds of people such technologies helped form.

## *Historia literaria* and the Sciences of Print

The unity of the empire of erudition was cultivated not only by lexica and scholarly journals but also by extensive forms of meta-commentary that organized a burgeoning range of printed material as scholarly knowledge. Over the course of the eighteenth century, these technologies came to be referred to collectively as *historia literaria.* An early form of literary history, *historia literaria*'s purpose, as one eighteenth-century scholar explained, was to "tell the history of erudition well and to link it together."[25] For its practioners, *historia literaria* recounted and organized the history of everything that had been printed and provided the raw material for scholars to advance knowledge. Although techniques for dealing with printed texts had a long history, dating back to at least the sixteenth century, *historia literaria* and its related bibliographic sciences organized, produced, and transmitted knowledge about books and other forms of print, as well as the methods and practices scholars used to interact with them.[26]

In the last third of the eighteenth century, these bibliographic sciences were increasingly dismissed by critics as false sciences of arbitrary aggregation beholden to an aimless curiosity. But they were actually crucial predecessors to science and disciplinarity. They helped assemble the material infrastructure necessary for science to emerge as a distinct cultural sphere.[27] *Historia literaria* organized the material stuff—journals, articles, books—into coherent and manageable scholarly traditions. It allowed scholars to differentiate their sciences from other types of knowledge. It helped construct the authoritative traditions from which the disciplinary self of the research university would eventually draw his standards and disciplinary models of excellence.

*Historia literaria*'s origins can be traced back to Francis Bacon, who hoped it would redress the lack of a "complete and universal history of learning."[28] It would be capacious in its reach and, initially at least, withhold judgment, because its primary purpose was not to produce new knowledge but

> to collect out of the records of all time what particular kinds of arts and learning have flourished in what ages and regions of the world; their antiquities, their progress, their migrations (for sciences migrate like nations) over the different parts of the globe; and again their decays, their disappearances and revivals. The occasion and origin of the invention of each art, the manner and system of transmission, and the place and order of study and practice . . . [the] history of sects, and the principal controversies in which learned men have been engaged . . . an account of the principal authors, books, schools, successions, and academies, societies, colleges, orders—in a word everything which relates to the state of learning.

As Bacon envisioned it, however, *historia literaria* was not simply to collect such facts. It was also to explain how these facts related to one another. Understanding how one discovery or book led to another would help scholars to advance the sciences. Knowledge about the causal relationships of past discoveries, publications, and insights was the necessary, if insufficient, condition for the advancement of knowledge.[29] In this sense, *historia literaria* was a "bridge" concept between Bacon's inductive, experiment-based science and humanism's textual traditions.[30] Although Bacon acknowledged that given the continued growth of print it might well be an "endless labor," he argued that extensive, consultative forms of reading, and not necessarily deep, intensive reading, were a necessary propaedeutic for the future growth of sciences.[31]

The most influential *historia literaria* text in the German tradition, although first published in Latin, was Daniel Georg Morhof's *Polyhistor*, published as four separate books between 1688 and 1704.[32] Surveying the humanist scholarly tradition, Morhof's text was a consummation of the scholarly reference work of figures such as Julius Scaliger and Justus Lipsius of the sixteenth century and Grotius, Alsted, and Kircher of the seventeenth century. As a distinct genre of text and a concept of knowledge, *historia literaria* can be traced in a direct line from these early modern figures through Morhof and then to a series of eighteenth-century German scholars, who not only cited Morhof's *Polyhistor* as an influence but also embedded it in their own work in a nesting effect. They would include portions of Morhof's text in their own and simply add to them with every new print edition. They built directly off of one another's texts as they accumulated an edifice of erudition over time.

Like Jöcher's lexicon or the earliest scholarly journals, Morhof's book was primarily a bibliographic account of knowledge. It was a book of books. He avoided more theoretical methods based on complex theories of memory or philosophical systems that filtered out too many books or other texts. Morhof's inclusive methods were based in part on a rejection of what he saw as the excessive efforts of early modern scholars, who also organized "all the sciences according to a secret and singular method."[33] Such methods were futile because "only the knowledge of libraries shows the way to the perfect knowledge of all things." Knowledge of the world of things was a product of knowledge of the world of books. In place of the early modern "speculative" methods, Morhof and later practitioners of *historia literaria* encouraged the more minute labor of working with particular print objects. They practiced a bibliographic empiricism intent on assembling the building blocks for an edifice of erudition.

Morhof's *Polyhistor* was also a guide to the emergent empire of erudition, which did not yet pose a threat to the world of nature or people themselves, as it would in the last third of the century when worries about book plagues and floods abounded. Like the natural world, it only required knowledgeable and able guides. If the world of nature would soon have its Linnaeus, the realm of print already had its Morhof.

Early in the eighteenth century, an article in Zedler's *Universal Lexicon* explained *historia literaria*'s basic logic. Any attempt to familiarize oneself with a science without knowing its history, it explained, would be like trying to get to know a person without knowing anything about his "curriculum vitae."[34] But *historia literaria* considered the history of a science to consist only of published work. Both a science and a human life were defined by a series of discrete events. In the case of *historia literaria*, the historical events were the publication of books and articles by scholars. Alluding to Bacon's description of the eye of Polyphemus, the article compared a science without a history to "the body of the enormous Polyphemus," the Cyclops who when blinded by Odysseus became bumbling and incoherent. *Historia literaria* was the eye of the sciences. It guided erudites by giving them a "genealogy" and a "coherent unity"; it constituted the empire of erudition and the history of knowledge by narrating and displaying it in print.

The German theologian and scholar Jakob Friedrich Reimmann embraced Bacon's vision and Morhof's model in the first German-language contribution to the genre, *An Attempt at an Introduction to Historia Literaria*, published in 1708. He explained that because he only wanted to prepare the way for future scholarship, he had refrained from offering "a heap of judgments about books and authors."[35] Comparing *historia literaria* to the cataloguing of plants, he contended that both provided a comprehensive, synoptic "view" of their particular objects of study.[36] Out of the "topsy-turvy" world of print, *historia literaria* created a bibliographic order by cataloguing printed items according to their external characteristics, as physical artifacts: "where the book was printed, with what type, on what kind of paper, in what format it is, how much it costs, whether it had been reprinted." These bibliographic details, Reimmann advised, should be collected and printed alongside the pertinent information about the author: "where he lives, his condition, and how old he is."[37] Just as naturalists like Linnaeus would decades later organize the world of plants, so too must scholars organize the world of books. *Historia literaria* was the bibliographic analogue of natural history. Different kinds of knowledge could be organized according to distinct categories. And printed books made this form of organization that much easier. If the books were organized, so too was knowledge—that is, knowledge could be controlled and mastered.

Many practitioners of *historia literaria* echoed Reimmann's contention that scholars should refrain from judging other books lest any printed text be relegated to oblivion.[38] "Who am I," wrote one scholar, "to appoint myself judge of texts and impose a hierarchy among authors? This right is the privilege of the public."[39] In the first half of the eighteenth century, however, this "public" was still assumed to be the same as the homogenous, generally educated, and capable class of erudites, who could judiciously select books on their own. They only had to be informed about which books had been published.

Despite the focus on accumulating and ordering a world of print and erudition, *historia literaria*'s efforts at a complete accounting of print material were continually frustrated. When confronted by the actual printed material, the project of relating one coherent story of erudition proved unwieldy. Nikolaus Gundling, a professor of jurisprudence at University of Halle, opened *Complete History of Erudition* with a complaint about the "many thousands and millions of books in the world."[40] Anyone who hoped to report on them all was fooling himself: "His life would not be long enough." Given these conditions, Gundling admitted that his "complete" history would in fact remain "incomplete."

In order to manage the perceived excess of print, early eighteenth-century scholars sought to make *historia literaria* a clearer and more efficient method through ever more detailed categorical or historical typologies, which proliferated just like Linnaean species categories. Johann Adam Bernhard, rector of a Lutheran secondary school, proposed a print typology organized according to authors' dates of birth and social status (such as "children of whores"), temperament, or religion. He also included separate categories arranged according to books that had been "printed by famous printers," "promised but never published," or, more simply, the rare, the curious, and the best.[41] Bernhard also proposed a historical typology that organized the practices and methods of erudites—what they read, how they studied, how they accumulated and transmitted knowledge—into distinct epochs. Echoing Reimmann's dictum about the preparatory purpose of *historia literaria*, he argued that it should not judge which forms of scholarship were better or worse; instead, it should simply gather together and organize them all into an accessible order that facilitated easy use.

Bernhard's typology also crafted a narrative of the ineluctable advancement of knowledge over time. He dubbed the seventeenth century the "critical age" in which scholars busily "purif[ied]" manuscripts from the "decay" that had befallen them in the Middle Ages. This was also the epoch when scholars first yearned for a modern form of scholarship which would enable them to do more than accumulate past knowledge. The "literary age,"

Bernhard's term for the eighteenth century, was set to achieve just this, he claimed. *Historia literaria* was in the midst of developing a coherent and distinct realm of literature, a *literarium*—a unified account of print. With its historical sensibility, *historia literaria* "told" the story of this emergent realm and its related practices—the improvement of libraries, the bibliographical critique of books, the biographical descriptions of scholars—which were designed "to better filter" and unify print.[42] And yet, Bernhard did not consider *historia literaria* the apogee of knowledge. It was "an instrumental science," he cautioned, that "clears the path so that the pace of advancement in other sciences can be that much faster."[43] It was a propaedeutic science that excavated the work of previous scholars in order to create the historical consciousness that was crucial for advancing knowledge.

In one of the last published examples of *historia literaria*, *Outline of a Universal History of Erudition* (1752), Johann Andreas Fabricius, who held positions at various universities and secondary schools, advised readers to use his text as the "basis for a commonplace book."[44] They could "interleave it with paper and thus add additional and curious items that they have found here and there and over time without much cost or effort collect good historical treasure." *Historia literaria*, suggested Fabricius, was a set of print works and practices which functioned like a commonplace book. It was a means for storing and organizing the raw material, the information, that scholars could retrieve for their own work. It accumulated knowledge for future generations. Such technologies were especially needed in the eighteenth century, because previous scholars, claimed Fabricius, had done lackadaisical bibliographic work. When he looked for a book that an author cited, he often "found nothing."[45] The original citation had listed the wrong author or omitted the publication date. Without accurate data, the world of print and thus the empire of erudition could not be navigated. It was just an incoherent mass of particular printed texts. Scholars needed a "complete index" to orient themselves in an ever-expanding world of print. Advocates of *historia literaria* tried to remedy this confusion by creating an objective bibliographic interface. Scholars conceived of it as the means by which they could efficiently and precisely locate and engage printed texts. *Historia literaria* stood for all the particular objects, designs, and systems that facilitated interaction between humans and the multitude of printed objects. But some scholars were never able to overcome fully the worry that *historia literaria* could not move beyond this preparatory orientation. Could such sciences ever become more than mere historical knowledge? Would they ever be able to produce new knowledge?

*Historia literaria* and its related technologies sought to assemble printed material as raw, potential knowledge. They also encouraged, facilitated, and trained scholars in the forms of scholarly communication and interaction which would orient them in an empire of erudition increasingly thought to constitute its own distinct sphere. In so doing, *historia literaria* played a crucial role in the gradual emergence of science as a distinct cultural and scholarly practice. It gradually formed scholars who conceived of the sciences as coherent bodies of printed knowledge. It outlined the content, boundaries, and relationships of various sciences. Under such conditions, scholars could more easily consider knowledge as consisting of distinct entities—the sciences—that had to be understood according to their own unique principles. The accumulating and organizational focus of *historia literaria* provided a basis for the expansion of erudition into a more research-oriented paradigm of science as it emerged over the last decades of the eighteenth century.[46]

*Historia literaria*'s pursuit of "completeness" also made it highly eclectic, but this eclecticism was undergirded by a basic, more theoretical premise: antidogmatism. The transmission of information and ideas unfettered by filters of good or bad (or dogma)—the refusal to judge the quality of what they were collecting—was considered by most erudites to be necessary for the advancement of knowledge. Scholars were anxious that overly selective filtering or search mechanisms would hamper the circulation of knowledge and block the unforeseen ends to which it might be put. *Historia literaria*'s accumulating tendencies, then, were indicative not only of a lust for information but of a particular epistemic hope that as knowledge was made more freely available, it might prove ever more useful. Furthermore, the notion of a unified empire of erudition was deemed necessary if scholars were to be able to distinguish their work from previous work. New arguments could be seen as new only to the extent that they could be related to older, more established arguments. Every argument, every particular text, had meaning only relative to a broader whole. This would become a basic premise of the concept of research.

And yet, even as some scholars appealed to an empire of erudition, others voiced doubts about the ultimate value of the humanist scholarly practices that *historia literaria* adopted. They doubted that print-focused practices and techniques alone could produce the types of knowledge that they deemed legitimate. Such diverse thinkers as the jurist and civic philosopher Christian Thomasius and the rationalist philosopher Christian Wolff leveled similar criticisms against *historia literaria*. Erudites, they argued, had failed to make ancient learning accessible to the "larger society."[47] More interested in rhetorical style, eloquence, and eulogizing "moral qualities," erudites eschewed what Thomasius considered the more basic task of distinguishing the true and the false and the good from the bad and accounting for the grounds of knowledge.[48] The scholars of the empire of erudition had lost site of the proper ends of knowledge.

Thomasius worried that *historia literaria*'s pursuit of comprehensiveness was driven more by the desire for a theoretical perfection of knowledge in some elaborate material form than by the practical relevance of knowledge. Most scholarly journals, he complained, published "merely excerpts" but "no judgment[s]" about the quality of books.[49] And yet *historia literaria*'s general refusal to separate good from bad, as Thomasius occasionally acknowledged, was premised on a more basic commitment to eclecticism, which encouraged its proponents to embrace the empirical influx of print. At its best, *historia literaria* could generate an ever-expanding repository from which scholars could draw as they saw fit without having to worry about whether something had been left out because another scholar had deemed it not worthy of being included. *Historia literaria*'s eclecticism also represented a challenge to what Thomasius considered the "ignorance" of an arcane medieval university structure that divided up knowledge into distinct and unrelated sciences and "disciplines."[50] In contrast to universities, *historia literaria* assumed that all sciences were related insofar as they had been printed. Print was a unifying agent. Thomasius's critique of the university resurfaced in the debates preceding the founding of the University of Berlin in 1810, but it highlighted a more basic feature of the eighteenth century—the university was not the central technology for the production and transmission of knowledge. Print and its related artifacts were. And any critique of the organization of knowledge had to begin with a broader critique of the material forms and technologies of knowledge.

Even though his rationalist, system-oriented philosophy subordinated practical, civil, and ethical concerns to theoretical ones, Wolff similarly contended that *historia literaria* amounted to little more than compilation—the ordering of titles under an author's name without attending to the logical relationships of ideas. It lacked a systematic organization that demonstrated how the ideas of one book related to the ideas of another.[51] The history of erudition was a useful technology for locating "what has already been found," actual print objects, but it needed rational guidelines to advance knowledge.[52]

While acknowledging the propaedeutic aims of *historia literaria*, critics like Thomasius and Wolff highlighted its limitations. However eclectic and open to various forms of learning its advocates claimed to be, *historia literaria* tended almost reflexively to defer to the presumed value of past, established learning over present learning. In the battle of the books and the quarrel of the ancients and the moderns, it was clearly on the side of the ancients and all forms of established authority. A further criticism, already voiced at the beginning of the century, but one that became more important over the course of the eighteenth century, was that *historia literaria* reduced knowledge to the history of erudition and ultimately to facts about the materials and technology of knowledge.[53] *Historia literaria* failed, that is, because it was never self-conscious enough about its propaedeutic character; instead, it became its own end. In its tireless effort to maintain the unity of the empire of erudition by providing a comprehensive account of it, *historia literaria* assumed that knowledge was not a problem to be labored over and constantly improved, but rather an already-accomplished story that simply had to be told as a singular, unified, and continuous whole.

## The Enlightenment and the Promise of Print

As the aids to erudition multiplied, the sciences grew as well, and the story of erudition became increasingly difficult to recount. "The great mass of sciences" that had emerged in the first half of the eighteenth century, as one anonymous lexicon article put it, now posed a threat to the unity of the empire of erudition. "In all parts of erudition the truth appears in such abundance, and the mediation that is thus required is so exhausting, that human life is much too short for anyone to attain even moderate levels of knowledge in several parts of erudition."[54] And it was the proliferation of print that had made this "abundance" of truth so visible. Every corner of the "empire" now seemed to have its own literature, its own method, and even its own "truth." The "empire" was expanding so rapidly that it was becoming difficult to maintain a view of the whole. As a consequence, warned the article, any erudite who "divides" his time and energy among a range of sciences will never become skilled in his own "discipline." The traditional erudite's pursuit of a comprehensive knowledge and disdain for specialization precluded him from knowing anything well. In an age of rapid growth in the "aids" and technologies of knowledge, the erudite—the polymath scholar and denizen of the homogenous and unified empire of erudition—was doomed to become an anachronism. In order to manage the fragmenting effects of information abundance, the scholar would have to make way for a new figure: the specialized university scholar engaged in research.[55]

As fears about the imminent fracturing of the empire of erudition continued to mount, however, the learned class faced another internal challenge to its unity and authority: the high Enlightenment and the influence of what came to be known as popular philosophy.[56] Beginning around the middle of the eighteenth century, scholars began to exhort one another to forgo their traditional scholarly commitments to the empire of erudition and attend instead to a wider public and make knowledge, culture, and philosophy "fundamentally available to all."[57] In 1754, Johann August Ernesti, a professor of ancient literature at Leipzig, summarized the new attitude: "Let us hasten to make philosophy popular!"[58] Setting a new tone for the culture of learning, Ernesti called for a philosophy that eschewed traditional forms of scholarship and scholastic philosophy and made knowledge, truth, and morality concerns for everyone, not just the guild of erudites.[59] Scholars like Ernesti called on fellow erudites to adopt a different set of intellectual and cultural norms and assume a different identity, namely, that of the "reflective writer" dedicated to educating a more literate public to be useful citizens.[60]

Over time the authority of erudition began to give way to the norms and legitimacy of a "popular philosophy" that valued the general writer and thinker over the more narrowly construed erudite.[61] Scholars, insisted the popular philosopher Christian Garve, had no reason to write in esoteric terms. "If something is true, it should be easily communicable to a wide audience."[62] "Popular" and "scholarly" should not be mutually exclusive terms. A "scholar," wrote Johann Georg Walch, was not only a member of the empire of erudition but "more accurately" anyone who possessed "knowledge of truths" of interest to all people.[63] Knowledge was a common possession to be used for the common good.

Writers and scholars like Garve and Walch were attempting to expand the parameters of what counted as real, worthwhile knowledge and who should have access to and produce it. Even Fabricius, the author of the last canonical example of *historia literaria*, defined erudition not as a capacity to interact with scholarly printed material and a devotion to a particular canon of learning but rather as a tool with which one could battle "inexperience, stupidity, ignorance, falsity, prejudices, sophistry, malice, atheism, naturalism, sectarianism, pedantry, holy simple-mindedness, hypocrisy, idiocy, superstition and other similar human monstrosities and defects."[64] He used "erudition" interchangeably with a host of other terms—doctrine, literature, polymathy, science, discipline, wisdom, and pansophism—to describe a knowledge at work in the world.[65] Erudition properly understood was Enlightenment, and its purpose was not to accumulate a comprehensive collection of established learning, but rather to facilitate the "perfection, true satisfaction, felicity and blessedness of the human."[66] Following his contemporaries, Fabricius enjoined his fellow erudites to integrate themselves into broader society and participate in the Enlightenment drive to educate more people and contribute to the good of the state.

Such efforts by scholars to recast erudition as popular philosophy or Enlightenment were attempts to make the requisite abilities to navigate the world of learning available, in principle at least, to everyone, regardless of knowledge of Latin and the methods of *historia literaria*. These criticisms gradually led to a drop in the prestige of erudites, who were increasingly dismissed as guardians of a traditional, guild-based knowledge unfit for the demands of modern societies.[67] Despite their criticisms of certain forms of scholarly ways of managing print, these attempts to reimagine the intellectual and cultural values of knowledge also challenged the empire of erudition's basic assumptions about print. Print technologies were not merely buttresses against possible catastrophe, depositories of established knowledge. They were agents of Enlightenment. Books and print more broadly were celebrated not only as the embodiment of all knowledge but also as a central agent of social and political progress.

The Swiss popular philosopher Isaak Iselin described "the printing of books" as "the most formidable foe of darkness and superstition."[68] Like many of his contemporaries, Iselin asserted a direct correlation between a society's ability to read and write and its moral development: as literacy rates improved and access to print expanded, so too would progress toward the good. As "the art of reading and writing, which were once a secret reserved for the clergy, . . . expanded to all classes," moral progress would become universal.[69] The correlate of this assumed moral progress was an abiding confidence in print to organize and facilitate the continual advancement of knowledge and maintain its unity beyond the limits of the empire of erudition. The technological and economic effects of the book market combined with the political, social, and philosophical program of the popular Enlightenment to accelerate the creation and dissemination of printed objects and their imagined analogue—a homogenous reading public or public sphere [*Öffentlichkeit*].[70]

The expanded claims made on behalf of print, the normative arguments about its inherent capacity to enlighten, however, created a clash of cultural expectations about what print technologies could or should do, especially among scholars. The crisis of the empire of erudition and the learned class was, as Steven R. Turner puts it, an internal one of "self-image and justification."[71] It was a crisis about what constituted real knowledge and how it should be generated and transmitted. On the one hand, even as Enlightenment scholars insisted that knowledge was a common possession to be used for a common social good, they remained committed, as Kant wrote quoting Bacon, to the "renewal" and "growth" of the sciences.[72] On the other hand, they had put themselves under increasing pressure not only to disseminate these advances but also to make sure that this new knowledge was socially relevant and practical. Many scholars gradually began to worry that the second imperative conflicted with the first. The production of new knowledge required greater familiarity with ever more specialized sciences. In order to count as new, it also had to distinguish itself more sharply from other sciences. This, in turn, made the task of sharing such knowledge with a broader reading public more difficult. The rapid expansion of knowledge and its fragmentation created a feedback loop that challenged the basic imperative that all new knowledge be broadly disseminated. Specialization facilitated the production of new knowledge, but it also widened the gap between a specialized knowledge and a universal knowledge that was assumed, theoretically at least, to be accessible to all scholars. In this sense, knowledge advanced in inverse proportion to its broader dissemination. Demands to make all knowledge accessible and comprehensible to nonspecialists threatened to impede the advance of knowledge. How could scholars push ahead with the creation of the new, if they were constantly compelled to distribute and make accessible work they had already completed? The injunction to explain and share would stall the advancement of knowledge. The interests of those who generated new knowledge increasingly conflicted with the interests of those who communicated it.

The popularizing and moralizing imperatives of the high Enlightenment represented an existential threat to the imagined unity and homogeneity of the empire of erudition, both because these new intellectual values undermined many of the traditional norms of the empire and because they acknowledged the worth of printed texts that did not neatly fit within the traditional categories of erudition. In the name of Enlightenment, scholars and intellectuals wrote and published in an increasingly diverse range of popular printed forms, from works of fiction and travelogues to newspapers and journals.[73] Writers and publishers juxtaposed these new print forms to an empire of erudition that they denigrated as irrelevant and incapable of addressing society's needs.

Above all, however, the proliferation of print, regardless of its source or cause, resulted in an uncontrollable diffusion of knowledge which scholars on both sides of the Enlightenment debate worried "overwhelmed" human capacities to deal with it.[74] The proliferation of popular works of print, as well as traditional "aids to erudition," exceeded the abilities of readers and scholars of all types to manage and interact with them. Although Fabricius longed for a comprehensive knowledge that could be controlled and made useful, he begrudgingly conceded that, given the new material conditions of print, scholars would have to devote themselves to "one science."[75] The only feasible solution to the proliferation of print, already anxiously anticipated in Zedler's lexicon article, was for scholars to abandon the empire of erudition's disdain for specialization and embrace it. Such a shift also meant that the very concept of the scholar and what counted as authoritative knowledge would have to change. The erudite who operated in a homogenous, unified empire of erudition would have to become the specialized scholar.

## The Flood of Print and the Differentiation of a Scholarly Public

The transformation and fragmentation of the empire of erudition over the second half of the eighteenth century involved both an external and internal differentiation. Externally, scholars began to distinguish themselves and their particular forms of knowledge from the more popular forms encouraged and supported by the high Enlightenment and popular philosophy. Scholars gradually began to distinguish themselves from what came to be known as the "gebildeten Stände," the cultured or educated classes.[76] Internally, scholars began to distinguish more sharply among themselves. They specialized.

"Sixty years ago," declared an anonymous author in 1780, "those who bought books were erudites; these days there is hardly a woman who doesn't read."[77] This expansion in kind and number of the reading public, continued the article, had been made possible by the sheer growth in publications. Over the course of 160 years, the catalogue of the Leipzig book fair, long the biggest in Europe, had expanded from five to seventeen printed sheets. Even the catalogue's established categories had changed over the course of the eighteenth century to accommodate these changes. Between 1740 and 1800, for example, theology titles plunged from over one-third of all advertised books to less than one-seventh.[78] And, perhaps most importantly, a new category of book was added: "books for pleasure reading [like] novels, comedies, and books for women." At midcentury, the types of books that could be categorized as "learned literature"—literature expressly addressed to a generally undifferentiated learned class and their professions (law, theology, and medicine), that is, the empire of erudition—still made up almost two-thirds of all texts published, but by 1800 this figure had declined to less than one-quarter of all titles.[79] In 1740, the book fair catalogues did not have much to offer a "larger public" except theology texts and texts addressed to erudites.[80] The number of texts devoted exclusively to the figure of the erudite, like Jöcher's lexicon of the learned, or periodicals addressed to the empire of erudition, such as the learned journals of the early eighteenth century, had declined dramatically by 1770 and all but disappeared by 1800, as had the "universal" lexica and erudite compendia.[81] The proliferation of print was thus an increase not only in the sheer quantity of printed texts but in types of printed material.

These initial challenges to the homogeneity and singular authority of the empire of erudition, however, came not, primarily at least, in the form of new lexica or books but from the explosion of periodicals. Whereas more widespread complaints about a "plague of books" did not surface until the last three decades of the century, concerns about an uncontrollable surfeit of periodicals came much earlier. In the period between 1700 and 1720, the number of periodicals in Germany, and not just scholarly journals like *Acta Eruditorum*, increased from 58 to 119; between 1720 and 1740, from 119 to 176; between 1740 and 1760, from 176 to 331; between 1760 and 1780, from 331 to 718; and between 1780 and 1790, from 718 to 1225.[82] The majority of these new periodicals were not addressed to the empire of erudition, but rather to the unlearned "public."

Like the lexica and books discussed above, the growth of periodicals in number and kind challenged the continued invocation of a singular, homogenous empire of erudition by scholars and writers. The periodicals that expanded most rapidly in the first third of the century were the so-called moral weeklies and their related print forms often referred to by contemporary scholars as "trivial literature."[83] Addressed to a broad range of readers including erudites, business people, and women, they were key instruments in the distribution of particular Enlightenment values to the broader reading public. They exhorted readers to civic engagement and Christian virtue. In contrast to the more internal focus of the scholarly periodicals, which were intent on maintaining the unity of the empire of erudition, these periodicals focused on a wide range of topics, including politics, the family, moral education, and civic life, and a broad set of genres, such as satires, moralizing tales and fables, allegories, and published letters.[84]

Reflecting on the expansion of the German book market over the course of the eighteenth century, the fictional book dealer Hieronymous, one of the central characters in Friedrich Nicolai's late eighteenth-century novel *The Life and Opinions of Sebaldus Nothanker*, described how the empire of erudition responded: "The class of writers in Germany addresses almost exclusively other writers or members of the learned class. Only rarely is it the case with us that an erudite is a man of letters [*Hommes de Lettres*]. With us an erudite . . . writes only for his audience and his subordinates. This small group of teachers and students, which is around 20,000 strong, disdains the other 20 million people who also speak German, so much so, that they don't even take the effort to write for them."[85] Until the middle of the eighteenth century, the German book market primarily addressed only erudite readers, who had always assumed that the entire print market was only for them. Whereas a member of the empire of erudition wrote for a public whose boundaries were generally marked by literacy, over the course of the century there arose another public that still had a "need to read," even though they were not, as traditionally conceived, erudite.[86] And, as scholars have recently detailed, they began to organize their reading habits and desires in various social forms as well by establishing reading societies and lending libraries that were organized by private citizens for public use.[87]

The empire of erudition's first response to the proliferation of print and the pressure to write for a broader audience was to differentiate itself from an external reading public—the one envisioned by the proponents of a more popular Enlightenment and cultivated by periodicals and moral weeklies. Erudites began to make sharper, even more moral distinctions between erudite and non-erudite. Johann Bodmer dismissed the "corrupt taste" of popular periodicals such as *Der Patriot* or *Die vernünftigen Tadlerinnen*, which encouraged "a reading addiction" that portended the emergence of a new reading public that represented "a new and dangerous power in the state."[88] Although such critics acknowledged that the state needed literate citizens, they also contended that the intention to expand vastly the reading public needed to be better managed. Erudites, they worried, were losing control over books and reading and thus their authority over the production and transmission of knowledge. "The condition of literature in Germany," lamented one scholar, "is such that publishers print no longer for erudites but for their readers."[89] The class of erudites and the empire of erudition were no longer synonymous with books and print. The categories of scholar and reader were gradually becoming distinct.

Even as scholars sought to differentiate the empire of erudition externally from other institutions and orders of knowledge, they also began to differentiate internally among fellow scholars and among sciences heeding the reluctant advice of Fabricius to specialize. Scholars increasingly wrote not just for a homogenous group of learned erudites but for specialized groups of fellow scholars. Internally, the empire of erudition began to fissure into the distinct domains of knowledge that many scholars earlier in the century feared were inevitable. In 1783, one author declared that Germany had succumbed to a "journal addiction" and wondered how any scholar could possibly keep up with every journal. The journal's editors responded that over the past twenty-five years scholars had already solved the problem of proliferation by establishing specialized journals. "Every science can profitably have its own journal," advised the editors, if it maintains the boundaries of its particular science.[90] Such internal differentiation helped avoid the "repetitions and monotony" that they claimed characterized earlier scholarly journals, which, in their efforts to account for the empire of erudition as a complete and unified whole, ultimately printed only superficial details. Describing changes in the Leipzig book fair catalogue, another commentator marveled at the proliferation not just of books and periodicals but of the very "categories of erudition." There were now categories, he observed—natural history, economics, physics, pedagogy—that had not existed two hundred years ago.[91] These forms of specialization were a means for managing the overgrowth not just of particular print objects but of the categories devised to contain it.

The internal differentiation of the empire into more specialized subgroups can be traced over the second half of the eighteenth century, especially between 1740 and 1760, to the proliferation of specialized periodicals, which for many scholars signaled the final disintegration of the empire of erudition.[92] The fate of one scholarly journal, *Göttingische Anzeigen von Gelehr-*

*ten Sachen*, was exemplary in this regard. When the journal was established in 1739, its editors explained that it reported on the "news" from an undifferentiated "empire of erudition."[93] By 1753, however, its new editor, Göttingen University professor Johann Michaelis, made it clear that the empire of erudition had already begun to fracture and that he should not be held responsible for providing a comprehensive account of erudition: "Concerning the disciplines that are so totally different from the part of the sciences to which I am devoted, may I just say: it is obvious that I could have nothing whatsoever to do with articles dealing with law, medicine or mathematics."[94] But, as the previous success of the journal demonstrated, it was not clear that such sharp distinctions in the empire of erudition could have been made earlier. Michaelis's admission represented an increased awareness of the emergent distinctions among, as he put it, "disciplines." Under his editorial direction, the yearly "index," which had previously printed reports on scholarship from the metropolitan centers of the empire of erudition, was replaced with an index that reported only on local events and scholarship from Göttingen. The empire of erudition was dissolving and the order of the disciplines was beginning to emerge.

Although the changes in the Göttingen journal were subtle, the changes in the broader world of scholarly journals, as evidenced by the explosion of specialized journals devoted to particular sciences beginning around midcentury, were clearer.[95] The number of periodicals dedicated exclusively to natural science, for example, increased sharply.[96] In the preface to the first volume of one such journal, *Physikalische Belustigungen*, the editor defended his exclusive focus on the physical world and particular interest in, as he wrote, "microscopic discoveries," by which he meant seemingly narrow, specialized topics.[97] "If something occasionally should appear that is outside its [the journal's] sphere," he wrote, "it will still always bear some relationship to the study of nature."[98] In order to carry out such a task, he called on "researchers of nature" to contribute articles that demonstrated this focus. One article in the first volume, for example, discussed a debate about whether a cochineal, a scaly insect, belonged to the animal or plant kingdom. In so doing, it cited an array of contemporary articles and books. The article devoted as much space to discussing the scientific literature as it did to the actual insect. Through its citations, it connected a range of published books and articles on the subject so as to facilitate further study. This journal assimilated many of the features of *historia literaria*.

Specialized scholarly journals such as *Physikalische Belustigungen* shared a number of common features. First, they undid the distinction between the history and practice of a science. They treated sciences as though they could not be detached from their history of practices, methods, and concepts, all of which were stored and connected in a network of citations and bibliographic information. Second, they recirculated and thus constituted a differentiated form of scientific knowledge. Journals reprinted translations, recommended texts, published bibliographies, and identified these print objects as science. Every article situated itself within an intellectual tradition that it helped construct and maintain. Third, authors and editors presented every article and the journals themselves as a supplement to science as a whole. Journals were oriented to the production of future work. Finally, none of these journals appealed to the authority of or invoked an empire of erudition. They identified with particular, differentiated sciences. In short, the objective, material conditions for a concept of a science as a modern discipline were gradually coming into place, as the homogeneity and authority of erudition were dissolving.

## Putting It Together Again, as Formal Unity

The print technologies of the early eighteenth century which reported on a unified empire of erudition were gradually superseded not only by the emergence of a more popular-oriented print market and the proliferation of specialized journals but also, in the last third of the century, by cultural anxieties about overload and fragmentation.[99] The erudite was being eclipsed by two distinct figures: the more generally educated reader as imagined by the Popular Enlightenment, and the specialized scientist—the botanist, philosopher, philologist, and theologian. By the last two decades of the century, many scholars and writers had begun to doubt the claim of Zedler's lexicon to account for a universal knowledge and make it available to all people. The purportedly homogenous world of the erudite had begun to give way to a complex of loosely related segments.

And yet various attempts to imagine an order of knowledge which could manage the desire to control and master knowledge continued. One such effort was the reinvention of the universal learned journal. Whereas the journals of the first half of the century reported on the people and events of a circumscribed empire of erudition, in the last third of the century a number of journals emerged as critical organs devoted not to erudition but to literature, that is, the whole range of printed forms. Another, related effort were highly formalized print projects that sought not just to report and display but to systematize all that had been printed. Although the empire of erudition had begun to fracture, the commitment to print and the belief in its capacity to unify knowledge on the part of scholars and writers remained just as strong.

In the last two decades of the eighteenth century, there was a resurgence of general scholarly journals devoted to addressing a broadly educated audience. One of the most well known of these was the *Berlinische Monatsschrift*, founded in Berlin in 1783 by members of the Wednesday Society, a secretive group of scholars and Prussian elites who met to discuss a wide range of intellectual and social issues. Echoing the purposes of early scholarly journals, the two editors Johann Biester and Friedrich Gedike sought to coordinate their journal with the larger realm of print. But they intended to gather "news from the entire empire of the sciences," not just the "empire of the erudition."[100] They self-consciously sought to keep up with the constant development of knowledge and not just report on what was already known. They collected not only information concerning scholars but also "descriptions of peoples and their customs" from foreign lands, news about the "most interesting people," excerpts and translations from foreign-language texts, and various other material that would contribute to a distinctly German language and literature. Whereas the scholarly journals of the late seventeenth and early eighteenth centuries had collected and transmitted knowledge of and about a more circumscribed empire of erudition, Biester and Gedike were more ecumenical in their interests. They sought any information that might contribute to the sciences. Despite the fact that the *Berlinische Monatsschrift* sought to address a wider audience, although one already distinct from the broader audiences of the moral weeklies, its two editors emphasized the difference between an "initiate" and a "non-initiate." They acknowledged that the empire of erudition had fragmented into specialized sciences and that this made the dissemination of scientific knowledge even more difficult. Journals like theirs attempted to translate specialized knowledge into a language that was accessible to a broader audience of generally educated readers. They tried to cultivate an audience that was neither simply specialized nor popular, but rather highly educated and interested in the various forms that knowledge had taken in the late German Enlightenment. They sought to create middle ground between simply popular and more specialized knowledge.

The second attempt to refigure a unity of knowledge was an ambitious effort to better organize the proliferation of print through highly formal, technological means. Whereas general Enlightenment journals pursued a unity of knowledge by cultivating a unified reading public, these other projects—complex organizational schemas, highly formal indexes—pursued a unity of knowledge by organizing print itself.

Surveying the changes that the print market had undergone since 1700, an article on "ephemera" in the *German Encyclopedia* from 1788 wrote that scholarly periodicals now appeared in such "enormous amounts, that if one wanted simply to list all the titles he could fill an entire book."[101] Perhaps one day, mused the article's author, someone would write an "ephemera of ephemera." A few years later, the editor Johann Georg Meusel called for something similar—a "universal index" that could account for the periodicals that had "grown to so many volumes."[102] Over the following decade, these periodicals of periodicals began to appear as part of a renewed effort to organize knowledge in one, comprehensive printed text. These print projects represented both the apogee of the bibliographic order as organized by *historia literaria* and its impending demise.

They had precedents in the journal *Learned Germany, or Lexicon of the German Authors Now Living*, which Meusel began to edit after the death of its founding editor, Georg Christoph Hamberger, in 1773. Like Jöcher, Hamberger had organized his volumes around the person of the scholar, but he excluded the curious details about their lives. He reduced his entries to only the barest of bibliographic information: name, date and place of birth, and position of the author, as well as a list of published works. Hamberger had initiated a formalization of the bibliographic order and a reduction of the empire of erudition to a set of information about print, what Meusel described as a "literary butterfly collection."[103] It was a highly static and structured display of print objects.

Even this formalization, however, lamented Meusel, could not solve the problem of information overload and provide a comprehensive account of print. "With the great abundance of our writers and their exceptional fecundity," he worried, ever more supplementary volumes and editions would be needed to track even just the basic bibliographic information.[104] And Meusel was right. The first volume of the fifth edition was published in 1796, and the final, twenty-third volume of that same edition was not published until 1834, fourteen years after Meusel's death. And upon publication, it was immediately out of date.

The interminable character of Meusel's lexicon did not deter a series of similar projects in the 1790s from attempting to reassemble the empire of erudition in such formal terms. In an article printed four years after Meusel's initial proposal for a "universal index," an anonymous author called for a repository that would provide information and a reference about all printed material, what he called a "book of books."[105] Considering the expanding bounds of "erudition, literature, culture and Enlightenment," he wrote, "it is impossible for even the most assiduous person of literature" to read everything that impacts his "job," much less hold it in his memory and organize it.[106] What was needed was a condensed index that reduced printed objects to their identifying elements—not their ideas or concepts but their bibliographic characteristics.

Some of the earliest scholarly journals and lexica of the empire of erudition had, of course, included bibliographic information along with terse summaries or announcements of particular books. But these new, highly formalized projects took the concern with a strictly bibliographic order to unprecedented levels. In *Universal Repository of Literature for the Years 1785–90*, for example, the German scholar Johann Samuel Ersch sought to realize Meusel's appeal for a "universal index," but one limited from the outset to a narrow time frame.[107] Ersch organized the first two volumes of the *Repository* according to what he termed "the systematic index" of printed texts, which, he boasted, provided "a convenient and illuminating arrangement of the classes, orders, and sub-orders" into which over 32,000 articles could be fit.[108] Ersch's primary objective was to organize book reviews of the late eighteenth century and thereby create an index of everything that had been printed in that period. Consisting of bibliographic information on both published books and their reviews, Ersch's text was based on individual books, not authors. The print object and not the person of the erudite was the basis of organization. The last volume was an alphabetical listing of articles according to title and cross-referenced with the first volume's systematic index. The content of the *Repository* was not particularly new; instead, its highly reduced form and exacting detail were what distinguished it from previous texts. Its level of formalization was new and striking. One sample entry, "A.L.Z. 85, I. 1," translates as "*Allgemeine Literatur-Zeitung* (journal title). 1785 (year). I (volume). 1 (page)." Even authors' names were abbreviated: A. = August; Abr. = Abraham; etc.

Although Ersch's index fit within the bibliographic tradition of *historia literaria*, it introduced a hyper-formalization that extended to the classes of the sciences. Ersch organized the first volume of the *Repository* according to sixteen categories, ranging from philology to theology. Following the title page was an encyclopedic table that divided each category into multiple Roman numeral divisions, which were then further divided into numbered subdivisions. The printed table goes on for twenty printed pages in the first volume and is followed by a four-page detailed explanation of the abbreviations and other guidance on how to decipher the entries.

Ersch also included a second-order category devoted to the study of the organization of printed material and the sciences. He referred to the category as *Wissenschaftskunde*, a term coined by the German encyclopedist Johann Eschenburg,[109] and it included reviews of books concerned with scholarly methods and the organization of the sciences. Such a category was necessary not only because of the proliferation of print but also because of the proliferation of the categories of the sciences. Several sciences, he wrote, had assumed a "different form in the recent past."[110] Some had expanded, others had been absorbed into other sciences, and some no longer existed. The "existing classification," according to which each science had its particular "position," could not account for the ways in which sciences developed over time. A fixed organizational order of the sciences was becoming impossible given how sciences were increasingly thought to develop over time. Sciences had a history. Under these conditions, any pursuit of completeness, concluded Ersch, would require a highly formal and integrated organization of print. In his own index, for example, he interwove volumes 1 and 2 (the systematic index) with volume 3 (the alphabetical index). Thus, in volume 3, under "Kant," one finds "—. Critique o. Pure Reason. 85. III. 41, 53, 117, 121, 125. SR. VI. 381." The "85" refers to the review of Kant's *Critique* published in the *Allgemeine Literatur Zeitung* in the publication year 1785, and the following numbers indicate the particular pages. The second abbreviation refers to the systematic index, sixth category (*Philosophie*), entry 381, where one finds listed the *Critique* and its reviews. The search of the systematic index is facilitated by the fact that the pages are numbered according to category and entries, so the page with entry 381 reads over the top "VI. 379–381." For scholars like Ersch, the growing consciousness of the impossibility of completeness led to formal projects that bypassed the problems of excess and the exigencies of categorization by reducing the empire of erudition to a bibliographic order.

Johann Heinrich Christoph Beutler and J. C. F. Gutsmuth published a similarly hyper-formalized solution to proliferation with their *Universal Subject Index* of all articles published in German between 1700 and 1798.

Like Ersch, they embraced what they considered to be the enlightening potential of print—its capacity to "purify and transmit scholarly knowledge broadly"—but they also acknowledged how difficult this had become.[111] "In the countless hoards of periodicals," they wrote, "the best fruit of scientific thought" lay hidden.[112] Their solution was to produce a "repository of human reason" which sought to make the products of the human mind accessible and ready to hand.[113] Organized according to subject or topical category, each entry gave a terse account of the scholarly knowledge "stored" in printed texts. But like Ersch's index, theirs too was hyper-formalized. The entry "Enlightenment," for example, read as follows:

> *Aufklärung—*
> *Gedanken und Fragen sie betreffend.*
> D. Mk. J. 79. B. 2. S. 94
> "*Was heißt aufklären*" v. Moses Mendelssohn. B.M. J. 84. Sept. S. 193
> und v. Kant. B.M. J. 84. Dez. S. 481.[114]

Designed to make visible the "advances" in the sciences,[115] Beutler and Gutsmuth's index excised the peculiar, the fascinating, and, ultimately, the human. They formalized the empire of erudition into bits of bibliographic information. The problem of excess and the hypertrophy of the sciences and print was technical, and so too was its solution. It was all a matter of print. The indexes of Ersch and Butler and Gutsmuth represented a science that was embodied not in a clear thinking individual—the person of Cartesian rationality—but in the ever-expanding interrelations of printed objects. Science was a bibliographic repository that humans need only manage.

These indexes represented the apogee of *historia literaria*, which, along with all the other technologies of print for which it stood, was the necessary if insufficient condition for the emergence of the modern research university. Before science could become an institutionally supported practice, it referred to an endlessly expanding material interface that allowed readers, publishers, and scholars to interact with print and each other. Printed texts pointed to, indexed, cited, and quoted other works in increasingly expansive feedback loops. All this pointing, citing, and accumulating, all this bibliographic work, helped constitute authoritative traditions of the sciences and, gradually, distinct communities of scholars working in particular fields or sciences. There was no biology, jurisprudence, theology, or philosophy without its literature. The sciences were inextricable from print. Wilhelm von Humboldt's formulation in 1809 of science as research—"as something that has never been nor will ever be fully discovered"—has its prehistory in the empire of erudition's conceptions of what constituted authoritative knowledge.[116]

The hyper-formalized indexes of Ersch and his contemporaries, however, represented not only the apogee of the empire of erudition and the bibliographic order but also their imminent collapse. For many late eighteenth-century critics, these "aids of erudition" embodied the empire's overweening confidence in print and reduced Enlightenment to information management. Humans had become detached from their print technologies. Out of the empire of erudition—the imagined community of learned persons who communicated through print—had emerged a bibliographic order that had autonomously developed into a second nature that obscured the very persons it was designed to connect. There were no human agents in these final projects, just desiccated bibliographic data. The empire's technologies of print had come to outstrip the ethical and normative capacities of humans to interact with them. The sharpest criticisms of the traditions of *historia literaria* were directed at projects like Ersch's which exemplified the empire of erudition's assumption that a well-organized world of print could produce and transmit broadly an authoritative knowledge. Toward the end of the century, however, these criticisms gradually began to yield to efforts to imagine different forms of epistemic authority, different ways of organizing the desire to know and master knowledge: a reimagined university and the institutionalization of science. The material forms of knowledge that characterized eighteenth-century scholarship and learning were ultimately an insufficient ground to support science as a practice, as a distinct culture. What was also required was an ethics of knowledge that could form people capable of making meaning out of this mass of material. What had in large part been a virtual arrangement of loosely related materials, practices, and persons was to be institutionalized in the research university, which, claimed its idealist and romantic champions, melded the objective, material forms of knowledge with subjective practices. It formed persons through the discipline of science. The gradual shift from material to formative technologies over the course of the eighteenth century was nowhere more evident than in the case of the encyclopedia, to which I now turn.

CHAPTER THREE

# Encyclopedia from Book to Practice

FOR MUCH OF THE eighteenth century, the erudite scholar, the general man of letters and not the expert scientist, reigned over an "empire of erudition," that virtual world of a unified, homogenous knowledge accessible to everyone who possessed the means and facility for interacting with print. In this imagined world, true knowledge was knowledge in and of print. As various forms of print proliferated, the "aids to erudition"—journals, lexica, and periodicals—came to stand in for knowledge itself. Many scholars saw these as indistinguishable from the sciences they were designed to organize and communicate. This conflation of knowledge with the technologies of knowledge was central to the success of the bibliographic order, which made the homogeneity of the empire of erudition and a "comprehensive" account of knowledge conceivable. As the empire gradually began to fracture under the expansion of print technologies, however, scholars began to seek out new ways to imagine the unity and authority of knowledge. They sought different technologies and new metaphors for conceiving of the norms and practices underlying authoritative knowledge.

One of the most important of these aids to erudition was the encyclopedia. Perhaps more than any other print technology of the eighteenth century, the encyclopedia stood in for a range of desires and ways of conceiving of how knowledge should be produced, organized, and shared. The very concept bore a range of aspirations and assumptions about what counted as real knowledge. From Ephraim Chamber's *Cyclopedia* (1728) and Denis Diderot and Jean d'Alembert's *Encyclopédie* (1751) to W. T. Krug's and A. W. Schlegel's lectures on encyclopedias in the 1790s, encyclopedias, both printed and conceptual, exemplified the shifts in form and function of Enlightenment print technologies. In particular, encyclopedias offer an example of how these technologies went from being merely a printed aid to erudition to an ethical technique that epitomized the purpose of a university in an age confident in the capacity of print to organize and transmit knowledge.

For most of the eighteenth century, encyclopedias referred to printed reference works arranged alphabetically and designed to organize the world's knowledge and manage the desire for a unified, authoritative knowledge in material form. By 1800, however, in Germany at least, many of these encyclopedic technologies began to break down. The dual Enlightenment imperative to produce new knowledge and to share this knowledge with an expanding reading public became impossible for any single project. In the last decades of the eighteenth century, writers like Herder, Goethe, and Humboldt used "encyclopedic" as a dismissive shorthand for the naïve and ultimately futile pursuit of a comprehensive knowledge. Eighteenth-century writers eventually used the term to refer not only to actual encyclopedias but to the whole set of technologies and practices of the bibliographic order, which they argued had failed to do anything more than accumulate information.

Encyclopedias did not disappear, however; instead, late eighteenth-century scholars reinvented them. Some were variations on the highly formalized indexes of Ersch and his colleagues, while others were more akin to philosophical undertakings obsessed with the very form of what became known as "encyclopedics"—a science of encyclopedias. Many of these later print encyclopedias, which focused on the character and person of science, began as lecture courses and had an explicitly ethical dimension, a deep concern for the habits and forms of life which printed technologies facilitated. These new encyclopedias referred not just to printed texts for accumulating information but to technologies for forming particular types of people. They represented what A. W. Schlegel referred to as the "encyclopedic revolution" that had overtaken German intellectual life and ushered in a new order of knowledge, one in which encyclopedias referred not merely to printed objects but to scientific institutions that embodied the "scientific spirit."[1] Universities were "encyclopedic institutions" that formed "thinking minds" into a distinct culture of knowledge defined by the virtues of science. As German intellectuals sought to reimagine the university, they did not leave print technologies behind; instead, they incorporated them and the norms and practices associated with them into the university.

## Universal Books

The concept of an encyclopedia has a long history that extends from antiquity through the Middle Ages and well into the eighteenth century,[2] but only since the Renaissance has "encyclopedia" referred to a work that contains the knowledge of "many books" in an accessible and organized material form.[3] "Encyclopedia" first appeared in the titles of sixteenth-century handbooks for the study of the liberal arts and referred not to knowledge that had been brought together in a printed book but rather to "the circle or course of learning," that is, to a curriculum.[4] In Hellenistic and Roman cultures, *enkyklios paideia* generally referred to the program of education that began with learning to read and write and moved on through the canon of Greek and Roman authors and then grammar, arithmetic, philosophy, rhetoric, and music. *Enkyklios* connoted a general or common education, everything that a free, educated man needed to know before dedicating himself to a particular profession. The actual word "encyclopedia" was first used toward the end of the fifteenth century to refer to such an ancient "circle" of learning and the *artes liberales* more particularly.

Not until the middle of the sixteenth century did scholars use "encyclopedia" in the titles of books to refer to a particular work that represented the unity of all knowledge organized in a certain order.[5] Characterized by a desire to provide a complete account of human knowledge, the ideal textual encyclopedia contained the "collective knowledge of a community which might be put together again if all other books were lost."[6] Encyclopedias were textual buffers against possible catastrophe, like the massive loss of ancient texts that had occurred in the Middle Ages.

The seventeenth-century German scholar Johann Heinrich Alsted conceived of his seven-volume *Encyclopaedia septem tomis distincta* (1630) as a unified account of all learning: the seven liberal arts, all the scientific disciplines, and the "circle" of everything that should be taught.[7] In this last sense, Alsted's encyclopedia was not just a collection of knowledge but a system that presented a method for comprehending "all things in this life."[8] Alsted organized the liberal arts and sciences into the proper place both in the order of knowledge and in the curriculum according to a set of timeless categories or topics. These commonplaces were also mnemotechnic devices designed to help readers retrieve information. For early modern encyclopedists like Alsted, an ideal encyclopedia recovered, safeguarded, and displayed the totality of knowledge and erudition. It was, as one seventeenth-century scholar put it, the "art that comprehends all others, the perfection of knowledge."[9] If successful, a true encyclopedia would render any future encyclopedia redundant, because a real encyclopedia would display and organize knowledge that was timeless and complete. It would provide a universal, unchanging knowledge that mirrored the perfect and timeless order of creation.

The compilers and authors of eighteenth-century encyclopedic works inherited some of these universal aspirations toward an exhaustive account of learning, but they reconceived of what a properly "universal" encyclopedic work might mean. This shift in the valence of an encyclopedic, universal knowledge was nowhere more evident than in Johann H. Zedler's *Grosses vollständiges Universal-Lexicon aller Wissenschaften und Künste*, which, although entitled a lexicon, was presented by its editor and publisher as a project of universal, encyclopedic knowledge. Originally planned as a twelve-volume work, Zedler's lexicon was published in sixty-four volumes and four supplements between 1732 and 1754.[10] By 1754, the *Universal-Lexicon* included over 62,000 pages and over 750,000 articles.

Zedler's *Universal-Lexicon* was no mere dictionary. It was an attempt to collate and display a comprehensive knowledge. It was designed, as the preface plainly stated, to "finally" put the "entirety of erudition" in a single work and provide the erudite with an easily accessible account of all sciences.[11] Publishers of similar works even feared that Zedler's work would eventually render their own lexica and encyclopedias irrelevant.[12] The articles were published anonymously; Zedler swore his contributors to silence. He printed only one edition and did not acknowledge any Latin antecedents. All of these elements were central in turning "Zedler," as the lexicon quickly came to be known, into an authoritative institution that exceeded any particular author or even its eponymous publisher and embodied the desire for an encyclopedic knowledge—that is, a comprehensive and universal account of knowledge in print. Zedler epitomized the desires and aims of the empire of erudition.

The "universal" character of Zedler's project operated on at least three distinct but related levels.[13] First, it was universal because it represented a Baconian expansion of knowledge and what counted as knowledge. It included not just established sciences, like theology and metaphysics, but "all sorts of mechanical arts that were not taught in the schools."[14] The traditional university faculties—philosophy, theology, law, and medicine—come only at the end of the long series of other types of knowledge. The lexicon's title page captures this universal character of its aspirations:

> Great Universal Lexicon, of all science and arts, which have heretofore been invented and improved through human reason. Including both a geographical-political description of the earth / according to all monarchies, empires, kingdoms / principalities / republics / independent do-

minions, lands, cities / ports, fortresses, castles / market towns, bureaus, cloisters, mountains, passes, forests, oceans, seas / islands / rivers / and canals; including the natural treatment of the realm of nature, according to all heavenly, airy, fiery, water and earthly bodies / and all stars, planets, animals, metals, minerals, salts and stones that can be found within. As well as a complete historical-genealogical account of royal and most well-known people in the world: the life and deeds of emperors, kings, and princes . . . / [including] all state, war, police, and domestic activities of the aristocratic and civil classes, of business, exchanges, / arts and trades . . . a complete account of all famous church fathers, prophets apostles, popes . . . as well as all councils, synods, orders . . . and finally, a complete embodiment of all learned men, famous universities, academies, societies, and their discoveries: furthermore mythology, antiquity, coin science, philosophy, mathematics, theology, jurisprudence and medicine, as well as all free and mechanical arts, including an explanation of all technical terms that are contained in them.

Similar to early modern encyclopedic texts, the underpinning assumption of Zedler's lexicon was that an already unified and ordered universe of knowledge existed outside the text. Scholars only had to collect, organize, and transmit its constitutive elements. The underlying order of the world only needed to be transmitted more broadly and remade as social knowledge. Zedler's lexicon was universal, therefore, in a second sense: it catalogued a universe of ideas. It sought to collect already-established knowledge, not produce new knowledge. The accumulative character of the title page, its enjambment of one area or type of knowledge after another in a series that goes on for a full page without any commentary on their relationship, continued in every volume. It was a monument to erudition. The only clear connection among these sciences and fields of knowledge was the fact that the different volumes were a compilation and a cooperative effort of various masters working in their own areas or fields. These "assistants and coworkers," as the preface put it, were constructing a giant building, an edifice of learning. Like Jöcher's lexicon, Zedler's was an additive project that assembled the raw material of knowledge for future work.

This leads to the third sense in which the lexicon was "universal": it aimed to expand access to knowledge. As Peter von Ludewig, chancellor at Halle University, wrote in the preface, "Once a truth is publically in print, everyone can make use of it."[15] After having secured a rare privilege to publish the *Universal-Lexicon* in all German states, Zedler used the dedication page to list all territories under the rule of the emperor and thus made a claim that his text would be universally distributed.[16] This expanded access was facilitated by the indexical bibliographic information provided by individual articles, which included citations not only to other texts but also to particular pages. The "*Polyhistorie*" entry, for example, reads, "See Gellim lib. 5. cap. 14, see Walchens *hist. critic. latin. Linguae cap. 4. S. 7.*" Each entry was designed to facilitate further interactions with other printed texts. The goal was a comprehensive account of knowledge that existed in print. And yet, Zedler made no claims about how all this knowledge was to be engaged or its proper study. It was not a "circle" of learning; it was display of learning. Even then, as we saw with the expansion of periodicals in the previous chapter, as knowledge grew with each new volume, scholars struggled to fulfill the Enlightenment imperatives to advance knowledge and increase access to it. The aspirations of a universal, printed encyclopedia, as understood by Zedler—one that represented all sciences and types of knowledge, a universe of knowledge universally accessible—began to seem out of reach given the continued growth in print. It was both a product of and monument to the empire of erudition and a sign of its impending fracture.

## The French *Encyclopédie* and Its Reception in Germany

Despite the monumental influence of Zedler's project in the first half of the century, it was the French *Encyclopédie, or Systematic Dictionary of the Sciences, Arts, and Crafts* of Diderot and d'Alembert that shaped German conceptions of the encyclopedic. It also set the terms for a debate about what constituted, as Diderot put it, "the unity of knowledge" in the second half of the eighteenth century.[17] And this was primarily because Diderot and d'Alembert took the question of the proliferation of print and its ramifications for a universal knowledge, or a universal book of books, seriously.

Published between 1751 and 1772, the *Encyclopédie* had twenty-one volumes of text and eleven volumes of engravings, something that Zedler had none of. In his *Preliminary Discourse to the Encyclopédie of Diderot*, d'Alembert wrote that the *Encyclopédie* "set forth as well as possible the order and connection of the parts of human knowledge," as both a dictionary and an encyclopedia. As a dictionary, it was to accumulate and connect the "general principles that form the basis of each science and each art, liberal or mechanical, and the most essential facts that make up the body and substance of each."[18] As an encyclopedia, it was to reduce the complexity of the world through a system that collated and condensed knowledge, thus making it miniature and manageable. The purpose of an encyclopedia was not to guide a young man through a "circle of learning" or produce new knowledge; it was, wrote Diderot, "to collect all knowledge that now lies scattered over the face of the earth."[19] The encyclopedic compiler gathered up knowledge that already existed and clarified its structure in order to share it with future generations. The universal project of the encyclopedia "enchained" or linked knowledge as well as people over time. The *Encyclopédie* did so not only by "collecting" knowledge but also by developing the print tools and formal elements that made encyclopedias the "book of books." These included the *Encyclopédie*'s renvois, or cross-references; its diagrammatic elements, such as the tree of knowledge and the *system figure*; and alphabetical order of the articles. By incorporating the paratextual and organizational elements that previous encyclopedias had used into their own text, Diderot and d'Alembert helped further establish an encyclopedic genre by embedding other encyclopedic attempts into their particular project.

Diderot and d'Alembert never considered the *Encyclopédie* to be universal in the sense that early modern encyclopedists like Alsted or early Enlightenment figures like Zedler had pursued. They did not presume a ready-made universal knowledge that simply had to be catalogued; instead, they cultivated a certain skepticism toward the universal "enchainment" of knowledge. In his article "Men of Letters," for example, Voltaire wrote that "the course of History is a hundred times more vast than it was for the ancients. . . . Men of letters are not expected to study all of these subjects in depth; universal knowledge is no longer within the reach of man. But true men of letters put themselves in a position to explore these different terrains, even if they cannot cultivate all of them."[20] Voltaire limited the purview of the encyclopedia by dispersing its organizational center to "different terrains" to be explored. In contrast to Alsted's aspirations for a timeless order of knowledge or Zedler's "complete" account of knowledge, these constraints lent the project a more provisional character, because the imperative to seek new facts, produce new knowledge, and explore "different terrains" meant that any encyclopedia would have to change constantly.

The entire project had a practical orientation. It was designed to help navigate a complex world of knowledge. D'Alembert described it as "a kind of world map that shows the principal countries, their position and their mutual dependence, the road that leads directly from one to the other. This road is often cut by a thousand obstacles, which are known in each country only to the inhabitants or to travelers, and which cannot be represented except in individual, highly detailed maps. These individual maps will be the different articles of the Encyclopedia and the Tree or Systematic Chart will be its world map."[21] D'Alembert used the map metaphor to spatialize the relationship between the particular and the general and thus make it visible. The metaphor assumed a cartographer who, through his epistemological gaze, discerns these complex relationships, but the territory represented by the map is knowledge itself. It is a "sort of labyrinth, a tortuous road that the intellect enters without quite knowing which direction to take." Expanding on these metaphors of encyclopedic organization, Diderot, in his entry "Encyclopedia," noted that the real universe and the world of ideas could be comprehended in an infinite number of ways. The number of possible systems of human knowledge, the number of ways in which knowledge could be represented, he wrote, is as large as the number of these "points of view." Thus, no human system is free from "all arbitrariness"; all systems of knowledge contain some element of individual choice.[22] And thus, concluded Diderot extending d'Alembert's map metaphor, there are an infinite number of maps. While acknowledging the limitations of "our human condition," Diderot still encouraged a striving toward a "more elevated point of view" from which a grand simplicity could become visible. From such a vista, the artificially produced order, not the terrain of knowledge itself, would become "clear and easy to grasp; not a tortuous labyrinth in which one goes astray and never sees anything beyond the point where one stands. No, it must be a vast, broad avenue extending far into the distance, intersected by other highways laid out with equal care, each leading by the easiest and shortest path to a remote but single goal."[23] The purpose of an encyclopedia was to simplify the "labyrinth" of knowledge by mapping functional paths through an infinitely complex world. The map is both an "arbitrary" ordering device and a crucial and practical epistemological tool. Thus, as David Bates argues, the epistemological functions of the map are tied to the complexity of the terrain it surveys. It is precisely the map's manipulation of scale and perspective that makes it not simply totalizing but useful.[24]

The self-consciousness with which Diderot and d'Alembert describe their encyclopedia lends it its practical value as a technology for orienting the scholar amid a surfeit of print. After all, a map scaled 1-to-1 with the world would be useless. The philosopher-encyclopedist's gaze was not a view from nowhere. Diderot and d'Alembert grounded their encyclopedia in a self-reflexive errancy that expanded its function from, as twentieth-century German philosopher Ernst Cassirer notes, the mere acquisition of knowledge to the formation of a particular "mode of thinking."[25] To think encyclopedically, as it were, was to reduce complexity to simplicity, to reduce "apparent" diversity to a manageable identity.[26] For Diderot and d'Alembert, the pursuit of encyclopedic knowledge should always be constrained by the awareness that a general system from which "all that is arbitrary would be excluded" might not be worth pursuing. "For what would be the difference," asked Diderot, "between reading a book in which all the hidden springs of the universe were laid bare, and direct study of the universe itself? Virtually none: we shall never be capable of understanding more than a certain portion of this great book."[27] Humans need reductive modeling technologies to make sense of a world that will always exceed their capacity to comprehend it fully.

Given such measured skepticism, Diderot placed the "thinking and contemplating entity man" at the center of all projects of knowledge because, without the perspective of the human being, "the sublime and moving spectacle of nature will be but a sad and silent scene. It is only the presence of men that makes the existence of other beings significant."[28] Without the human, empirical reality is silent. Thus, he continued, the human being should not only be introduced as the meaning-making agent into the *Encyclopédie* but also be made the "center" of all that is. According to Diderot and d'Alembert, the *Encyclopédie* was radically new because it produced meaning by placing the human being at—and thus displacing God from—the center of knowledge.

Such an encyclopedia project was needed in 1750 Europe more than ever, explained Diderot, because it was the only appropriate response to a specific historical reality: "As long as the centuries continue to unfold, the number of books will grow continually, and one can predict that a time will come when it will be almost as difficult to learn anything from books as from the direct study of the universe. It will be almost as convenient to search for some bit of truth concealed in nature as it will be to find it hidden away in an immense multitude of volumes. When that time comes, a project, until then neglected because the need for it was not felt, will have to be undertaken."[29] Looking into the future, Diderot anticipated that the printing press, which "never rests" and has filled "huge buildings with books," would produce a world of its own. In this print-filled future, both the natural world and the world of books could become obscure, distinct, and closed in on themselves.

And although natural historians had begun to map nature, the new world of print would require maps of its own. It would have its own complexity that would require particular ways of knowing and particular forms of perception.

As we saw in chapter 2, this is precisely what happened over the course of the eighteenth century. As print continued to proliferate, scholars began to regard the empire of erudition as its own realm, a bibliographic order that required complex technologies to guide readers through a world of print: more complex indexes, more journals, more encyclopedias. Before late eighteenth-century German intellectuals invoked the specter of "book floods" and "book plagues," Diderot feared that the advent of such a world would eventually come untethered from the world of humans. He anticipated late eighteenth-century German anxieties about a printed nature separated from empirical reality, an anxiety about an anonymous and abstract print world detached from nature.

Diderot oriented the *Encyclopédie* toward a future in which its goal of collecting and relating was deferred. The interminable nature of the encyclopedic endeavor made a certain Enlightenment exhaustion inevitable. As a map, the encyclopedia dealt with the increasing complexity and dimensions of knowledge.[30] Its provisional character was motivated by an overburdened consciousness of its own historicity brought on by an Enlightenment vertigo. The perceived progress of knowledge threatened to make any organizational efforts obsolete and thus threatened to undermine encyclopedic thinking. What good, asked Diderot, would Alsted's "exhaustive" encyclopedia or Zedler's "complete universal" lexicon be now?[31] Diderot's basic skepticism and insistence on the practical use value of any encyclopedic technology detached the encyclopedia from any normative claims about the order in which the sciences and other forms of knowledge should be studied. He made no explicit claims about what type of person should study what. Like Zedler, he divorced the encyclopedia from the ancient encyclopedic notion of a "circle of learning."[32]

One of the first German scholars to take up Diderot and d'Alembert's revised conception of the encyclopedia was Johann Georg Sulzer. In the first edition of his encyclopedia, *Brief Account of All Sciences and Other Parts of Erudition*,[33] which actually preceded the French *Encyclopédie* by almost a decade, Sulzer expressed confidence that a "natural order" of the sciences could be adequately represented. The pursuit of an encyclopedic order of knowledge which mirrored the order of nature was not possible. With each new edition, however, the skepticism that had characterized the *Encyclopédie* became more evident, such that in later editions Sulzer described any attempt to represent the unity of sciences as arbitrary. Already in the second edition of 1759, just as the *Encyclopédie*'s influence was spreading throughout the German-speaking lands, Sulzer articulated one of the core conclusions of the French *Encyclopédie*: an encyclopedia should aim not to accumulate all possible knowledge, but rather to distinguish the significant from the insignificant.[34] For Sulzer, earlier encyclopedias could afford to be indiscriminate because they could accumulate as much information as possible without worrying about subsequent threats to their basic organization and functionality. Prior to the eighteenth century, thought Sulzer, knowledge was limited. There just wasn't as much of it. Modern encyclopedias, in contrast, had to be more selective because they were being written in the ceaseless "inquiry" of a "modern age" that produced an endless stream of new knowledge.[35] Given such unprecedented expansion of new knowledge, so visible in print, modern encyclopedias had to filter out the "vain, useless, or outrageous."

Sulzer's anxiety about the "ever-increasing" material forms of knowledge, traceable with each new edition of his encyclopedia, was also evident in one of the most common metaphors for encyclopedias, the tree. Whereas previous encyclopedias appealed to the tree metaphor for its ordering capacities, Sulzer thought that such trees had grown out of control. In the "modern age," he wrote, the branches of knowledge had grown so much that they obscured the metaphor's ordering capacity. In Sulzer's use, the tree metaphor figured overgrowth, confusion, and disarray, not, as in Diderot and d'Alembert's use, a visualization of a manageable organization of knowledge that moved from general forms of knowledge (the trunk and main branches of the tree) to more particular types of knowledge (the ever smaller branches). It failed because it tried to represent the "most exact relationships and lines of descent" among proliferating forms of knowledge which were becoming unrepresentable. In place of the tree metaphor, Sulzer adapted Diderot and d'Alembert's geographic and spatial metaphors and their related technologies, especially the map. Such a geographical depiction of knowledge as spatial terrain, he argued, renders knowledge mappable and thus surveyable. Echoing Diderot and d'Alembert's insistence on the errant but practical encyclopedic map, Sulzer argued that given the "expansive state of erudition," it had become impossible to represent the disparate elements of learning "in a natural relationship" free of constraints.[36] Any attempts to connect the order of knowledge to the order of nature were, he concluded, bound to fail. Future encyclopedias would have to recognize the gaps between representation and the world of things and be satisfied with a useful, if not metaphysically accurate, encyclopedia.

These concerns, however, did not hamper Sulzer in his insistence that a "true scholar" should not be "completely ignorant" in any part of knowledge. "There is no part that does not either require or make use of all other parts."[37] Despite his suspicion of the possibility of a universal knowledge, the polymath erudite of the empire of erudition remained for Sulzer the norm. Throughout the eighteenth century, scholars like Sulzer continued to insist on the normative, if not actual, unity of knowledge. Sulzer's encyclopedia exemplified eighteenth-century efforts to reconcile the fragmentation of knowledge with the continued aspirations for a unity of knowledge.

Following in the tradition of the French *Encyclopédie*, Sulzer clarified the two forms that an encyclopedic knowledge had come to refer to in the last third of the eighteenth century in Germany. On the one hand, an encyclopedia could provide the fullest, most complete presentation of knowledge, as exemplified in its lexical order. An encyclopedia, as Diderot put it, "collect[s] all the knowledge that lay scattered over the earth."[38] On the other hand, a formal encyclopedia could provide a systematic account of knowledge which offered a theory of the whole, as visualized in its famous tree of knowledge. The former provided no account of the whole, while the latter gave no account of particulars.[39] The expectation was that both aims—to collect particular material knowledge and to theorize a universal formal order of knowledge—could be held together without one overwhelming the other. In the German reception of the *Encyclopédie*, these two aims were separated and came to represent two different types of encyclopedias: a material encyclopedia that aimed for completeness, and a formal encyclopedia that aimed for systematic unity.[40] Whereas the former sought to collect as much information as possible, oftentimes for didactic purposes, the latter sought to provide a more formal or discursive schematization of all possible knowledge. As the eighteenth century wore on, these two types of encyclopedic projects became increasingly opposed to one another.

## Material Encyclopedias and the Excess of Knowledge

The fate of Friedrich Martini's *Universal History of Nature in Alphabetical Order* exemplified the hypertrophy of late eighteenth-century German encyclopedics in general. It began as a translation of Valmont de Bomares's natural history dictionary [*Dictionnaire raisonné universel d'Histoire Naturelle*], with its first volume being published in 1774. But whereas Bomares's *Dictionnaire* covered the entire alphabet in fifteen volumes, Martini only managed over a period from 1774 to 1793 to produce eleven volumes, the last of which ended with the letter C. One reviewer noted in 1790 that "since the appearance of the first volume 15 years have passed and two letters have yet to be completed. If this continues, there will be over 100 volumes and the last one will appear sometime in the middle of the twentieth century."[41]

For Martini, the difficulty of publishing a natural history encyclopedia lay not in the infinite complexity of nature but in the emerging complexity of print. Writing a natural history, he lamented, required "looking up an incalculable mass" of related texts.[42] His encyclopedia made it difficult to tell whether the object of natural history was actually the living organisms or the printed, material objects that described the living organisms. Almost every page of the *Universal History of Nature* is half filled with footnotes and citations pointing to other printed works. It was less an inventory of natural objects than of print objects.[43] As was the case with *historia literaria*, however, these material, accumulative projects went a long way in constituting sciences as distinct sciences. Their accumulating and cross-referencing work helped construct a networked canon of print authority.

Martini's was not the only German encyclopedic project to collapse under the saturation of print. In 1778 a group of scholars began work on the *German Encyclopedia, or Universal Dictionary of All Arts and Sciences* [*Deutsche Enzyklopädie, oder Allgemeines Real-Wörterbuch aller Künste und Wissenschaften*]. This project ceased publication in 1804 at the letter *K*. In the preface to the first volume, the editors described the goal of the project as transforming the sciences and other useful arts from a "possession and exchange among a few initiated" to a more universally accessible form of knowledge.[44] One consequence of these explicitly popularizing ends was that the editors saw no need to provide a theoretical, systematic account of knowledge (no map, no tree); instead, their encyclopedia was only to be an inventory. They unapologetically described their encyclopedia as a dictionary and not a system. The entry "Encyclopedia," which itself extends from page 372 to page 396 of a double-columned text, expresses deep skepticism toward previous encyclopedic efforts to provide a genealogy or system of sciences. It described the *Deutsche Enzyklopädie* not as a "system of all human knowledge" but as a "mere aggregate."[45] This self-consciously less ambitious aim nevertheless failed to confront its own limitations in aggregating knowledge, that is, the fact that it never made it beyond the letter C.

The first volume of Johann Samuel Ersch and Johann Gottfried Gruber's *Universal Encyclopedia of the Sciences and Arts in Alphabetical Order* appeared in 1818 and, as its editors put it, sought to "completely encompass" all categories of human knowledge.[46] The project did not end, however, until 1889, having progressed through just over half of the alphabet. One early review mocked the persistent desire for completeness: "It must be immediately apparent to even the ignorant that with such accumulation of materials that either the objects will be treated only in a facile and fragmentary manner or that the extent and conclusion of the work cannot be calculated."[47]

In the last three decades of the eighteenth century, these material encyclopedic projects were increasingly dismissed for their purportedly arbitrary organization. They lacked, as Christian Heinrich Schmid, a professor of rhetoric and poetry, put it, a clear "concept of erudition."[48] "When will people stop squandering the name encyclopedia," he wrote, "on an arbitrary number of alphabetically organized articles, in which half-baked ideas, which one would be ashamed to present to scholars, are offered to the masses as a prize."[49] Schmid's dismissal of encyclopedias designed on the French model as "arbitrary" anticipated a wave of criticisms that peaked in the 1790s when the very term "encyclopedic" connoted not a universal organization of knowledge but rather a reduction of knowledge to mere information.[50] Goethe chided the encyclopedists for reducing knowledge to historical facts; Herder mocked their inability to produce "original work";[51] Lichtenberg dismissed them as purveyors of "dictionary learning."[52]

These late eighteenth-century accusations of abstraction were broad indictments of an encyclopedic tradition associated with the French *Encyclopédie* and Sulzer. Critics claimed that such encyclopedias were organized around superficial characteristics or elements that had no real connection to underlying processes in the real world. Why, they wondered, organize an encyclopedia alphabetically? The errant, and thus more pragmatic, orientation of Diderot and d'Alembert and Sulzer had become a liability. These worries were part of a broader anxiety concerning the radical materialism many suspected lurked behind the French project—a materialism that denied purpose, order, and thus a meaning external to material processes.

## Formal Encyclopedias and the Excess of Categories

Despite the collapse of many material encyclopedia projects, the broader genre of the encyclopedia was not simply discarded. A range of more formal encyclopedic projects were published over the last two decades of the eighteenth century. These projects focused not only on collecting as much material as possible but on articulating the formal characteristics of the sciences: how they were organized, how they developed, how they related to one another, and how they could be represented in print. They also displayed an increasingly self-referential and systematic character. At the University of Mainz, for example, encyclopedics was even recognized as a distinct science when a chair in "general encyclopedia" was established in 1784.[53]

As we saw in chapter 2 with the emergence of periodical indexes—books of books—toward the end of the century, the organization of knowledge had become an increasingly self-referential undertaking. Many of these new encyclopedias elevated philosophical or theoretical questions about the organization of knowledge, systems, and the character of science over efforts to collect and organize "the world's knowledge," as Diderot had put it, and display it in print.[54] These new projects sought to avoid the tendency toward assembly and display which had led material encyclopedias to collapse by formalizing knowledge and focusing on the very concept of science.

One particularly influential example was Johann Joachim Eschenburg's *Textbook for a Wissenschaftskunde: A Compendium of Encyclopedic Lectures*.[55] Eschenburg, a professor of philosophy in Braunschweig, distinguished his "universal formal encyclopedia" from previous encyclopedias and their pursuit, as he wrote, of a "material completeness."[56] And yet, in what would prove to be a crucial qualification, he claimed that "systematic rigor" was not his only concern. His more basic intention was to outline in accessible language the history of the sciences and their relationship to the university.[57] As indicated by the title, his encyclopedia was not just a system. It was also a textbook [*Lehrbuch*] designed to introduce students to what he termed the "scientific culture" of the university.[58] Eschenburg's attempt to use an encyclopedia to inculcate students into a "scientific culture" represented a crucial shift in the history of the genre. It identified science as a distinct culture and tied the culture of science to a particular institution. The textbook, as one review tellingly put it, was designed as an "isagogic"—from *isagogics*, meaning "introductory study"—text to supplement "isagogic" lectures.[59] Instead of leading students into the study of the Bible and exegesis as had traditional forms of isagogics, however, Eschenburg's lectures led students into the study of science. Its goal was to train them in the proper modes of interacting with its canon of texts and practices.

Eschenburg coined a term for such an encyclopedic introduction to the university's scientific culture—*Wissenschaftskunde*, a systematic description of all sciences. The term quickly became a standard element in German encyclopedics. It was, as one reviewer noted, a "fitting expression" that subsumed a century's worth of efforts to organize the sciences and conceptualize science. Eschenburg envisioned *Wissenschaftskunde* as a science of sciences which would outline the content, extent, methods, and supplements of the sciences, as well as the previous historical efforts to organize them. It also organized eighteenth-century encyclopedias into a clear typology: material and formal, internal and external, historical and philosophical. And like any other science, reasoned Eschenburg, *Wissenschaftskunde* would require its own bibliography, so he concluded the section of his textbook on *Wissenschaftskunde* with a bibliography of books on the very concept of the encyclopedia.

Eschenburg's *Wissenschaftskunde* was a conceptual breakthrough for German scholars. The term, wrote one scholar, "encompasses the study of something and science in itself; in this respect, [it] represents the unity of cognition. It is this, *Wissenschaftskunde*, or a study that has been potentiated through science."[60] Eschenburg's neologism was not just an abstract idea. It was a particular material form for thinking and writing about science. *Wissenschaftskunde*, wrote Eschenburg, treated science "objectively"—that is, as the embodiment of organized truths themselves, "which have been lifted out of the entire mass of recognizable objects and organized into a unified whole." This allowed scholars to consider science as a distinct and observable whole that existed apart from the subjective cognitions of individual minds. Considered "objectively," science could then be ascribed its own properties, like other objects in the world, and thus made an object of analysis. Echoing other attempts to reconceive of science discussed in chapter 1, Eschenburg suggested that science was not merely a cognitive capacity or a category of knowledge; it was the embodiment of interrelated cognitions, observations, explanations, and principles, all of which were organized in a purposeful relationship that constituted an "edifice of teachings or system." As the objective "embodiment" of particular truths, science existed in time. It was not static. It developed over time. Given the failure of grand rational typologies, future encyclopedias would have to treat sciences as distinct, objective entities with their own particular histories that could not be fully fixed by a timeless rational taxonomy. Encyclopedias would have to embrace a more historical approach that traced, for example, the "origins and advances of the literature" of the sciences.[61] These were some of the first anticipations that sciences could constitute distinct disciplines and could be arranged into a system of disciplinarity.

Eschenburg, however, did not fully forsake the bibliographic traditions of *historia literaria* and Zedler's commitment to the empire of erudition. His textbook covered not only the basic content of the sciences—relevant discoveries, inventions, arguments—but also the "means of transmission," as well as the "published texts and various editions" of scholars' work which would have to be studied. Echoing descriptions of *historia literaria*, he concluded that one of the most important tasks of an encyclopedia was to coordinate the printed material of science into coherent traditions. Every section of Eschenburg's encyclopedia bore out this theoretical description. The first section on "Philological Sciences," for example, begins with a short paragraph on the concept of philology, followed by a paragraph of nothing but citations: "See. Guil. Budaei de Philologie Libri II. Parif. 1536" or "Compare. Herder's *Ideas for a Philosophy and History of Humanity.* V. II. P. 233 of the Quarto edition, P. 269 of the Octavo edition." This format was repeated for all eight of the encyclopedia's categories. The point is not that (or whether) Eschenburg was arguing anything new, per se, but rather that he presented his encyclopedia as a print interface that constituted a different unity of the sciences. He described himself as a "literary geographer," who mapped the different areas and boundaries of literature. The world of science was the world of print. Unlike *historia literaria*, however, Eschenburg suggested, although he never fully elaborated what he meant by it, that his encyclopedic project was related to the university, whose unique task was to train students in how to engage the material of the sciences. Universities were needed to form students and encourage virtues particular to science, such as scientific "private industriousness."[62]

For many critics and fellow encyclopedists, Eschenburg's attempts at a formal encyclopedia were not systematic enough. As one blunt reviewer put it, Eschenburg was not Kantian enough. After the failure of the French *Encyclopédie* and its German successors to provide more than a "merely arbitrary order" of the sciences, Kant's critical philosophy, wrote one scholar, guaranteed that "the structure of human knowledge" could finally be "properly ordered."[63] Any encyclopedia that failed to embrace fully a Kantian schema was doomed from the start. The distinctions that Kant had drawn in his so-called critical philosophy—between mathematics and philosophy, pure and empirical knowledge, propaedeutic and metaphysics, speculative and practical use of reason—provided a framework for a systematic organization of knowledge which eclipsed the older, arbitrary orders. Central to this reconceptualization of an encyclopedic order was Kant's definition of system: "the unity of the manifold under one idea."[64] The difference between a mere catalogue (or aggregate, index, etc.) and a system was that, in the former, the parts precede the whole, whereas in the latter, the whole, as the organizing idea, precedes the parts. The idea of the system was an a priori determination of the whole and its organizing purpose.

As scholars pursued what they considered a Kantian encyclopedics, they did so in necessarily imaginative and inventive ways, since Kant's own discussion of a systematic encyclopedia was cursory at best.[65] One of the most extensive attempts to develop an encyclopedia based on Kantian principles was undertaken by Wilhelm Traugott Krug, successor to Kant's philosophy chair in Königsberg. For Krug, Kant's transcendental philosophy made possible a "science of the sciences," or what he simply called an "encyclopedic,"[66] which could finally manage the growth of the modern sciences.[67] Krug organized sciences according to common elements that were preestablished as the basic characteristics of a science: their basic concepts, content, boundaries, relationships to other sciences, current states, and the methods and literary aids for engaging them.[68] By analyzing sciences according to common traits, he treated them as though they were natural objects to be classified. Like a botanist organizing plants according to a Linnaean taxonomy, the systematic encyclopedist would observe common traits and then place similar sciences into established classes.[69]

Whereas the material encyclopedias had been overwhelmed by the sheer amount of printed material, Krug's encyclopedia succumbed to a "classificatory fury."[70] He began his book with a detailed typology of encyclopedias and then proceeded to a basic classificatory scheme for the sciences, which he based on nine classes as opposed to Sulzer's eight. He continued to divide this scheme into more subcategories: form, material, purpose, determining ground, and external form. (In a later version, he added even more divisions.)[71] Krug's self-described "systematic" organization of the sciences repeated in its very form the hypertrophy of Enlightenment taxonomy projects that he mocked. His discussion of the problems of natural history taxonomies reads like a description of his own attempts to order the sciences. The classes of natural history, he mocked, "run together or seem gradually to lose themselves in one another," such that a complete classification of natural objects in accordance with their natural order is nearly impossible. From this Krug concluded that as new natural objects were discovered, human knowledge of nature would undergo "constant growth," which would render previous classificatory schemes "insufficient and imperfect."[72] Krug could just as well have been discussing his own encyclopedia.

Just as the material encyclopedias collapsed under their own accumulating efforts, formal encyclopedias such as Krug's collapsed under their own systematizing impulse, what one fellow encyclopedist described simply as Krug's "too much"—too many categories, too many classes, too many subdivisions. Despite the common critique that previous classificatory schemas were arbitrary, encyclopedias like Krug's were no better than their predecessors. They continued to offer an analytical account of the sciences, even though they recognized, to varying degrees, the inadequacy of these forms of organization. They ultimately undercut the broader desire for unity and obscured the relationships among sciences.[73] The embrace of the organizing potential of reason may have alleviated the pressure of the "growth" of the particular, but it also distanced encyclopedic projects from individual persons and undercut encyclopedias' appeals to the ethical and formative potential of the university. The same tension repeated itself—that of abstract classes versus material, particular interactions. All of these efforts to offer a theory of science ultimately exposed the contingency of science itself. But they also revealed the problem of encyclopedics to be the same problem as science more generally: how to relate the universal form of science to particular sciences? How could a formal account of the sciences and its embrace of the relationship of all sciences be reconciled with a material account and its embrace of the particular?

For many of the Kantian-inspired projects, including Eschenburg's and Krug's, the idea of a systematic order replaced the Enlightenment pursuit of material completeness. Since the advent of early modern encyclopedias, a comprehensive or complete knowledge had been understood by most scholars as a representation of a natural order; the proper measure of completeness of a project was the extent to which it reflected an order in nature (be it the order of plants or the order of books). Previous efforts—from Alsted to Zedler—were in their own terms systematic, but they had conceived of their systems as reflecting a natural order. Kant's critical philosophy envisaged a different systematic order, one projected onto the world by a rational subject.

"Human reason," wrote Kant, "is according to its nature architectonic, that is, it considers all knowledge as belonging to a possible system."[74] Therein lay the potential for encyclopedics that Eschenburg, Krug, and others saw in Kant's transcendental philosophy; it promised to disentangle the order of knowledge from the ever-proliferating things of the world. As we saw in the case of Krug in particular, however, these projects, like their material cousins, encountered their own forms of uncontrollable proliferation and excess and never achieved their universal aspirations.

## The Ethics of Encyclopedias and the University

Over the course of the 1790s, scholars continued to produce Kantian-inspired encyclopedias, but many began to relate their emphasis on the character of science and system to the institution of the university, thus recovering in new ways the ancient and early modern conception of encyclopedia as a "circle" of learning. Around 1795, scholars at the University of Jena, a hotbed of post-Kantian thought, added another dimension to these projects when they began to offer annual, encyclopedic lecture courses for students of all faculties.[75] Their courses were designed to introduce students not to a specific science, or even the relationship among various classes of sciences, but rather to the particularly scientific character of what they considered authoritative knowledge. The main conclusion of these encyclopedic lectures was that the character of science could not be detached from the character of the scholars.

German universities had offered encyclopedia lecture courses throughout the second half of the eighteenth century. Beginning around midcentury, professors at the University of Göttingen, for example, were required to offer a "*Collegium Encyclopaedicum*," a lecture course designed to introduce first-year students to a particular field of study.[76] In the winter semester of 1756/1757, faculty members offered encyclopedic courses in medicine, theology, mathematics, and philology. Whereas earlier print encyclopedias attempted to cover, as Göttingen professor Johann Gesner put it in the middle of the century, "the whole world of learning," these university-based encyclopedia courses focused on a particular science.[77] Göttingen law professor Johann Stephan Pütter defended his course on what he termed *Rechtsgelehrsamkeit* on the grounds that "completeness" could not be achieved through the "connection" of all sciences, but only through the cultivation of particular sciences.[78] Pütter's notion of "completeness," however, was primarily bibliographic. Following the bibliographic imperatives of *historia literaria*, he listed as one of his main course objectives helping students develop their own "law library," which he facilitated by providing an extensive "directory of several useful law books."

Similarly, scholars at several universities, including Jena, gave introductory, encyclopedic lecture courses that sought to introduce students to all sciences. Many of these university encyclopedia courses used common texts, such as Johann Georg Heinrich Feder's *Outline of the Philosophical Sciences* (1767), which, as Feder put it, provided the "content" and "book knowledge" necessary to introduce students to all "philosophical sciences."[79]

Scholars and faculty members conceived of both the specialized and universal encyclopedic lecture courses in terms of the empire of erudition and its bibliographic order. They emphasized the content of particular sciences and the accumulation and transmission of print knowledge about them. Like *historia literaria*, they were propaedeutic in function and exposed students primarily to the literature of the sciences. They were the oral analogues of the specialized periodicals and aggregating scholarly journals that were proliferating around the same time, and once published as texts, they fit right into the logic of the homogenizing aims of the empire of erudition.

Whereas the purpose of these earlier encyclopedic lecture courses, whether they were highly specialized or more synoptic, was primarily didactic and accumulative, the encyclopedia courses that scholars began to offer in Jena in the 1790s had both epistemic and ethical aims. The lecturers did not just present the scientific character of their encyclopedias in propositional terms; they emphasized their formative potential as well. Initially delivered as university lecture courses, most of these encyclopedic projects appealed to the university as a unique institution in the history of knowledge and described the encyclopedic lecture as a distinct mode for interacting with the proliferation of print. Material encyclopedias were limited not just by an epistemic failure to establish the right categories of knowledge but by an ethical failure as well. They formed intellectual dolts. The failure to produce a well-organized encyclopedia was both a symptom and consequence of the broader culture's incapacity to help people orient themselves amid a surfeit of information.

One of the first scholars to offer such encyclopedia lecture courses at Jena was the philologist Christian Gottfried Schütz. Between 1783 and 1787, Schütz lectured several times on "An Encyclopedia according to Sulzer's Concept of All Sciences," which was based on Sulzer's encyclopedia and its attempt to provide an overview of all sciences in the tradition of the French *Encyclopédie*.[80] As we saw above, although Sulzer was skeptical of any attempt to systematize all sciences, he did continue to uphold the polymathic erudite as the norm for what counted as real knowledge. Embracing these tendencies, Schütz described the purpose of the philosophy faculty, and thus his own lecture course, as covering "all those studies that can truly be considered *studia humanitatis*, that is, all those sciences in which one cannot be ignorant if he wants to be a cultivated person and not a mere handworker or illiterate."[81] Beginning in 1793 and continuing to 1800, however, Schütz based his lecture course not on Sulzer's encyclopedia but on Eschenburg's textbook.

Soon thereafter in the winter semester of 1796/1797, Johann Heinrich Gottlieb Heusinger offered a lecture course in Jena on the encyclopedia of philosophy which epitomized the new concept of an encyclopedia as an ethical technology designed to form students into particular types of scholars. The systematic aims of a philosophical encyclopedia, he wrote, should not be separated from the ethical task of transforming students' relationship to science.[82] The subtitle of the published version of the lectures described the book as a "practical guide" or code of practice for studying at the university.

Heusinger compared the young university student to a traveler who, unprepared and ignorant about what he is about to encounter, sets out in his "carriage" to explore the world. Lacking any "basic" knowledge, the student has no way to discern what is new, old, or even useful and tends to conflate "erudition" with mere "memory." He presumes that patients can be cured, cases argued, and sermons written simply by consulting a "catalogue" of established knowledge. To know is to consult compendia and make use of the "aids of erudition." These youthful misconceptions are rooted in an ignorance of what "science is, and how the man of science (the scholar) distinguishes himself from other people, who don't at all lack knowledge."[83] This common misconception leads the student to mistake the appearance of learning for real knowledge.

It is in this context of confusion about the sources of epistemic authority that the encyclopedic lecture was necessary, because its basic purpose was to distinguish university-based knowledge, science, from other forms of knowledge and inculcate students into the life, practices, and principles that undergirded it. The initial task was to develop the "character" of the student by introducing him to a particular way of relating to science as a unified, distinct realm of knowledge. For Heusinger, such a process of inculcation involved two basic steps. First, the student had to be brought to recognize the excess of material forms of knowledge and realize that no one person could master them all. Second, the student had to be trained to imagine science as a unified "field"—a more general domain of knowledge within which the multitude of sciences could communicate with one another. Science, wrote Heusinger, was an arrangement of particular sciences or "discrete possessions" of the general field of science. What a university student needs—and what Heusinger promised to provide—is not a taxonomy, but rather the ability to distinguish between scientific knowledge and common, popular knowledge. Whereas the latter consisted of "a mere index of names of sciences," the former consisted of the "proper concepts" that underlie individual sciences. The authority of scientific knowledge rested on the integrity of underlying principles, not on a complete or exhaustive accounting of what has been known. "Whatever it may consist of," wrote Heusinger, science "distinguishes itself primarily from that which is never science."[84]

Heusinger sharpened the basic binary between science and nonscience by tying real knowledge to particular characters, whom he then juxtaposed to their opposite: the thinking and the nonthinking person, the enlightened and the unenlightened, the scholar and the common man. Within the realm of science, he explained, the scholar is not simply opposed to a nonthinking man. One could be a "thinking man" but not a scholar. With respect to law, for example, what distinguishes a common man, who may very well be familiar with the laws of a particular country, from a learned jurist is that the latter bases his judgment on basic principles that are rationally communicable. The scientific character [*Wissenschaftlichkeit*] of a particular science is a function of the extent to which it pursues its particular "material" from objective principles that are theoretically available to all and can be disseminated and replicated—that is, what made science was a commitment to method. Each individual science, as Heusinger put it, was the "embodiment" of method that rested on common principles. Method stood in for unity and the relationship of knowledge and thus dispensed with the need to pursue a material completeness, which inevitably led to mere accumulation.

The actual content of Heusinger's encyclopedia offered nothing new, no real advances on Kant's critical philosophy. Most of the chapters introduced or recapitulated elements of Kant's critical system, focusing in particular on the *Critique of Pure Reason*. What was different, however, was how Heusinger framed the purpose of the lecture course in terms of producing what he termed a "critical spirit [*Geist*]" or the "character" of the true scholar.

Such a scholar would be critical, precise, and dedicated to strenuous labor. The goal of the lectures was not simply to introduce students to the basic principles and concepts of Kantian philosophy, or provide them an overview of the sciences, but rather to teach them how to think, which Heusinger characterized as the exercise of "self control."[85] Amid all the discussion of Kantian principles, Heusinger was sketching a vignette of the ideal scholar, the disciplinary self.

Between 1804 and 1812, Carl Christian Schmid, author of a famous Kant lexicon[86] and professor of philosophy in Jena, lectured on "Encyclopedia and Methodology of the Sciences." The purpose of an encyclopedia, he argued, was not merely to provide an overview of the sciences but to cultivate a particular disposition within students toward science. "The study of the science of sciences," he wrote, "excites, enlivens, and maintains the drive to infinite expansion, as well as perfected unity and the organic mutual determination of all knowledge. It gives all students a purposeful direction. It removes vacuity [*Geistlosigkeit*], pedantic one-sidedness, liberality and plan-less indeterminacy from the form of thought and from the life of the scholar" and maintains "the drive to unity, system, the all-round permeation of all knowledge."[87]

Schmid connected the study of science directly to the university, whose task was to give these efforts institutional form. "The singular purpose of life at a university," he wrote, was to form the mind to science. "Even in academic lectures, the primary purpose is to present the spirit and essence of science in its totality, to enliven the idea of science, and to awaken the scientific sense, a purpose to which the transmission of particular facts, as mere means and subsidiary ends, must be subordinate." "The condition of possibility of human science," wrote Schmid, "is university study, that is, the step-by-step development and exercise of the faculty of knowledge." The university organized knowledge by forming students, who were critical, thorough, and industrious. The well-formed student would see himself in relationship to the "whole" that was the broader university community, while simultaneously committed to his particular "discipline," where he actually is formed into the broader "scientific spirit."[88]

Even Krug, who had studied in Jena in the early 1790s, published a revision of his highly formal encyclopedia along the lines developed by Jena scholars in hopes, as he put it, of contributing to contemporary debates on instituting a "better organization of universities."[89] The organization of universities, he argued, should reflect the organization of the "scientific system itself." In order to flourish in the modern age of print, the university had to become a system, but a system not just for organizing sciences but for organizing people. The purpose of a university was to cultivate a "mind for science, for the higher insight into learned study, for self-thinking." University teachers were to inculcate a set of virtues unique to science: "industriousness, discipline, and order."[90] Krug's attempt to combine encyclopedics and questions about the university echoed not only his Jena colleagues but also an increasingly public debate that was beginning to embrace the university as the ultimate technology for organizing the sciences, print, and knowledge. Eschenburg, Krug, and the Jena scholars pointed to the university as the print encyclopedia's true successor—an institution that could subsume a bibliographic order and embody a new disciplinary order dedicated to the culture of science. Most importantly, they argued that the character of science should not be detached from the character of scholars.

## The End of the Bibliographic Order and the Promise of a Living Encyclopedia

The authors of the Jena encyclopedias cast themselves in opposition to the empire of erudition and the pursuit of a material completeness. They considered the objective, material accumulations of the "empire" a necessary but insufficient condition for science. They sought to outline a culture and character of science as not only a way to systematize knowledge but an ethical solution to the proliferation of print. And they tied that culture and character to the institution of the university.

Lecturing in Berlin in 1803 after having spent several years in Jena studying and working with the likes of Fichte, Schelling, Hegel, and his brother Friedrich, A. W. Schlegel gave a biting resume of German intellectual life at the end of the eighteenth century. Although the "Germans are one of the primary writing powers of Europe," he claimed, they have yet to consider "to what end" they were reproducing so many texts. Before the invention of the printing press, manuscripts and texts had been scarce. They were a "precious possession that was passed on from generation to generation. It was a Romantic poverty." But given the ubiquity of printed texts, their seemingly infinite reproducibility, people no longer read them with "devotion but merely for mindless distraction." The only thing that kept pace with the endless reproduction of texts and distracted reading was the "useless writing" that supplied seemingly infinite amounts of words. The combination of too many books and too much writing produced an age in which the "spawns of immature minds" had "infested" the world. He wrote,

> When one observes the external framework of our so-called literature, the great institutions and the slight and unedifying effect on the mind, one is overcome by disgust and displeasure. Twice a year the great book fair flood (not even counting the smaller, monthly floods with which the journals are washed up) casts on land the newly birthed books in great balls from out of the great ocean of writerly shallowness and banality. These are then ravenously devoured by the great mob of the reading world, but without affording any nourishment. And they are just as quickly forgotten and fade away into the filth of the reading libraries and with the next trade fair the whole cycle begins again.[91]

In a world of easily accessible books and distracted reading, what was needed was a radical cultural change, a new way to conceive of what counted as real knowledge in a world awash in so much print. The technologies of the bibliographic order were of little help. In fact, they contributed to the confusion. They simply tried to keep up with the printing presses and invented new categories or bigger books that catalogued a world that now exceeded the capacity of humans to control and order it.

The proliferation of print could not be solved by the cataloguing and aggregating of "mindless pedants." New technologies and methods were needed to distinguish valuable books and knowledge from among the "raw aggregate of books."[92]

In his *Lectures on Encyclopedia*, Schlegel offered a possible solution: the new German conception of an encyclopedia and its connection to the university. Over the past decade, he wrote, a "fortunate revolution"[93] in encyclopedias had swept over Germany, offering a way out of the "shallowness" that had befallen eighteenth-century readers. On Schlegel's telling, what was once a printed reference book that collated unrelated facts had become a locus for ethical practice. Invoking the work of Eschenburg, Krug, and other late eighteenth-century German encyclopedists, Schlegel hailed the encyclopedic "revolution" as heralding a change not just in a scholarly technology but in the very "morals, the general mode of thought, and the condition of society." The new formal encyclopedias and their emphasis on the character of science and the person of science had awakened a unique "historical and philosophical sense" that would reinvigorate the sciences and thinking more broadly.[94] The encyclopedic "revolution" represented a revolution in the ethics of knowledge. Whereas the empire of erudition and *historia literaria* had stood in for the desire to control and master knowledge for almost a century, the "encyclopedia" had, for idealist and romantic German thinkers, supplanted these Enlightenment metaphors for authoritative knowledge and stood in for a different order of knowledge.

Looking back over the various conceptions of "encyclopedia" over the past century, Schlegel noted how more recently most "moderns" used the term either positively to refer to a formal, systematic "concept of all sciences" or negatively to refer to a material accumulation of facts. While the latter led to an "empty spouting of knowledge," the former led to abstract theorizing about sciences that could never be taught "fully" in one book. Both the formal and the material encyclopedia projects had failed to deal adequately with the contemporary proliferation of print. Either their formalism led to abstract theories of the whole without much actual content, or their materialism devolved into unorganized "masses of facts."[95]

The encyclopedic "revolution," suggested Schlegel, was incomplete, but it had produced new ways of figuring the unity of knowledge, new metaphors that could stand in for the desire to organize and master knowledge in an age of excess. D'Alembert's geographical metaphors had made the "relationships and points of contact" of the different sciences visible,[96] but they reduced the sciences to ahistorical, fixed categories. On a map, sciences were separate lands connected, if at all, by stable, seemingly timeless borders. D'Alembert's *mappe du monde* was predicated, as Schlegel put it, on the "fiction" that particular sciences were organized according to timeless mental faculties. Similarly, Kant's architectonic represented the sciences as one "building" in which the "foundation stone" carried the weight of everything else—in Kant's case, a purified philosophy. While Kant's architectonic metaphor provided a clear account of the objective order of the sciences, it neglected their subjective element—that is, the way in which knowledge was always bound up with the particularity of people in time, the ways in which knowledge was a product of actual human beings and their interactions with the world of things. Both metaphors reduced encyclopedias to static containers and thereby described knowledge as itself static and inert. Neither could give an account of how knowledge developed over time. In this sense, there was little difference between the purportedly modern encyclopedias of Diderot and d'Alembert and the post-Kantians and the premodern encyclopedias of Alsted and Zedler.

The only way to combine the universality of the formal encyclopedia and the particularity of the material encyclopedia, argued Schlegel, was to conceive of the encyclopedia not in terms of the static categories of previous metaphors but in terms of the "method and spirit of the sciences," which Schlegel defined as the "internal life principle of all human science." The "method and spirit" of science was something that all sciences shared; it was their common life. Without it sciences desiccated into a "dead carcass" and ceased to be a "living organization." To account adequately for the dynamism of human knowledge, "one must take refuge in another metaphor, that of an organization that is both cause and effect of itself," that is, the organism.[97] Neither the map metaphor nor the architectonic metaphor could fully capture this dynamism, both failing to articulate "the manner of connection and unity" of human knowledge, which developed over time.

Previous encyclopedists, such as Bacon or d'Alembert, had used metaphors of nature before, when they represented the unity of knowledge as a tree with a central trunk and branches that grew out of it, but they imagined a tree that never developed, as well as branches that could never have an effect on the whole. The tree was just another representation of a fixed order of knowledge that needed to be visualized in a clear image, but once represented, the image would need to change. In contrast, Schlegel used the organism metaphor to describe an organized being that developed over time and whose parts were both cause and effect of the whole. An encyclopedia was the analogue of an organic being. It embodied the same "spirit" [*Geist*] or "life principle" that coursed through all sciences. It shared in the same life force.

With such a vitalist conception of knowledge, a scholar could observe the various sciences as though they were interdependent organs with their own "processes of distribution, the sicknesses and by-products, the tumors that threaten to take from other parts and even the death of certain parts."[98] The advantage of thinking and observing the sciences as though they were parts of an organic whole, what Schlegel called a singular, unified *Wissenschaft*, was that one can observe not only their fixed "limits and points of contact" but their constantly changing "universal inter-linkage and reciprocal influences." Sciences were bound together over time because they shared a living, life-giving force. On Schlegel's account, the organism's defining characteristic was its internal teleology. It was goal directed and its purpose was internal to itself, organized as it was by a life-giving "*Geist*." Describing the sciences in these terms, Schlegel imbued them with purpose and order as well.

The organism metaphor also allowed the scholar to conceive of his own relationship to the sciences differently. It represented an integration or grafting whereby whoever partakes of the "method and spirit" of the sciences participates in the growth of a living, dynamic knowledge. It described the harmony of the subjective (the person) and objective (the material) aspects of knowledge. As a living being, the scholar shared in the same life force that bound the sciences together. The task of the real scholar, then, was not simply to compile bibliographic facts but rather to process actively the material of knowledge just like a living organism consumed and processed its nutrients. The material of erudition sustained a scholar, but only if the scholar consumed it as though it were nourishment. In this way, the scholar was productive.

By inserting the language of purpose and teleology into discussions of the encyclopedia, Schlegel distinguished encyclopedias from their print predecessors and tied them to the ethical and rational lives of humans. Encyclopedias were not simply static storehouses, curiosity cabinets designed to alleviate the burden of memory. These organic processes describe both products of nature and products of the mind. Encyclopedias and other technologies of print were the material objectivations of human labor and thought and thus could not be detached from the humans who designed and used them. They were both elements in something that exceeded their own particularity. Whoever or whatever was imbued with the "method and spirit" of science could participate in the historical development of knowledge over time. According to this description, sciences are observed not according to a logical typology or classificatory schema, but according to the spatial and temporal relationships of their own forms as the move toward an internal but distinct end. Sciences are thought of as objects that change and develop over time. Such an encyclopedia would provide a narrative of how sciences develop "genetically" from one another, as "members" of a "chain" of knowledge.[99] And human beings who participate in this common life participate in this development. They become part of the historical development of sciences.

The organic metaphor was also a crucial concept for connecting the encyclopedia to the practice of science and, ultimately, to the institution of the university. What the organic metaphor described was the way in which the individual scholar was, in his individual acts of observations and scholarship, part of the larger whole of science. On such an account, knowledge could no longer be taught as mere erudition, as the accumulation and display of already-established knowledge; instead, it had to be approached as something still in the making. Thus, the metaphors describing the relationship had to change to account for this dynamism.

Far from intimating some ethereal, otherworldly realm, the "method and spirit" of science intimated the living reality of science, its embodied reality in material objects, persons, and, ultimately, the practice of science. This is why Schlegel used the "method and spirit" of science interchangeably. The method of science was just this life-giving force, that which bound the sciences to one another, scholars to the sciences, and scholars to scholars. Method or "spirit" was the basis of scientific community and interaction. When the Jena encyclopedists insisted on the importance of the "principles" of science, they were referring to these underlying elements that bound sciences and scholars. The organic metaphor figures shared principles and common methods as a participation in something that exceeded individual scholars. Whoever was committed to a method or the life of science could participate in the progress of science.

The organic metaphor entailed a different pedagogy and conception of science. The sciences were no longer fixed categories, individual capacities, or canons of literature. They were, as one early nineteenth-century scholar put it, the "laws of formation of life."[100] They were authoritative laws of formation, laws for harmonizing the individual person with the "method and life" of science. And, as Schlegel put it, universities were the "encyclopedic institutions" that embodied this "scientific spirit" and formed "thinking minds" into a common life organized around science.[101] Following his Jena colleagues, Schlegel used the "university" as an institution to stand in for authoritative knowledge. The "encyclopedic" university was committed not simply to transmitting knowledge but to developing in students the capacities, habits, and virtues of science. Its purpose was the "universal dissemination of a scientific sensibility [*Sinn*],"[102] and the encyclopedic university was an ethical technology. According to this new metaphor, the encyclopedic university, the objective forms of knowledge could not be detached from the character of those who produced and interacted with them. The authority of knowledge, its legitimacy and proper organization, depended on the people who produced it.

Schlegel's reformulation of the "encyclopedia" metaphor, from a book of books to a formative institution, was another proposed solution to the proliferation of print, but it was one that changed the terms of the debate. By qualifying the university as encyclopedic, Schlegel suggested that the university was not supplanting the print encyclopedia, but rather integrating certain elements into an institutional structure that could sustain and cultivate them by forming people. The university was a living encyclopedia that reproduced itself by producing certain types of people. And this was particularly necessary in an age of easily accessible information, when it was difficult to determine, wrote Schlegel, what was "worth being known because of the imperceptibly vast extent of the external framework of erudition through the multiplication of books."[103]

In the remainder of this book, I consider what Schlegel referred to as the "life" of science and its relationship to the research university. Just as Schlegel was delivering his lectures in Berlin, his colleagues there, in Jena, and elsewhere, including Fichte, Schleiermacher, Schelling, and Kant, were beginning to imagine the university as the singular institution that could sustain and cultivate this grafting of the person into the living whole of science.

CHAPTER FOUR

# From Bibliography to Ethics

THE PREVIOUS TWO CHAPTERS have brought us to the cusp of that absent concept and its institution: disciplinarity and the university. I say "cusp" because the objective, primarily material aspect of science outlined in the previous chapters was a necessary but insufficient condition for the emergence of science as a cultural practice and social system. As we began to see in chapter 3, the turn by intellectuals and writers like Schlegel and the Jena encyclopedists to science as a form of life was driven in part by anxieties that the material technologies of the bibliographic order had begun to outstrip subjective capacities to interact with and make meaning of them. Throughout most of the eighteenth century, authoritative knowledge was inextricable from, if not interchangeable with, the material forms of communication, the "aids of erudition," which constituted the bibliographic order. The increasingly common and nervous complaints about too many sciences and too much print were manifestations of a deeper worry about a gap between this objective material realm of knowledge and the capacity of people to navigate it without being overwhelmed or lost.

In their zeal to make all knowledge "public through print" and "universally comprehensible," as Schlegel dismissively described the goals of the Enlightenment, scholars had created profoundly productive technologies that organized and produced more knowledge but relatively few norms and practices that could help human agents engage and make meaning of all this material.[1] Over the course of the eighteenth century "science" increasingly referred to a process of objectivation that produced a highly self-referential material infrastructure, which was necessary for science to emerge and establish itself as an authoritative tradition, but it was only when science was institutionalized as a practice guided by a stable ethos, a new form of consciousness into which persons could be formed, that it truly became modern science [*Wissenschaft*]. None of this is to suggest that *historia literaria* and related sciences lacked practices and formative traditions. The previous two chapters have described some of these in detail. They did lack, however, a clear ethical account of the ends and purposes of all this collecting and organizing and a common institution to ensure their continuity and regularity.

Another way to understand the difference between the previous chapters and the ones to follow is to consider them as tracing different sets of value-laden metaphors for the unity of knowledge, for the desire to control and manage knowledge in particular ways and to particular ends. The previous chapters describe Enlightenment metaphors from the "empire of erudition" and the "bibliographic order" to the "book" and the "encyclopedia," as well as "Enlightenment" itself. The following chapters describe the emergence of a new set of metaphors, already anticipated in the Jena encyclopedia lectures and Schlegel's vague intimations of a "life" of science. Romantic and idealist scholars invoked "the university" and "science" as stand-ins for a particular order of desire to maintain and control knowledge. Just as the erudites of the empire of erudition appealed to print and its particular manifestations to stand in for real knowledge, late eighteenth-century scholars invoked their own metaphors, ones both more ethically oriented and emphasizing the structures that could guarantee the stability and continuity of those formative practices. The encyclopedia, the university, erudition, the ethos, and, even more broadly, the Enlightenment (and, as we shall see, even Berlin) stood in for particular knowledge orders, for particular ways of organizing and conceiving of what counted as real, authoritative knowledge. I am describing a shift in Enlightenment metaphors of print to romantic metaphors of the disciplinary self, a shift in the ways the desire to know are organized and cultivated.

One possible counterclaim to my argument that the preconditions for disciplinarity emerged around 1800 is that scientific disciplines are not uniquely nineteenth-century phenomena. As we have seen in the previous chapters, scholars have referred to distinct sciences and categories and systems of knowledge for centuries. But in almost every case these references were to fixed, objective categories of knowledge and had little connection to the practices of science which I am emphasizing. A discipline was simply another name for a stable category of knowledge and not a practice that formed people. Discipline in this earlier sense was always bound to external ends. Furthermore, earlier scientific systems—from the sciences organized in a Lullian combinatorics system to the more institutionalized *artes liberals*—were not sharply differentiated from other institutions or structures (e.g., the church or the state) and thus did not entail the relative autonomy that modern, scientific disciplines developed. One of the key preconditions for the emergence of science as a differentiated form of knowledge and as a distinct cultural sphere was the emergence of these new communicative structures made visible and more effective through print technologies. This distinction was most evident in the proliferation of discrete, material forms of communication, which made it possible for science to communicate with itself and, increasingly, for particular sciences to communicate independently.[2]

Internally, like the early modern scholars who devised a range of textual techniques to organize knowledge, late eighteenth-century scholars continued and expanded on these technologies, and they increasingly turned to one technique in particular to manage the proliferation of texts: specialization. Specialization functioned as a filtering mechanism that produced meaningful distinctions and helped sort what was relevant from what was not. In order to discern something meaningful from the surfeit of printed material, a scholar, as the physicist and intellectual Georg Lichtenberg wrote, had to work solely on "the small parts" of a particular science.[3] If scholars were not simply to reproduce what had already been printed, if they were going to advance knowledge, they had to break the empire of erudition down into more manageable parts. But in order for particular, specialized work to be recognized as authoritative and not merely a meaningless fragment, the various sciences had to be conceived of as constituting a unified whole of which they were a part. The external differentiation of science from other social and cultural spheres was repeated internally in the specialization of science itself: disciplinarity. The reduction of complexity externally corresponded to an increase in internal complexity. That is, as science gradually emerged as a distinct realm, it underwent an internal differentiation of its own. The external and internal differentiation of science was a mutually reinforcing condition.

My description of the functional differentiation of science is indebted to the systems theory analysis of Niklas Luhmann and Rudolf Stichweh and, although to a lesser degree, the functional analysis of Talcott Parsons, but my account differs sharply in one respect.[4] Whereas Luhmann's and other functionalist accounts assume that science and the university can be explained without norms and values, my account of the university emphasizes just these normative aspects. The assumption that the modern university and modern science are systems that can do without norms and values is, at best, premature. The forms of differentiation which characterize modern science and the research university happen in an institution that "still holds norms and structures together."[5] This is Luhmann's blind spot. His radically functionalist account of science reduces human agents and their orienting norms to mere elements of a system, but, as we shall see in the following chapters, actual humans were constitutive agents of these systems, such as science.

Theories such as Luhmann's which focus on social systems assume from the outset that all spheres of social action are held together by organizing mechanisms such as money or administrative power which operate well below the level of norms and values.[6] In the case of the university, the normative ends of science are subordinated to the administrative rationality of the university as a social system.[7] Strictly functional accounts ignore, if not deny, the power of ideas, norms, and values to bring people together and form communities and institutions. Norms are merely an ideological edifice built upon the ground of communication processes and administrative mechanisms. In contrast to these assumptions, I argue that however differentiated modern social systems and the university itself may be, the norms and values that course through them are not so easily differentiated and cannot be dismissed as epiphenomena of more basic administrative structures.

Accounts such as Luhmann's are also beset by an asymmetry between a system's structure and its function.[8] They cannot fully explain how social systems, such as science, reproduce themselves. Luhmann and his acolytes consistently appeal to organic metaphors to describe the complex forms of causality undergirding modern social systems. Unlike organisms, however, social systems lack universal conditions for their continuity. They do not have the organs and genes that allow living organisms to reproduce themselves. In contrast to Luhmann's appeal to organic metaphors, German romantics such as Schlegel appealed to organic metaphors precisely because they were attempting to describe these conditions of continuity and reproduction, to explain how science could develop and maintain itself over time, and they claimed that it did so through processes of formation. As we saw in the previous chapter, sciences reproduced themselves by forming people, and the organism metaphor was used to conceptualize these processes. It described the reproduction not of some "airy," insubstantial system of differences, as Luhmann described science, but of particular types of people. The continuity and stability of science necessarily included historical individuals. The recursive processes of science were a function of the formation of a particular self.

Science as a distinct culture as it emerged around 1800 was not only a material interface or an inevitable rational system; rather, it was a meaning-making, ethical system that emerged out of the late eighteenth-century collapse of encyclopedic learning and the proliferation of print. Scholars and intellectuals sought to design practices and institutional structures that would recognize science as authoritative. They sought to link the objective, material elements described in the previous two chapters to a specific institution, the university, which could embody and stabilize a unique ethic and notion of authority. The modern practice of science which emerged around 1800 was not just a pedagogical tool that transmitted knowledge efficiently, however. It was a social and intellectual structure that formed particular types of scholars, defined their labor and social identity, and legitimated a particular type of knowledge and way of knowing as authoritative. As the authority structures undergirding knowledge—both ecclesial and juridical—began to collapse, the system of disciplinarity afforded a different authority structure that allowed for a new way of coping with a deluge of print and the epistemological anxieties that came with it. It was a technology that adjudicated the value, worth, and meaning of knowledge by forming individual persons, and its institutional home was the modern research university.

In science and disciplinarity, the two traditional notions of discipline as behavioral modification and a category of knowledge come together. In order to grasp the social import of disciplinarity, it is crucial to recognize it as the formation of individual persons as actors, who often precede large-scale institutional changes. Particular actors are necessary, and cultural resources have to be available for them to draw on.

The scholars and intellectuals who held up specialized science as the solution to eighteenth-century information overload sought to replace the Enlightenment desire for a universal and complete knowledge with an ethos of research—the endless, unceasing pursuit of knowledge. As an epistemic and ethical idea, research gave individual, increasingly specialized scholars the sense that they were participating in something larger. The ethos of research was predicated on the historical progression of human knowledge, and it consoled scholars, whose own work could seem fragmented and isolated, that they were contributing to a broader historical enterprise. The burden of knowledge—the responsibility to know and produce knowledge—shifted from the individual, ahistorical man of erudition to the disciplinary self as researcher operating and living within a disciplinary or scientific community of fellow scientists. It is this normative aspect that more radically functional accounts cannot register, because normative accounts presuppose particular moral agents. In the following chapters, we will familiarize ourselves with this disciplinary self, his desires, habits, and virtues, as well as the community to which he devoted, as Fichte put it, his every waking minute—the university.

First, however, we need a clearer sense of the broader epistemological and cultural anxieties and concerns about, as Schlegel put it, the "contemporary state of German literature."[9] Much to Schlegel's chagrin, around 1800 "literature" still referred not to a particular type of writing but simply to everything that had been printed. Just as some scholars were busy creating a new bibliographic order and producing ever more complicated indexes and encyclopedias, others were growing increasingly anxious about the reduction of knowledge to print. "In the last thirty years," wrote the German book dealer and publisher Johann Georg Heinzmann in *A Plea to My Nation: On the Plague of German Literature* in 1795, "no nation has printed so much as the Germans."[10] For Heinzmann, late eighteenth-century German readers suffered under a "reign of books" in which they were the unwitting pawns of ideas that were not their own. Nevertheless, readers continued to demand more books: "New, new. Anything more than a year old, is deemed vulgar goods."[11]

As with other historical complaints about too many books, increased chatter about book plagues, floods, and the oversaturation of books in the last few decades of the eighteenth century was inseparable from worries about their deleterious effects on readers, who, critics like Heinzmann claimed, had been infected by a reading addiction or madness. These complaints were never merely descriptive. Every lament entailed a normative claim, either implicit or explicit, about how books should be read. As one commentator put it, in a book about the "current reading addiction," "The reading addiction is a foolish, detrimental misuse of an otherwise good thing, a really great evil that is as infectious as the yellow fever in Philadelphia. It is the source of moral degeneracy in children. It brings folly and mistakes into social life. . . . Nothing is achieved for reason or emotion, because reading becomes mechanical. The mind is savaged instead of being ennobled. One reads without purpose, enjoys nothing and devours everything. Nothing is ordered; everything is read in haste and just as hastily forgotten."[12] The perceived excess of books was not just a managerial problem; it was an ethical one. With their similarly stated complaints about the endless reproduction of print and distracted reading, the Jena encyclopedists and A. W. Schlegel were responding to these pervasive cultural anxieties.

Herein lies the basic difference between earlier eighteenth-century debates about too much print and later eighteenth-century anxieties about reading madness and book plagues. Whereas the former were framed primarily in terms of the possibilities and limits of material technologies, the latter were framed in moral terms. For critics like Heinzmann, print threatened the character of eighteenth-century readers. Whereas scholars in the first half of the eighteenth century sought to manage the proliferation of print through objective means—the publication of a more complete encyclopedia or specialized journal—many scholars in the last third of the century sought to manage excess through disciplines and technologies that formed particular types of people.[13]

Complaints about too many books and the increasing worries about their ethical effects—about the types of people that the perceived excess in print was producing—were in part a function of the impending collapse in the Enlightenment's overriding confidence in the capacity of print to spread knowledge more broadly and thereby improve people. Before worries about a flood of books became common in the late 1770s, books and print, as we saw in chapter 2, were considered the central medium for social and political progress.[14]

Against this cultural background, Heinzmann's seemingly hyperbolic plea exemplified a gradual decline in an Enlightenment confidence in print. Worries about book plagues and reading diseases were not simply irrational fears, however. They were, in part, cultural responses to actual increases in the production of printed texts over the last three decades of the eighteenth century. The catalogue of the Leipzig book fair, the center of Germany's book trade, gave a sense of this rapid growth with an increase in annual titles of more than 125 percent between 1770 and 1800.[15] This growth corresponded to the broader proliferation of printed texts in Germany in the last third of the eighteenth century.[16] It also corresponded to the proliferation of bookstores during the same period, at least a threefold increase in Prussia from the 1760s to 1802.[17] Collectively these trends point to strong growth in the German book market in the last third of the century. The complaints and anxieties registered by Heinzmann and others are discursive traces of the cultural effects, reactions, and catalysts of these material shifts in the book market.[18]

Like other critics of the eighteenth-century book market, Heinzmann contended that the rapid expansion of print had turned books into commodities. And this shift was changing not only the ways in which people interacted with books but the ways in which they thought. Over the course of the eighteenth century, critics extended their worries about a surfeit of books to critiques and anxieties about their deleterious effects on reading and thought. Many of these critics pointed to what they considered to be an inherent conceptual paradox between books as commodities and books as mediums of thought. The profit-maximizing logic of the book market undermined the idea of books as expressions of original thought.

The "double existence" of the book as a commodity and as a medium of thought was repeated in Enlightenment notions of the person. On the one hand, the Enlightenment confidence in print was predicated on the assumption that individuals were capable of making judgments and distinguishing among the surfeit of circulating books. On the other hand, the culture of the Enlightenment posited such an individual as its end; that is, the autonomous subject was the goal of a process of enlightenment which required print. But the market produced books in accord with the actual needs and desires of its consumers.[19] The liberal, autonomous individual of the market was presumed to have always already been capable of discerning among wares, while the Enlightenment model of cultivation toward the good assumed that such abilities and such an individual had to be formed. The book market encouraged the production and circulation of books as commodities and was not, primarily at least, concerned with questions of epistemic authority and the cultivation of subjects capable of navigating their way through so much print. This tension became especially acute in the 1780s and 1790s in Germany.[20]

The underlying worry of critics like Heinzmann, then, was that the commodification of print brought on by the expansion of the book market could not be regulated by either market mechanisms alone or improved material technologies, that is, better encyclopedias, dictionaries, or review journals. The unregulated expansion of the book market threatened to replace the sovereignty of reason with the sovereignty of print. By the end of the eighteenth century, the terms of the debate had shifted: "The damage that the universally torn factory-like writing exerts on the public spirit is incalculable. And still says Kant . . . 'Humankind progresses ineluctably ahead in Enlightenment, virtue and wisdom.' We do not want to argue and dream sweet dreams. We want to hear facts. And the facts tell us that as the reign of books increases, in just the same measure does practical reason and the power to act diminish."[21] The proliferation of print, the ascendance of the bibliographic order, and the attendant fragmentation of knowledge are registered here as direct threats to practical reason and thus to human autonomy, that is, moral agency. As the Jena encyclopedists intimated, late eighteenth-century scholars described the problem of information overload as not just a managerial problem but an ethical one that threatened the integrity of the person.

Friedrich Nicolai, Berlin author and publisher, presaged these cultural anxieties in narrative form in his three-volume satirical novel *The Life and Opinions of Sebaldus Nothanker*, published between 1773 and 1776. The master mocks Sebaldus's claim that an author is simply a scholar who is attempting to share "useful knowledge" with the world. The vast majority of authors, insists the master, "carr[y] out a trade" and regard the "few actual scholars as meddlesome, non-guild tinkerers."[22] Books are like any other commodity, and authors are like any other tradesmen. Books are manufactured for profit and, as commodities, are indistinguishable from one another. "Erudition," mutters Sebaldus incredulously, "a craft? Writing books a business?" The consummate eighteenth-century writer, suggests the master, is not the novelist, poet, or reviewer, but the professor, who writes books only because he teaches courses that require students to buy his books. Professors peddle their books as compendia for "an entire science" under the assumption that these tools make learning more efficient and almost effortless. Professors write books for "prestige" as well, even though, laughs the master, in the new book market, "whatever can be printed can become a book." The young Sebaldus worries that the master's account of print as a mere business threatens the very authority of knowledge. "What is to become of German erudition if the majority of authors seek not the advancement of erudition but the advancement of fame and use?"[23] The master's only response is to repeat his suggestion that scholars and universities are complicit in the degradation of print. Anxieties about information overload—too many books—were worries about the sources of epistemic authority. The purported excess of print threatened to undermine prevailing notions of what counted as real knowledge.

Amid these debates, Johann Gottfried Herder offered a brief history of print which culminated in a bleak account of the contemporary print market.[24] The printing of books produced a world in which "cast letters" (modeled metal letters) "speak and scream" about that which human beings had been silent.[25] The printing press created not just books but an independent world of printed voices, a world inhabited and animated by printed letters, both pieces of type and the inked images made from them. In this other world of print, "thoughts of all nations, old and new, flowed into one another" and made possible a giant, borderless market of ideas. In this bibliographic world, increasingly larger feedback loops emerged in which books "reproduced themselves." In this "printed Babel," all thoughts and voices "entangle[d] themselves" and the "original" voice or thought was lost in a cacophony of disembodied voices. "Everyone reads everything whether they can comprehend it or not."[26] Lost in this print vertigo, modern humans are "confused letter-men and themselves printed letters" who have been infected by "book madness"—a pathological assumption that books constituted a distinct realm with its own logic and organization. Print had made possible that abstract, homogenous non-thing: a public that seemed to transcend the limitations of space and time. The German physicist and satirist Georg Lichtenberg summarized the critique in this way: "There are scarcely in the world any stranger wares than books—printed by people who don't understand them; sold by people who don't understand them; bound, reviewed and read by people who don't understand them; and now even written by people who don't understand them."[27]

The commodification of books rendered them into abstractions, circulating wares untethered from their human authors. For critics like Herder and Lichtenberg, nothing exemplified the abstraction of print more than the encyclopedic endeavors of Diderot and d'Alembert: "Now encyclopedias are being made, even Diderot and d'Alembert have lowered themselves to this. And that book that is a triumph for the French is for us the first sign of their decline. They have nothing to write and, thus, produce *Abregés*, vocabularies, *esprits*, encyclopedias—the original works fall away."[28] Encyclopedias stood in for a whole range of print technologies that were abstractions of original books. They took what were once individual voices and recirculated them as a mass of printed material.

Contemporary scholars of the book and bibliography have argued that the kinds of reading described by Heinzmann and Herder were part of a reading revolution in eighteenth-century Germany.[29] Until the middle of the century, a more intensive, repetitive form of reading was dominant in which only a small selection of texts (the Bible, for example) was read repeatedly throughout the course of a lifetime. After printed texts became more available, modes of reading became more extensive and allowed for more contact with a wider range of texts. The spread of newspapers over the course of the eighteenth century was especially important, because people would read them only once and attend primarily to the most recent edition. Subsequent revisions to German book historian Rolf Engelsing's thesis have emphasized that the reading revolution of the second half of the eighteenth century involved less a change in the modes of reading for a broad population—a true democratization of reading in terms of literacy rates did not occur until almost a century later—than a change in the modes of reading in an already highly literate class.[30]

All of this recent work on book history provides convincing descriptions of how practices of reading shifted, but they tend to obscure the place of these shifts in the broader organization of knowledge and their relationship to what Robin Valenza recently described as "a core paradox of the Enlightenment project." On the one hand, Enlightenment figures saw themselves as carrying out Francis Bacon's imperative to produce new knowledge across all the sciences. On the other hand, they increasingly saw this first imperative as conflicting with the imperative to disseminate knowledge to an increasingly broader readership. The presumed homogeneity and common store of knowledge that Kant termed "the entire public of the reading world" began to fracture toward the end of the century.[31]

These tensions emerged in part because of the proliferation of printed texts and the expansion of cultural norms and assumptions about print over the course of the century: increased circulation of texts made more and different kinds of texts available to a reading public that itself was beginning to differentiate. The imperative to produce new knowledge led to greater specialization, which meant that ever more distinct forms of knowledge were increasingly difficult to communicate to a reading public as it was expanding and differentiating. One consequence of this paradox was that the popularization of knowledge coincided with the disciplinary differentiation and specialization of knowledge. Rising literacy rates, the increased publication of new print titles, and the spread of newspapers put pressure on traditional scholarly forms of knowledge to distinguish themselves from broader, more popular forms of knowledge. In the German context at least, the dual imperative of the Enlightenment became increasingly unsustainable. This presented a fundamental challenge to the logic of Enlightenment, whose confidence in social progress was in large part a function of its confidence in print.

Complaints that a book plague had infected German readers and brought with it an unmanageable growth in knowledge, then, were widespread. And the solutions and therapies proposed to treat such a pathology, in the German context at least, were multiple and various. There was no one solution, but the solutions that were offered focused increasingly on ethical technologies, technologies oriented toward forming the subject of knowledge.

The German poet and philosopher Friedrich von Hardenberg, a.k.a. Novalis, offered an exemplary account of eighteenth-century information overload and proposed a solution as well. In a series of fictional dialogues, Novalis describes the conversation of two interlocutors, who touch on questions concerning encyclopedias, the mental faculties, and the modern media ecology. While both interlocutors assume that the modern age is marked by an overabundance of books, they react differently. The arrival of the latest catalogue from the Leipzig book fair prompts Interlocutor A, who is a figure of the late Enlightenment's worry about information overload, to lament the modern "book plague" and, especially, those marks of black that fill their pages: "What burden are these letters!" The modern human being, he complains, is characterized by his "fatal habituation to the printed nature."[32]

Interlocutor B, in contrast, embraces a German print market, whose "articles of trade"—books—have put into circulation "more honest and worthy ideas" than all of "our neighbors combined."[33] But he goes on to introduce a different metaphor for books: "Everywhere we are bringing together the crude ore or the beautiful molds—we are melting down the ore, and have the skills to imitate and surpass the molds." Books are figured here as dense physical things. They are "mined" and refined like raw materials that "spur us to activity."

Interlocutor A counters that to compare books to minerals and ore is to reduce them to physical, fragmented objects. He then appeals to an Enlightenment encyclopedics as the more appropriate way to conceptualize books. In a true whole, he says, every book would be a "member in the educational chain." Every book would fill a hole in the system. Books are not simply physical objects; they are elements of a timeless, systematic order. Interlocutor A even envisions a "systematic catalogue" or encyclopedic system in which every book would have its appropriate position within a seemingly infinite chain of books. This homogenous world of print, the bibliographic order, described in chapter 2, facilitates the advancement of knowledge and maintains its homogeneity. But it also abstracts ideas from their material particularity as printed words are projected onto an idealized system of thought.

For Interlocutor A, the promise of Enlightenment is predicated on this dematerialization of thought, but—and this is the cause of Interlocutor A's anxiety—the increased production of books threatens to destabilize the systematic identity of such a *Bildungsideal*.[34] Interlocutor B's claim that every material book is a "spur to thought" is a direct threat to the Enlightenment promise of a homogenized system of books of a seamless bibliographic order. "There is not a book in the catalog," continues Interlocutor B, "that has not borne its fruit, even if all it did was fertilize the ground on which it grew."[35] And, for Interlocutor B, the potential of print lies not in some abstract realm of meaning but in a book's "armor of typography." It is with the density of their printed typography that books act upon their readers.[36] Interlocutor B counters the Enlightenment promise of a timeless bibliographic "chain of knowledge" with the sheer materiality of books. What if there were nothing behind the book, no abstract realm of ideas? What if there were just the materiality of impressed letters?

Interlocutor B's rejection of such an Enlightenment conception of the book suggests that the causes of the book plague lie not in the print medium itself but in "the unavoidable weakness of our nature, its propensity to habituation and indulgence." The "printed nature" is not responsible for the fact that, ultimately, "we will only see books, but no things. [And] that we will no longer even have our five bodily senses."[37]

As a way to cope with the cultural power of print, its ubiquity and authority, as Interlocutor B put it, "we cling so strangely, like thin and meager moss to the printer's vignette" (those elaborate vine-like designs impressed on title pages). Perhaps longing for an unmediated experience of things—one without the abstraction of modern print culture—people began to cling like sickly moss to what they could touch: books and, more precisely, those elements that highlight their crafted quality. Suffering from the plague of books, the self-reproduction of books, we "cling" to that moment of print's materiality: the contact of the printer's woodcuts or copperplates to paper.

Emphasizing this materiality of print in a letter to Friedrich Schlegel concerning his encyclopedia project, Novalis wrote that books are the "body" of science and, as such, cannot be immediately subordinated to human reason. Books are not separate from the empirical world; instead, they lie opposed to the reader as material objects. They are not simply gateways to other abstract worlds. They are things made at particular historical moments by particular persons with particular technologies. For Interlocutor A, however, the sheer number of books represents a threat to the organizational monopoly of reason over what are perceived to be its own products. The proliferation of books and the reliance on them constitute a book plague, a contagion, in the sense that books beget more books and thus threaten the organizational reach of reason over nature itself.

The irony of this anxiety should not be overlooked. As an advocate of a certain Enlightenment notion of progress, Interlocutor A is forced to renounce the very products thought to make Enlightenment possible: books. Interlocutor B, the romantic, evinces an ironic matter-of-factness when he replies simply, "There are only books." We find ourselves already, he suggests, in a world populated with rocks, flowers, and printed matter. There is not a world distinct from the "thingy" reality of books. But why was Interlocutor B not worried about the "plague of books"?

Novalis opened his *Blütenstaub* with the claim that we "search everywhere for the unconditional and always find only things." Perhaps in terms of his *Dialogues*, this statement could be rewritten as follows: we search everywhere for the absolute and always find only books. By reminding a culture given over to the idea of the universality of books that books are material objects, Novalis undermined the Enlightenment assumption of a homogenous reading public in which ideas flow freely, unencumbered by material exigencies. The smoothly interlinked whole imagined by Interlocutor A is predicated on a dematerialization of print and ultimately thought itself. The Enlightenment's abiding confidence in print ultimately fetishized print and obscured its relation with humans. It concealed the interactions between human and print technologies and thus granted these technologies a power they never had. Books were human technologies, interfaces for use and the production of meaning, not timeless containers of knowledge.

Recognizing books as technologies with which humans are inextricably bound, suggests Novalis, is the solution to perceptions of overload and excess. The key to dealing with the proliferation of books lies not in slowing down the printing presses but in becoming better readers. "Reading books" requires "practice," and "practice makes perfect, even in the reading of books." Instead of producing a book of books, an encyclopedia that would contain all encyclopedias, Interlocutor B suggests that we think of books as technologies that facilitate the practice of reading.

For Novalis, part of the practice of reading was an attention to a book's particular elements. Disciplined reading was bound up with the very technology of books: a title "can be read like physiognomy," a preface can be understood as a subtle measure of a book, and an introduction can be seen as the book's "root and square." These unique elements mark the printed book as something made and intended for use by a reader. As organs of "reading," books are not simply the ossified relics of thought. The specific form, the very medium of the book, implies an agent of use, who must be trained into reading and thus into interacting with the technology of the book.

Novalis's call for a well-disciplined reader echoed the Jena encyclopedists and their attempts to inculcate students into a scientific character. There were myriad other proffered solutions to cultural anxieties about excess and the dissolution of epistemic authority. Herder modeled a more dialogical, even esoteric, form of writing.[38] The philosopher and publicist Johann Adam Bergk offered detailed advice on how to approach books physically: advice on proper reading postures, where and when one should read, and the general disposition required for reading.[39] And, as I describe in the next chapter, Immanuel Kant offered a critical philosophy. Implicit and sometimes explicit in these proposed solutions was always a cultural critique of what the authors saw as the reduction of knowledge to a bibliographic order and the fetishization of print technologies. These critics not only voiced frustrations with the proliferation of print but claimed that the most appropriate response was ethical. The excesses of print required not more objective technologies but a new ethos, a new type of person.

In the following chapters I describe the debate around 1800 in Germany about the future of the university not only in the context of the decline of the empire of erudition or a perceived information overload but more particularly in terms of the epistemological and cultural anxieties that precipitated and shaped the university's formation and evolution. None of these technologies, including the reimagined university, solved the problem of information overload, but the research university came to stand in for them all. It represented a new conception of epistemic authority grounded in the formation of a disciplinary self and the practice of science. The research university came to represent not just another content delivery device, another more efficient technology for disseminating information, but rather an institution and community that bestowed epistemic authority and legitimated knowledge.

# Kant's Critical Technology

One of the first German thinkers to make an explicitly ethics-focused argument about the purpose of the university in the new media environment was Immanuel Kant. Like Herder and Novalis, Kant worried that eighteenth-century scholars and intellectuals had conflated knowledge with print technologies. The surfeit of print, he contended, required not more comprehensive books but different types of people. Echoing the claims of the Jena encyclopedists, Kant eventually embraced the university as the only institution that could sustain and form people in accord with science. Kant represented, then, not only a shift to technologies of the self but a direct effort to tie them to the university and its epistemic authority. Every discussion of a reimagined university which followed Kant's had to begin where he ended—the university was an institution devoted to science and the particular kind of authoritative knowledge that it made possible.

In the *Critique of Pure Reason*, Kant claimed that a critical philosophy could manage the "growth of the sciences" by "reigning in" metaphysical questions.[1] While this formulation of critique as transcendental philosophy is familiar, elsewhere Kant described the task of critique in explicitly bibliographic terms. One day, as he put it, critique would make the "great multitude of books dispensable."[2] This was the underdeveloped promise that Krug and other post-Kantians seized upon when they wrote their formal encyclopedias intended to organize what they considered a fragmented world of books and sciences. For them, Kant's philosophical project was another technology—in the same lineage as *historia literaria*, encyclopedias, indexes, and specialized journals—for managing print and the cultural anxieties that surrounded it.

Like Heinzmann and other late eighteenth-century writers, Kant complained that the overabundance of books encouraged people to "read a lot" and "superficially" and even fostered a pathological condition, *Belesenheit* (the quality of having read too much), which he considered a potential liability for the sciences, especially philosophy.[3] Extensive reading exposed one to a bibliographic "waste" or "literature," a term Kant used to refer to the totality of what had been printed.[4] Although he generally wrote about "literature," Kant's complaints about the mass, excess, and superabundance of circulating texts implicated the specific technology that had made such large flows possible—print. And, like so many others of his day, he worried that modern print technologies had outstripped the broader culture's capacity to distinguish good books from bad ones, to regulate its reading. Kant was concerned about epistemic authority in a modern media environment, and he was searching for new technologies with which to secure it.

Enlightenment named, then, not only, as Kant famously put it, an "exit from a self-incurred immaturity" but also an attempt to deal with a bibliographic vertigo brought on by the increased circulation of knowledge through print.[5] Recalling Plato's anxieties about writing, Kant worried that modern print technologies indiscriminately scattered knowledge without any consideration of quality or who might actually read what had been published. The printed word circulated without constraints, but its ubiquity and easy accessibility lent it the illusion of real knowledge. "It is so easy to be immature," he wrote in his famous essay on the question of Enlightenment, "if I have a book that thinks for me."[6] It was the book, and not the priest, the doctor, or the tyrant, that posed the first threat to enlightenment. And while Kant never claimed that critique would make books as such dispensable, he did cast critique and his philosophical system more generally as ethical therapy for an Enlightenment information overload. The task he set for his critical philosophy was not only to secure the ground of reason by preparing the way for a science of metaphysics but also to discipline philosophy by disciplining critique's relationship to the unchecked proliferation of print as literature.[7] This was only possible by forming disciplined thinkers committed to the project of critique.

Kant's systematic philosophy was a tool for distinguishing philosophy as a rigorous science from more bookish or bibliographic forms of knowledge which he thought had come to predominate over the course of the eighteenth century. *Historia literaria* and its related technologies were mere "tools of erudition" which valued the circulation of ideas over the discrimination among them.[8] Most importantly, scholars' overreliance on print obscured the place of the human in the production of knowledge. Kant's worries about the apparent excess of print were not about the sheer quantity of books, but rather about how changing modes of communicative technologies determined the conditions in which people thought. He feared that print technologies were creating the conditions that made "thinking for oneself" increasingly difficult. His solution was a disciplined form of philosophical thinking—a rigorous form of thought that he contrasted to the kinds of superficial and derivative ways of thinking he associated with the modern book market. Most importantly for our story, his cultural critique of eighteenth-century information overload and proposed solution culminated in an appeal to an institution that could serve as the bulwark of rigorous, critical, and disciplined thinking and a community for those who practiced it—the modern university.

## Kant on the Communicability of Reason and the Concept of the Book

Kant's concerns about modern print technologies were informed by a basic assumption that knowledge not only was possible but could be communicated to other persons.[9] This assumption is crucial to understanding just how entangled Kant's explicitly philosophical project was with his more empirical and historical concerns about books, the modern print market, and the authority of knowledge.[10]

In order to avoid the specter of solipsism and radical skepticism, Kant insisted that real thought, as opposed to fantasies or chimera, had to be communicable. Knowledge that was tied simply to a subjective "mental state" or an individual cognitive process was no knowledge at all.[11] In order for individual cognitions [*Erkenntnisse*] to qualify as actual knowledge [*Erkenntnis*], they had to be "universally communicable." Otherwise, they would be "a merely subjective play of the powers of representation, just as skepticism insists. But if cognitions are to be able to be communicated, then the mental state, i.e., the disposition of the cognitive powers for a cognition in general, and indeed that proportion which is suitable for making cognition out of a representation (whereby an object is given to us) must also be capable of being universally communicated; for without this, as the subjective condition of cognizing, the cognition, as an effect, could not arise."[12] A "mental state" is the subjective condition from which "knowledge as such" springs forth. Rational knowledge is an "effect" of a cognitive process that exceeds individual mental states. It is by definition communicable by means of the faculty of reason. If knowledge is to count as real knowledge, it must be in some way sharable. It is not reducible to an individual's capacity for thought or the mental state of a subject at a particular time. Alluding to the organic metaphors that he elsewhere used to describe the relationship between particular cognitions and the whole system of knowledge, what he termed "science," Kant describes knowledge as always more than the sum of its parts, which in this context means always more than individual cognitions.

Kant primarily discusses the "universal communicability" of reason in terms of the transcendental, a priori conditions under which aesthetics judgments qualify as objective and universal.[13] But even Kant's transcendental account presupposes that rational knowledge is "in some form made objective, that is, communicatively objectified."[14] How this actually happens in experience—that is, how mental states are actually communicated in particular objective forms—does not interest the Kant of the so-called critical system, because such questions would veer into empirical anthropology and history. In his *Anthropology from a Pragmatic Point of View*, however, Kant discusses the communicability of reason in more empirical and less categorical terms. If one is to avoid subjective delusions, he writes, then one needs to share one's thoughts publically. The "greatest characteristic of insanity" is the "loss of a common sense (sensus communis) and its replacement by a logical private sense (sensus privatus)."[15] It is therefore necessary to test one's private judgments and "the health of one's understanding" against the ideas of others "publically."[16] A common sense helps us test our private thoughts against a more common or intersubjective perspective. If someone shares his private thoughts by making them public, he can compare and test them against other peoples' ideas and recognize when his own ideas are false. Real, rational knowledge is in this sense objective knowledge, that is, knowledge that cannot be reduced to a private knowledge that is peculiar to one person's mind. "What cannot be expressed publically—with the openness to argument and revision that such openness implies—cannot belong to the rational" and therefore cannot count as real knowledge.[17]

Kant goes on in his *Anthropology* to make an explicit connection between the importance of such communication and one particular objective form of communication, namely, books. "The prohibition of books, which are merely concerned with theoretical opinions," he writes, "is an insult to humankind. We are thereby deprived not of the only but still the greatest and most useful means of correcting our thoughts. This happens by us making them publically in order to see whether they accord with the understanding of others, otherwise something merely subjective (habit or inclination) could be easily taken for something objective."[18] In order for private knowledge to be shared and tested, it must assume some objective communicative form. It needs to be shared in particular material form. All this is to say that although Kant never fully identifies real knowledge or science with particular concrete forms of communication, such as printed books, he presupposes their necessity.[19]

And on occasion he even considered books in detail. In "On the Unauthorized Reproduction of Books" (1785), for example, he distinguished between a book as a material object and a book as an action. Both communicate speech, but in different ways. As an object, the book is a commodity, a "mute instrument for the delivery of a speech to the public."[20] It is, as Kant puts it, an "*opus*" (Latin for "object"). In this sense, "he who has printed the book," the publisher, speaks in the name of the author. As an action, in contrast, a book is "not a thing" at all but a "speech." It is an *opera* (Latin for "exertion, labor, work"), an action whereby an author speaks to his readers. In this second sense, a book is not reducible to its material form. Kant based his dual conception of the book as *opus* and *opera* on the priority of an oral act of communication, the author's speech act.[21] According to this logic, the authority and integrity of a communicative form degrade the further it gets from a particular act of speech. As we shall see, this is the basic logic through which Kant would eventually evaluate the modern book market and its endless circulation of commodified books. And it would drive him to turn to the institution of the university as a bulwark against such commodification.

## The Style of the *Critique of Pure Reason* and the Proliferation of Print

Kant's dual conception of the book and his concerns about the culture of modern print can be understood in the context of his attempts to respond to the initial reception of *The Critique of Pure Reason*. Compared to some of his earlier and more popularly received texts—such as "Dreams of a Spirit-Seer Elucidated by Dreams of Metaphysics" (1766)—which were characterized by a freer, more open style, prone to imaginative digressions, the *Critique* is characterized by a rigid precision.[22] The first reviews of the *Critique* complained that its "terse writing style" required great effort on the part of the reader, that it would be "incomprehensible to the greatest part of the reading public," and that it hovered "in the clouds."[23] The popular philosopher Christian Garve worried that the *Critique*'s esoteric style betrayed the

Enlightenment imperative to educate the reading public and cultivate taste. A philosopher, he wrote privately to Kant, could surely "render more easily comprehensible (to readers not wholly unaccustomed to reflection) the truths that are supposed to bring about important reforms in philosophy."[24] A "useful" philosophical system must be expressed in a more "popular manner."[25] If a proposition were true, he concluded, it should be clearly and easily communicable to a relatively broad reading public.

Garve's criticisms of the *Critique* echoed a particular conception of philosophy common in Germany in the 1760s and 1770s. So-called popular philosophers insisted that true knowledge should be both useful and widely accessible.[26] Such a conception of philosophy also represented a shift away from traditional concerns with epistemology and metaphysics (theoretical knowledge) toward ethical and political philosophy (practical knowledge) and its popular effects. In this sense, Garve focused less on the content of Kant's philosophical arguments than on their form and what he considered Kant's false assumptions about the proper ends of philosophical inquiry.[27] Popularity, wrote Garve, concerned the "manner in which one dealt" with the objects of philosophical inquiry, and Kant had disregarded the public's capacity to deal with his new work.[28]

Garve based his commitment to a "popular" form of philosophy on the contention that truth claims should be derived from common experience. The task of reason was to chart a middle path between skepticism and dogmatism, a path best marked by common sense which should orient the speculative use of reason. Popular philosophers, like Garve, however, never offered a thorough defense of such a reason. They were more interested in implementing what they presumed to be the principles of reason than in articulating them.[29] For Garve, reason simply clarified propositions that were gleaned from experience. He argued further that because reason was common to everyone and its truth claims were derived from common experience, philosophical arguments should be broadly communicable. All readers should be able not only to follow but also ultimately to accede to the clear, rational force of good arguments.[30] True, authoritative knowledge was grounded in a common, accessible reason. Philosophers simply had to shine a rational light on this common ground and share this common knowledge with the broader public. The conclusion that Garve drew from these arguments was that esoteric forms of argumentation masked bad philosophy. Complaints that the *Critique of Pure Reason* was "a sealed book," then, were implicit criticisms of its underlying epistemology.

In a second, revised review, Garve expanded on his criticisms about what he termed the *Critique*'s two "literary" problems. The first concerned the abstract nature of its inquiry. It was "too removed from sensibility and intuition for the reader to approach [it] even with the greatest of efforts."[31] The second concerned its "new language," which Garve compared to mathematical terminology. The *Critique*'s language was an intricate, interdependent system that lacked clear points of access: "The terminology is like the thread of Adriane,[32] without which the most clever head would be unable to lead his readers through the dark labyrinth of abstract speculations. Even when the reader cannot see clearly he feels to his comfort that he still holds the thread in his hand and hopes for an exit. Regardless of how much one tries to shed the daylight of the common understanding in these dark, lonely passageways, the common understanding is rarely able to illuminate the path sufficiently, the path that one previously found through some type of feeling."[33] The *Critique* required the reader to proceed from point to point such that "no human being can comprehend the tenth member of this chain without having gone through the nine preceding members in their order."[34] The terse terminology was the thread that the reader had to grip tightly to find his way to the end. For Garve, this was the form not of philosophers but of mathematicians, who spin "artificial webs" that can only be entered by other mathematicians. Such systems are concerned less with truth as correspondence with things in the world and more with their own self-enclosed integrity. Had the *Critique* been a mathematical treatise, concluded Garve, its obscure, esoteric language and forms of thought would have been expected.

Kant readily conceded that the language of the *Critique* might "be damaging to the book itself" and thus the popularity of the system, but he also lamented that his "philosophical exposition" could not be as "fully armored as a mathematical treatise."[35] The unity of his philosophical system could only be guaranteed by its internal integrity and resistance to external influences. For Kant, the nearly impenetrable language of the *Critique* safeguarded the system, while for Garve it condemned it to irrelevance. But what was Kant protecting his philosophical system from? What were the external threats that he feared would undo his system?

In a letter to the German philosopher Moses Mendelssohn in 1766, Kant pointed to the dangers of philosophical chatter. Decrying the pitiful state of German philosophy, he confessed that he could not conceal "how much I regard the inflated arrogance of entire volumes full of such insights as they are now available with repugnance, even with a bit of hate. Even the complete eradication of these imagined insights could not be as harmful as the dream of science itself with its accursed proliferation."[36] A "waste" of books circulates methods, questions, and purported insights that do not deserve the name "philosophy." The Enlightenment dual imperative to advance knowledge and disseminate it as widely as possible in print had led to a confused and superficial public discourse on philosophy, characterized by the "dallying" and "tiring chatter" of contemporary "writers." The "battlefield" of metaphysics took material form in the pages of an increasingly profligate literature in which metaphysical questions "never end" and combatants never gain "an inch of territory."[37] Kant prophesied an end to this battle, however. Philosophy as literature, as popular printed discourse, would ultimately prove to be the "euthanasia" of "false philosophy." As he put it to Mendelssohn, "before true wisdom can come to life, it is necessary that the old one destroy itself. And just as putrefaction is the total dissolution that precedes a new creation, so does the current crisis of learning give me great hope that the long-awaited revolution is not far off."[38] The compulsion to distribute everything indiscriminately would ultimately undermine the Enlightenment project. It simply produced too much and made it increasingly difficult to distinguish good from bad thought.

Kant's dissatisfaction with "the prevailing culture of philosophy" was based in part on what he saw as the conflation of philosophy with literature.[39] The Enlightenment imperative to make philosophy immediately useful and popular reduced it to mere literature, to an easily accessible and communicable form of information. For Kant, the pursuit of popularity had given rise to the spread of bad philosophy. So, how did he envision philosophy's future role in an Enlightenment awash in print and under the demands of the market?[40]

The *Critique* represented one possible way forward. In this sense, the initial reception of the *Critique* and Kant's defense of its style offered a debate not only about particular philosophical propositions but also about the proper ends, form, and audience of philosophy. For Kant, the *Critique* represented a uniquely *philosophical* investigation. Its shift in tone and style was deliberate. In composing it, he wrote, he "chose the scholastic method" over the "free movement of the mind and wit" that had characterized his previous work.[41] And he did so even though he knew that the "dryness of this method would scare away readers" who only wanted an accessible account of moral issues of daily conduct. Even had he been capable of "wit and eloquence," he continues, he would have eschewed them because he wanted to "leave no suspicion that I might want to engage and convince the reader; instead, I would expect no other ascendance from the readers than through the mere power of the insight."[42] In contrast to the "false philosophy," true philosophy, as Kant envisioned it, shunned the ornaments and excesses of rhetoric and relied instead on the immediacy of pure "insight." The *Critique*'s "broodingly dry and constricted" form was an explicit rejection of what Kant saw as the excesses of the "dallying" and "tiring chatter" of philosophy as literature circulating in print. Its impenetrability was a bulwark against what he considered to be the Enlightenment's tendency to reduce philosophical inquiry to popular literary musings about the good. The truth value of a proposition or even the ultimate value of philosophy, suggested Kant, bore no necessary relationship to the ease with which it could be communicated. Although Kant advocated the "necessary communicability of reason," he gradually refined his understanding of it in the age of print and addressed questions concerning to whom and how it should be communicated.

## A New Order of Philosophy

Two years before the publication of the *Critique of Pure Reason* in 1781, Kant wrote his friend Marcus Herz that while composing the *Critique* he had struggled to articulate "the principles needed to achieve popularity in the sciences generally . . . especially in philosophy."[43] Despite his embrace of an impenetrable language and complex system, Kant had not completely abandoned the idea of popularity. In fact, throughout the *Critique* and elsewhere he insisted that the end of philosophy was not simply the construction of theoretical systems but ultimately a practical undertaking, concerned with the conditions of moral action. Philosophy, he wrote, ought to address the "interest of all humans" and cultivate their autonomy.[44] There was a role for "popular" concerns, but Kant had not yet been able to clarify it. In order to reconcile the Enlightenment's interest in popular ethical and moral questions with what he considered the necessity of a rigorous, scholastic method, a "wholly different order" of the sciences was required. Kant would have to reconceive of the unity of philosophical inquiry and its relationship to questions of communication.

Kant's intimation of a "different order" anticipated a shift in his thinking away from an encyclopedic description of the unity of knowledge toward what he would later term an architectonic order. Before the *Critique*,

Kant had considered an encyclopedia the best technology for organizing the burgeoning "expansion" of knowledge.[45] And, like many of his German contemporaries, Kant thought that Diderot and d'Alembert's *Encyclopédie* was the model. For d'Alembert, the encyclopedia's task was to "arrang[e] materials that for the most part have been furnished in their entirety by others."[46] An encyclopedia collated, distributed, and managed knowledge. In his lectures on a philosophical encyclopedia, however, Kant envisioned a different form.

In these lectures, which he delivered regularly from 1756 to 1796, Kant alluded to the encyclopedia's philosophical potential.[47] A philosophical encyclopedia would, he claimed, allow one to "find anything we desire" without having to "burden memory."[48] The proper response to information overload was to survey and map the very elements of knowledge. The task of an encyclopedia, according to Kant, was "to determine in advance the horizon of the entire human species (both past and future), . . . to determine the place of a science in the horizon of all knowledge (and to attempt to unite it with other sciences)."[49] Although appropriating d'Alembert's map metaphor, Kant described this encyclopedic task as an a priori determination of the limits of knowledge and the appropriate place of science within it. Aside from introductory comments on encyclopedias and the task of philosophy, the lectures consisted of various sections on the history of logic, including concepts and judgments, and a section on metaphysics. In its printed form, it reads as a brief introduction to philosophy and not a coherent system of sciences.

Kant's comments on a future philosophical encyclopedia remained incomplete, however.[50] In the so-called critical canon, there was a conspicuous absence of almost any mention of an encyclopedia.[51] By the time Kant had published the *Critique of the Power of Judgment*, he had stopped appealing to an encyclopedic unity and appealed instead to what he termed an architectonic unity of knowledge.[52] The architectonic was a concept through which one "derives the manifold from the whole."[53] He suggested that human reason was by its very nature "architectonic" and thus prone to consider all "cognitions as belonging to one possible system."[54] The architectonic, then, was an elaboration of what Kant considered the a priori task of a true, critical encyclopedia. It was the idea of the unity of the sciences. Kant also defined the architectonic as the "art of systems" that "out of a mere aggregate" created a system or unity of cognitions under one idea.[55] In an aggregate, the various parts precede the idea of the whole, but in a system the idea of the whole precedes the parts.[56] The unity of knowledge is not something one accomplishes through the busy labor of endless accumulation or an appeal to a common reason available to everyone; instead, it is the product of an act of thought. The unity of knowledge is an idea, not something that can be accomplished through the aggregation and accumulation of a material. It was discursive and not something to be achieved in print. Kant used the concept of the architectonic to separate his philosophical project from Enlightenment knowledge projects that engaged in what he considered futile pursuits of material completeness.

As Kant had intimated in his letter to Herz in 1779 and clarified in the *Critique of Pure Reason*, the "wholly different order" of the architectonic was ultimately an order of the sciences which would enable him to account for and manage the "growth of sciences."[57] For Kant, a science was not simply a taxonomic category or, as the term was used for most of the eighteenth century, a mental capacity; instead, it was itself a systematic unity of manifold cognitions. And the architectonic was the "doctrine" of how such "ordinary cognition"—disparate ideas, principles, and concepts—was made into science, i.e., how an aggregate of thoughts or ideas becomes a system. The unity of the sciences was achieved not by comprehensively collecting but by systematizing and organizing particulars according to an idea of the whole.[58] And science—not an encyclopedia, not a complete index, not a book of books—was the name of that whole that could not be found in empirical reality. Science was a new technology for "processing and treat[ing] scientific cognitions."[59] Science as system was a filter-and-search technology, and the architectonic was a second-order account of how science organized knowledge.

Kant described the organizing capacity of such a science in organic terms. Like a living organism, science "grow[s] internally, but not externally."[60] It grows not simply by adding each new limb but rather by integrating it into the whole and thus making both the part and the whole stronger without altering the integrity of the body. This integrated body, this whole, writes Kant, lies like "a seed" in reason itself. The unifying potential of a science is a product of reason itself. The unity of knowledge is not to be found in a divinely ordered cosmos or gradually gathered up like so many dispersed pieces of truth. It is a product of human reason. Science is a self-organizing, dynamic system. Herein lies one of the basic differences between Kant's architectonic concept of science and Diderot and d'Alembert's encyclopedia. Whereas for Kant the unity of science has a distinct center to which all particulars can be related, for Diderot and d'Alembert the encyclopedia had no distinct center.[61] Their encyclopedia had a central stem (reason) and a complex root system, but neither was reducible to a "seed" from which a self-organizing whole emerges. The best they could hope for was an errant representation of knowledge. For Kant, this unity is grounded in an idea of reason which cannot be represented in any material form. It can only be maintained as a concept.

Kant's contention that the idea of unity of knowledge corresponded to nothing in experience and could not be fully embodied in a particular communicative form conflicted with the assumption held by the empire of erudition that print technologies would organize knowledge. Enlightenment conceptions of the unity of knowledge as a material completeness fit well with the endless production of print. As scholars accumulated ever more information, they could just publish revised lexica, expanded encyclopedias, or simply more books. But as some of these projects began to collapse under their own hypertrophic growth, the pursuit of a complete account of knowledge in print struck Kant as futile, conceptually ill-conceived, and ethically irresponsible. The pursuit of "completeness" was the source of the increasingly anxious lamentations about an information overload. As a technology—a way of interacting with the world—the idea of a unified knowledge operated outside of time and space. It filtered the "expanse" of knowledge in a universal present tense that operated without delay, interruption, or mediation.

## A System without Print

But how could Kant's "different order" be shared, much less communicated in print? "What sort of treasure," he asked in the preface to the second edition of the *Critique of Pure Reason*, had been left behind "in the form of a metaphysics that has been purified through criticism but thereby also brought into a changeless state"?[62] And to whom was it communicable? Could a system of pure reason or the idea of the unity of knowledge be printed and distributed to an anonymous reading public and maintain, as Kant put it, its "utility," that is, prove a useful science?[63]

Kant acknowledged that the *Critique* had certain "difficulties and obscurities" that had adversely affected the "judgment of this book"[64] and would have to be removed. He cautioned, however, that any alterations to the published text should have no effect on the system itself, because "any attempt to alter even the smallest part [of the system] directly introduces contradictions not merely into the system, but into universal human reason."[65] The system would have to maintain itself in its "inalterability." Kant's solution, then, was to detach his system from the printed text and tie it to reason itself, which he considered unencumbered by technologies and even space and time. Like the pure reason from which it emerged, his system was timeless, disembodied, and thus necessarily technologically unadorned. The printed text was merely a derivative, material trace of the system itself.

Elsewhere, however, Kant acknowledged how difficult it was, even for a philosopher, to conceive of a disembodied system, to maintain the idea of the unity of knowledge without the aid of communications technology. In order to maintain the absolute integrity of a system, he wrote, the philosopher of pure reason "must hold his object hanging in midair before him, and must always describe and examine it, not merely part by part, but within the totality of the system of pure philosophy."[66] The unity of the system could not be written down or printed on a page; to preserve its integrity, it had to remain an idea. But the system's disembodied character, warned Kant, posed health risks. When the philosopher attempted to give the system a form, he risked a "pathological condition" that

> accompanies and impedes his thinking, insofar as thinking is holding firmly onto a concept (of the unity of ideas connected in his consciousness), produces a feeling of a spastic state in his organ of thought (his brain). This feeling, as a burden, does not really weaken his thought and reflection itself . . . ; but when he is setting forth his thoughts (orally or in writing), the very need to guard against distractions which would interrupt the firm coherence of ideas in their temporal sequence produces an involuntary spastic condition of the brain, which takes the form of an inability to maintain unity of consciousness of his ideas, as one takes the place of the preceding one.[67]

To think philosophically is to concentrate on a single concept. Any attempt to give the concept a communicative form threatens its unity. Whenever a philosopher tries to speak or write out the whole, he risks losing his focus. The concept seems to fade away. He then compensates by trying to focus even more intently, but these efforts to avoid "distractions" and redouble his attention only make things worse. He ends up with mere parts. The whole dissolves, the concept slips away, and the disoriented philosopher ends up mumbling, "Now where was I? Where did I start from?"[68] The entire endeavor turns him into "an invalid," incapable of concentrating on anything.

Since Plato's *Phaedrus*, one of the underlying oppositions in Western thought has been between speaking and writing, but Kant radicalized this ancient worry about the dissemination of thought. He suggested that any communicative form degrades the timeless unity of thought. His anxieties about modern print concerned the relationship between humans and their communicative technologies more generally. He worried that communicative technologies mired thought in the exigencies of time and place. As demonstrated by his emphasis on "concentration," Kant equated philosophy with absolute presence of mind, an unwavering focus on a unifying idea. In the context of an Enlightenment information overload and the ascendance of print's cultural authority, Kant's emphasis on "concentration" was understandable. Philosophy was not a set of axioms or arguments printed on a page; it was an ideal system that, in order to be accessed, required a new way of thinking. But as Garve put it, how could such a system and its organizing idea be transferred from the head of the author to the head of a reader without suffering "intolerable alterations"?[69] Kant was struggling with the effects of technological change on how people thought, but he was also struggling to hold together a theoretical concept of unity and an embodied, material account of the particular. He was struggling to imagine an account of the whole amid the profusion of particulars. He was attempting to imagine an authoritative form of knowledge in a media environment that seemed to lack any notion of epistemic authority.

## Two Discourses of Philosophy

The Enlightenment imperative to produce new knowledge and disseminate it to a broader public only exacerbated these concerns about thinking in an age of the proliferation of print. Kant embraced the need for an esoteric, systematic philosophy, but he remained committed to the Enlightenment vision of sharing the fruits of reason with all of humankind. The central question in a modern age of print, then, was how the benefits and insights of reason could be communicated while maintaining the quality and authority of rigorous, real knowledge. As he observed the emergence of an expanded reading public, Kant began to doubt the public's capacity to "concentrate" and engage in complex thought and the market's interest in cultivating anything like real knowledge. Unlike his systematizing predecessors, such as Descartes, who suggested that any clear-thinking person might understand and could conceivably undertake his method, Kant gradually distinguished between those who might enjoy some mediated benefit of his system and those who might actually grasp and work within it. He eventually rejected Garve's popular Enlightenment claim that the clear light of reason would be equally evident to all over time.

In order for "real" philosophy to flourish in the modern age of print, concluded Kant, it could no longer be invoked as an all-encompassing term for thinking. Throughout his various lectures and in his "critical canon," he increasingly distinguished between two types of philosophical discourse, two ways of talking and thinking about philosophy: philosophy as a scholastic system and a cosmopolitan attitude, or what he termed the scholastic and the cosmopolitan concepts of philosophy.

According to its cosmopolitan concept, the purpose of philosophy was to relate "all cognition to the essential ends of human reason."[70] It should not be reduced to maintaining the coherence of a philosophical system. "Until now," wrote Kant, philosophy had only been scholastic. Metaphysicians of all sorts—mathematicians, logicians, and other "craftsmen of reason"—had limited the ends of reason to constructing and maintaining rational systems. But philosophy was not just a science of "representations, concepts and ideas or a science of all sciences," it was also a science of "the human . . . its thinking and action."[71] It addressed what the human "is and what it ought to be." It was an explicitly ethical undertaking concerned with the shape of actual human lives.[72] The ideal "teacher" of a cosmopolitan philosophy sought to advance the "destiny" of the human being.[73] Echoing the ethical claims of popular Enlightenment philosophers, Kant contended that all humans should recognize the self-legislating potential of reason and that philosophers should never forget that "all interests are ultimately practical. Even speculative reason is limited and only complete in practical use."[74]

According to its scholastic concept, in contrast, the purpose of philosophy was limited to securing the ground for the "future system of metaphysics."[75] Kant employed the term "scholastic" in subtle and sometimes conflicting ways, however. When used in contrast to "cosmopolitan," it oftentimes denoted a truncated, solipsistic form of philosophy. When used in other contexts, however, it denoted a rigorous and necessary way of thinking. Depending on the context, Kant alternately expresses an admiration for the rigor and integrity of scholastic philosophy and a disdain for its solipsism and disinterest in ethical and social questions. When used positively, "scholastic" denoted not simply a descriptive claim about different methodologies but an evaluative distinction between his own de facto rigorous system and lesser attempts at philosophical thinking. In particular, he opposed the purported rigor of scholastic philosophy to the unprincipled methods of those whom he referred to as polyhistorians—practitioners of *historia literaria* who were driven by an all-consuming curiosity, an uncontrollable lust for information.[76] Echoing Thomasius and Wolff's early eighteenth-century critiques of *historia literaria*, Kant wrote that without rational limits and clear ends, historical knowledge, or *Polyhistorie*, "expands" uncontrollably and exceeds rational control.[77] Kant tied historical knowledge to book learning and the collection of mere facts. This association not only served as a convenient trope for distinguishing literature from philosophy but also marshaled broader cultural anxieties about the deleterious effects of print in order to sharpen the distinction with his critical philosophy.

In the *Critique of Pure Reason*, Kant appealed primarily to philosophy as a scholastic discourse. Here, philosophy was a system to be tinkered with and refined but not fundamentally altered by a circle of "deserving men, who combine well groundedness of insight so fortunately with the talent for a lucid presentation (something I am conscious of not having myself)."[78] Intent on maintaining control over his system so that it could be communicated in exactly the form that he intended for it, he summoned an elite band of scholars to ensure its integrity in print. These future scholars would not alter the system itself but "refine" its presentation in print and lend it the "necessary elegance."[79]

But how were these two philosophical discourses to be reconciled? In the age of print, how could philosophy combine the scholastic "precision of concepts" with the cosmopolitan interests of reason?[80] With respect to the medium of print, Kant's more popular essays—such as "An Answer to the Question: What Is Enlightenment?"—offered one solution. Although they were addressed to a broader reading public, they assumed the entire conceptual structure of the *Critique* but never directly mentioned it. The conceptual system hovered behind these essays.

This was because the cosmopolitan concept of philosophy, as Kant explained, consisted "not in setting aside scholastic standards, but only in not letting the form [of scholastic philosophy] be seen as the framework (just like one draws a penciled line, on which one writes, and later erases it). Everything scientific must be according to rules; but the technical [quality] of popular philosophy should not be seen, rather [the cosmopolitan philosopher must] condescend to the power of comprehension [of common people] and the typical expression."[81] Even if its exposition is accessible to a popular audience, a "scientific" text—that is, a rigorous and technical one—never abandons its systematic underpinnings.

A cosmopolitan, or what Kant calls a properly popular, philosophy necessarily functions on two levels of discourse. Implicitly, it adheres to the logic of its organizing system. Explicitly, it does not directly communicate in the system's idiom but rather in the idiom of the broader public. Some texts would operate on these two levels simultaneously, but others, like the *Critique*, were never intended to be popular. They were grounded "in the science itself and not in the exposition [or] the book itself."[82] By distinguishing between two philosophical discourses, Kant redefined Enlightenment notions of popularity not as an epistemological assumption but as something to be achieved. Garve assumed that truly rational thought should be readily communicable, but Kant suggested that it might only become communicable under certain conditions and constraints. "Popularity," he wrote, "cannot be the beginning point" of any science.[83] It is only possible, if at all, after the completion of the system, that is, after science has unified all parts into a whole. In this sense, Kant suggested that philosophy could have a cosmopolitan intent and yet eschew "a common language."[84]

## The Discipline of Reason

Kant's solution to the problem of philosophy in an age of print did not end simply with the distinction between two levels of discourse, however. He took efforts to protect what he considered to be philosophy's uniquely rigorous form of thought. In the preface to the second edition of the *Critique of Pure Reason*, he summoned his select group of editors not just to clean up his turgid prose but to become, as he put it, "members" of his system.[85] Those who had "mastered the idea of the whole" would become prosthetic extensions of the system. Kant's scholastic summons, then, was not only a bibliographic one—edit my book!—but also an ethical one. It required an existential decision to meld objective rational principles and definitions, the elements of the system, with subjective experience. The supposed author of the system, Kant, and a select few of its readers were interchangeable. The system could incorporate them both. Whereas the specialized journals and encyclopedias that had been common since midcentury focused on organizing objects of print, Kant extended the reach of his organizational intentions to persons. He framed specialization as a solution to information overload in explicitly ethical terms. "A good book," he wrote, was not one that sold well or that people read repeatedly, but one whose "particular style" could be made one's own.[86] A reader or a student could only really access the philosophical system and thus true philosophy if he were transformed.[87]

But what becomes of philosophy when it is refigured as an ethical transformation of its reader or subject? How can an ethical summons be communicated and shared? For Kant philosophy would cease to be embodied in and thus constrained by books; instead, it would be incorporated in the self-legislation of reason or, to put it in a less Kantian idiom, the discipline of philosophy. Philosophy would be disseminated not merely through print but through the discipline of philosophy as exemplified in the philosopher who, claimed Kant, "remains the exclusive trustee of a science that is useful to the public even without their knowledge, namely the critique of reason; for the latter can never become popular, but also has no need of being so."[88] The system is propagated by the philosopher himself, to whom a gift—the system—has been entrusted. Philosophy can become a science when it is grounded in the discipline of a science as embodied in a discipline or exercise that forms a self-thinking student.

In the *Critique of Pure Reason*, Kant defined a discipline as a "correction" through which "the propensity to stray from certain rules is limited and finally eradicated."[89] He went on to outline a "discipline of pure reason" in which a student of philosophy is trained in the proper use of reason by moving from dogmatism through skepticism and finally to critique. He termed this philosophical education the "tribunal of reason" through which the student of philosophy becomes a critical thinker. Kant detailed a method, replete with rules and advice on the proper forms of training, for the appropriate use of reason. He called it a "system of caution and self-testing":

> However a cognition may have been given originally, it is still historical for him who possesses it if he cognizes it only to the degree and extent that it has been given to him from elsewhere, whether it has been given to him through immediate experience or told to him or even given to him through instruction. Hence he who has properly learned a system of philosophy, e.g., the Wolffian system, although he has in his head all of the principles, explanations, and proofs together with the division of the entire theoretical edifice . . . still has nothing other than a complete historical cognition of the Wolffian philosophy; he knows and judges only as much as has been given to him. If you dispute one of his definitions, he has no idea where to get another one.[90]

"Historical" here signifies heteronomy—a form of knowledge whose source is external to the knower and particular, unified knowledge. Even if a student could articulate the particular definitions and propositions of a system, unless these "arise from reason" he "has formed himself according to an alien reason." "The faculty of imitation," Kant continued, "is not that of generation, i.e. the cognition did not arise from reason for him. . . . He has grasped and preserved well, i.e. he has learned, and is a plaster cast of a living human being."[91] On this account, philosophy is not merely a set of principles objectified in print or some other communicative form. It is rather a series of steps that every student of philosophy must "reflectively reenact."[92] A critique of pure reason is a form of philosophical therapy through the discipline of reason. Kant envisions a discipline of philosophy which harmonizes objective rational principles with a subjective experience and grounds the authority of knowledge in the thinking subject.[93]

Kant was attempting to undo the gap between what the encyclopedists had termed the objective and subjective aspects of science, the gap between the observer and what is observed. This harmony is "the only cure" for the "thirst for knowledge of the human," for the information lust that Kant claimed plagued literature.[94]

## The Book as Speech Act

Despite his general suspicion of communicative technologies, Kant did not leave such a science or discipline of reason completely disembodied. In various lectures, he described philosophy as a disciplined science characterized by a "form of rigor appropriate to the university."[95] It was in the university that philosophy could be "methodologically learned." He associated scholastic forms of thought, and thus science, with the university as a particular social institution. With the fracturing of the empire of erudition, scholars needed more than a hovering, disembodied system or an esoteric discourse to distinguish and identify real knowledge; they needed a structure or technology that could guarantee the continuity of science as a distinctive form of thought. The university both preserved scholastic methods and "disciplined" scholars into recognizing the differences between scholastic and merely popular forms of thought. But it also trained scholars to recognize the limits of scholastic discourse. Just as a mere popularizer ignores, or doesn't recognize, the borders of science, a pedant ignores the borders of the broad, popular reading public. Each social sphere had its appropriate discourse.

Implicit in Kant's discussion of philosophy as a distinct mode of thought, as an ethical transformation, was a cultural critique of the impact of print in modernity. This critique and his association of scholastic and cosmopolitan concepts of philosophy with particular social spheres became clearer over time, especially as Kant began to express his disdain for what he considered to be the market-driven, capricious tastes of an ever-expanding reading public. One of Kant's earliest defenders, Johann August Eberhard, directly addressed the issue in 1788. The reading public had failed to recognize the profundity of the *Critique of Pure Reason*, he contended, because the surfeit of print had made them poor readers. Because the *Critique* presented a complex system, it required an "industriousness" and "attention" that most modern readers lacked. "Whoever studies a speculative system with the same level of cursoriness with which he reads a novel or a newspaper," wrote Eberhard, can never hope to comprehend the whole that is so essential to grasping the system.[96] In the "superficial epoch" of modern print, readers' capacities to read closely and with deep concentration had atrophied.

The failures that critics attributed to the *Critique* lay neither in the text nor in the system, but rather in modern readers of print. These failures were subjective, not objective. Eberhard blamed the modern reader's "inattention, distraction, idleness, preference for comfort, hastiness, or even sometimes in the lack of exercise in thinking."[97] Kant's system required a completely different type of reader, one defined by "well-exercised thinking, a severe, tireless attention," in short, a "systematic mind." Kant's "discipline of pure reason," concluded Eberhard, was thus an intervention not only in the history of metaphysics but also in the cultural crisis defined by the proliferation of print and its deleterious effects on thinking itself. Kant's system required, as Eberhard put it, an entirely new person—"the scientist." "The scientist" was characterized not by his facility with organizing print, but rather by the "liveliness" of the act of thought.

Kant's concerns with communicative technologies were products of his anxieties about the increasing predominance of the objective aspect of science. If detached from individuals, objective communicative forms—oral, written, or printed transmissions—were derivative, faint imitations of a true "thinking for oneself." These technologies, just like the objective rational principles of a system that could be read in print, had to be harmonized with a subjective experience of reason. The "discipline of reason" required less a change in bibliographic technologies than in technologies of the self. Kant's attempts to reflect on printed literature, however—evident in his repeated comments about *historia literaria*, his frustrations with the reception of the *Critique*, and his distinction between a printed book as "corporeal artifact" and speech—suggested that printed texts posed a particular problem for him.

In the *Metaphysics of Morals*, published in 1797, Kant returned to the distinction he had made in "The Unauthorized Reproduction of Books" between a book as a speech act (*opera*) and a book as a "mute instrument for the delivery of a speech" (*opus*).[98] A book as an *opus* exists, he wrote, as a "thing . . . on its own," whereas an *opera* is an "affair" [*Geschäft*]—the free, intentional activity of a particular individual who occupies a particular social position.[99] As a speech act the book should not be conflated with the book object, which Kant feared risked commodification as it circulated in an ever-expanding book market. Kant reasoned, for instance, that were an author to die after he had given his manuscript to his publisher, the publisher would have no right not to print it. He would be ethically bound to publish it because the manuscript was "an affair that the author wanted to carry on with the public." The legal obligations of the publisher are derived from a prior affair, the presumed speech event, between the author and the public. The book-qua-opus, in contrast, was "mute" because it was merely an unauthorized, inert object that lacked the authority and value accorded to the book as the speech act of a person. It is in this sense that Kant considered *historia literaria* and related technologies derivate forms of knowledge. They treated print objects merely as "tools of erudition" which could be detached from their authors and thus their source of authority.

In the decade following the publication of the *Critique*, Kant made these arguments as he also began to echo broader criticisms, such as Herder's, of the commodification of print and the commercialization of the book market. On these accounts, books, book writing, and print scholarship had devolved into a less than prestigious subsidiary of the book market. In a biting satirical discussion, published in 1798, of the Berlin publisher Friedrich Nicolai, Kant mocked how the big business of print relied on the "factory-like" publication of books.[100] The successful publisher—and Kant mockingly extolled Nicolai as the greatest of them all—had to ignore the "inner worth and content of the commodities that he publishes." He could attend only to the fashionable tastes of the market, for which books were merely "the ephemeral products of the printing press" which circulated "like currency."

The circulation of books as commodities created a reading public characterized less by its public use of reason, as Kant had so famously envisioned in his 1784 essay "On the Question of Enlightenment," than by its capricious tastes as dictated by the market. The commodification of books threatened the enlightening potential of print. Although Kant worried about the theoretical implications of print as a medium (how to communicate a system), he was also increasingly concerned about the degrading impact of print on eighteenth-century readers—"the mass of books" that critique might one day make dispensable. Reason, for Kant, was the capacity to judge freely, that is, according only to rational principles, and he increasingly saw the circulation of print as an unregulated, unauthorized commodity as a threat to this capacity.[101] Conceiving of books as actual speeches allowed Kant to distinguish them from the "mass" of circulating books and link them to the authority of a particular rational being—the philosopher.[102]

But what was the nature of the authority and prestige that Kant ascribed books as opera, and wherein lay their potential to arrest the indiscriminate circulation of print objects? Why did Kant try to ground their authority in the book as opera, as opposed to the book as opus? In a discussion of how one learns and the authority of particular forms of learning, Kant complained about "the great mass of books" and then sketched a brief theory of different media and their relationship to thought: "Thinking for oneself is good, but learning on one's own is not. An oral presentation, even when it is not completely worked out, is very instructive. One does not hear something fully worked out or thought out, but rather one sees the natural way in which one thinks. And that is much more useful. . . . One thinks more when listening than with reading." Kant was concerned not with the relative autonomy of pedagogical practices—whether book-based learning was more or less autonomous than listening to a lecture—but with the performative value of a particular mode of transmission. Kant tied the act of presenting something orally to the act of thinking which the student observes in an oral performance by an authorized person, namely, one whom a community has deemed capable of such an action. Considered in this performative light, Kant's notion of the book as a speech act makes more sense, especially within the context of his cultural critique of modern print. The commodification of books as mere objects detached them from any particular person and thus deprived them of any performative context. Books as mere things circulated outside any signifying context that could authorize them as good or bad examples of thinking. To claim, as Kant did, that "thinking for oneself" is best facilitated by an oral presentation was not an abstract argument about the theoretical hierarchy of media—the oral lecture over the printed text—but more about the commodification of books and its potential effects on readers in late eighteenth-century Germany.

Kant's reconceptualization of the book as opera was a more basic attempt to redefine the normative understanding of scholarship, learning, and knowledge around 1800. By subordinating the material nature of the book—and with it the sciences of books that had emerged over the previous century—to the book as opera, Kant outlined new criteria by which to distinguish true learning "from the appearance of learning."[103] He tied knowledge and learning not to books as things in this world but to the authority of a communicative event laden with performative meaning. The solution to Enlightenment information overload was to ground the authority of print in the speech act of a philosopher.

## Institutionalizing the Discipline of Philosophy

An individual speech act, however, could not sustain philosophy as a discipline, so Kant ultimately returned to what he had previously only intimated. Whereas before he had only loosely associated scholastic thinking and true philosophy with the university, in the *Conflict of the Faculties*, published in 1798, he explicitly made the case that the university was the only technology capable of sustaining the continuity of philosophy, and thus scientific thought, as a practice. What was needed, as he put it, was a philosophical faculty, a "department" dedicated to the "testing and critique" of human knowledge.[104] As an institution the university could sustain the practices and organize the objective material necessary for a philosophical or scientific community. In this text that would become a touchstone for debates about the future of the university immediately after its publication, Kant argued not only for the autonomy of the philosophy faculty—not to be confused with contemporary philosophy departments, but rather one of the four basic university faculties housing all fields but law, theology, and medicine—but more basically for the role of the university in the modern ecology of knowledge.

In the first section of the *Conflict*, Kant differentiated the university's higher (law, medicine, and theology) and lower (philosophy) faculties on the basis of their relationship to reason, which, he argued, was determined by how they interacted with texts [*Schrift*]. Kant used the German term *Schrift* to refer not simply to particular texts or writing, but to the very process of how knowledge was communicated or transmitted. The fundamental difference between the higher and lower faculties lay in the fact that the teachings of the higher faculties were derived not directly "from reason," but rather from statutory texts "entrusted" to them by the state, like the Bible, the Code of Law, and state medical ordinances.[105] Each of the higher faculties represented a particular interest that the state had in its citizens: theology (eternal well-being and inner life), law (civic and social order), and medicine (maintenance of a strong and [re]productive population). Their authority was based on the particular teachings as embodied in the statutory texts. These texts ensured that the "norms" of the state were broadly accessible, durable, and stable. Members of the higher faculties could publish their own books and even endorse the teachings contained in them as "norms," argued Kant, but they would only have a "pedagogical" function. They would have no authority and thus no relevance outside the university. The higher faculties' reliance on government-sanctioned "texts," then, was a sign of their heteronomy, of the fact that their authority was grounded not in the self-legislation of reason but in the "command of an external legislator," the state.[106]

In contrast, the authority of the lower, philosophy faculty, contended Kant, was based on reason itself. The members of the philosophy faculty were interested not in how their teachings might influence people and institutions outside the university, but only in securing the "interests of science."[107] Their interest in what Kant called the "truth" of doctrines was a concern about the sources of the authority of knowledge and the capacity to distinguish truth amid the availability of unexamined, tradition-bound claims to truth. Whereas the state maintained the "right" to sanction the teachings of the higher faculties, it "left" the teachings of the lower faculty to the "reason of the scholars," who, wrote Kant, were not interested in exercising direct influence over "the people." One central qualification to Kant's distinction was the fact that the authority of the philosophy faculty was not ultimately self-given; instead, the state allowed the philosophy faculty to appeal to the reason of scholars. Its authority was thus ultimately granted by the state. The very organization of the university, including the relative autonomy of the philosophy faculty, was a function of the state's authority as manifested in the different faculties' relationship to texts. The university occupied a position between the life of science and the state and broader culture.

Within the university the philosophy faculty was free to raise objections against the teachings of the higher faculties, but only internally, argued Kant. Members of the various faculties could address their objections to particular teachings and doctrines "only to one another as scholars" lest they transgress the boundaries of the university established by the state.[108] Kant limited the pursuit of the pure ends of science (and thus the truth) not just to scholars in general—that is, as members of an empire of erudition loosely connected through print—but to those within the university. He limited, and thus endowed with higher prestige, the category of the scholar by tying it to the university. He distinguished the scholar from the "people" external to the university. "Practically speaking, the people takes no notice of" these internal arguments."[109]

Whereas the term "erudite" had referred for most of the eighteenth century to any member of the empire of erudition, Kant argued for a further distinction between

> actual scholars and members of the intelligentsia (university graduates or literati) who, as instruments of the government, are invested with an office for its own purpose (which is not exactly the progress of the sciences). As such, they must indeed have been educated at the university, but they may well have forgotten much of what they learned (especially concerning theory), so long as they retain enough to fill a civil office. While only the scholars can provide the principles underlying their functions, it is enough if they retain empirical knowledge of the statutes relevant to their office (hence what has to do with practice). Accordingly they can be called businesspeople (clergymen, magistrates, and physicians), they have legal influence on the public and form a special class of the intelligentsia, who are not free to make public use of their learning as they see fit, but are subject to the censorship of the faculties, so the government must keep them under strict control, to prevent them from trying to exercise judicial power, which belongs to the faculties; for they deal directly with the people who are incompetent (like the clergyman in relation to the layman).[110]

Kant goes beyond his previous divisions between university and general publics and implicitly identifies three distinct social spheres: the intelligentsia [*Literaten*], the general public, and university scholars. Kant divided the people [*Volk*], all those not directly associated with the internal workings of the university, into the intelligentsia and the non-intelligentsia. The distinction between these two groups was based on graduation from a university; those who had graduated, the intelligentsia, constituted the first public, and those who had not were simply the—"incompetent"—public at large. As graduates of one of the higher faculties, the intelligentsia, however, were "instruments" and "tools" of the government. They were "businesspeople," bound by their civil office. The third type of public was that of a "learned corporation that concerns itself with science," which Kant tied and limited to the university.[111] About this third, learned public, "the people," both the intelligentsia and non-intelligentsia, "knows nothing."[112] The scholarly public was protected by the university and was separate from both the general public and even the more educated public. It was three levels removed from the broader public. It could question the higher faculties that then trained the "businesspeople" who then either advised or actually served as government officials.

In making all of these distinctions, both among the university's faculties and among the three social spheres or publics, Kant cast the university as the medium between the interests of science, the interests of the state, and an increasingly complex and distinct extra-university public. The university was to be a "bulwark" against government control and the popular demands of the print market.[113] It would mediate the interactions of "actual" scholars and a broadly educated, literate public.

The metaphor that Kant used to describe the university's mediating work was not organic, as one might expect given German idealism's stock of metaphors, but industrial. The university was a factory for the production of knowledge. "It was not a bad idea," wrote Kant, "to treat the entire content of learning (actually the minds devoted to it) in a factory-like manner through a division of labor."[114] As a division of *labor*, this process focused on organizing the persons who produced knowledge. With this plan, there would be "as many public teachers, professors, trustees" as there were "categories of science." Taken together, he concluded, these teachers would constitute a kind of "learned community called a university."

Kant's use of the division of labor concept drew on Scottish philosopher Adam Smith's use of the term but differed in important ways. In *The Wealth of Nations* (1776), Smith had extended the concept of the division of labor from the more efficient production of pins to the more efficient production of intellectual goods:

> In the progress of society, philosophy or speculation becomes, like every other employment, the principal or sole trade and occupation of a particular class of citizens. Like every other employment too, it is subdivided into a great number of different branches, each of which affords occupation to a peculiar tribe or class of philosophers; and this subdivision of

> employment in philosophy, as well as in every other business, improves dexterity, and saves time. Each individual becomes more expert in his own peculiar branch, more work is done upon the whole, and the quantity of science is considerably increased by it.[115]

For Smith, the division of intellectual labor was a twofold process: professionalization and specialization. Professionalization indicated the possibility of making a living and developing a group identity with others of the same form of labor, that is, pursuing a career. Specialization, in contrast, indicated the extent to which one could pursue their scientific work within the constraints of a particular field.

In the *Groundwork of the Metaphysics of Morals*, Kant had adopted the division of labor concept to describe the organization of knowledge. "Where work is not so differentiated and divided, where everyone is a jack-of-all-trades," he wrote, "there trades remain in the greatest barbarism."[116] But whereas Smith claimed that the "progress of society" led to such a division of intellectual labor, Kant claimed that it was the "nature of science" that required it. Smith emphasized the "quantity of science" which could be produced by such a division, whereas Kant contended that it could lift the sciences out of their "barbarism." Kant imbued the concept with a moral force. Whereas Smith's division of intellectual labor focused on the more efficient production of intellectual objects, Kant's focused on the production of persons. The notion that the "entire content of learning" could be treated in factory-like terms was, as Kant put it, "really" an idea about how to handle "the thinkers devoted to it" in terms of the division of labor. In the age of the proliferation of print and the fragmentation of knowledge, the university should be thought of as a factory that produced not just books and articles but the producers of such knowledge, that is, teachers. By reproducing itself through persons, the university differentiated itself from other social groups and publics, like the literati described above. In a university, what is produced and transmitted is the university itself.

In Kant's use, the division of labor described an internal differentiation of the university as well. In the factory-like university, all "categories of the sciences" were entrusted to a particular professor with certain skills and a certain set of knowledge. This division not only of the categories, which preexisted this division, but of the scholars themselves allowed for greater efficiency in intellectual labor. "Where work is divided," one can perform it with greater facility. The university's "factory-like" division of labor granted it a certain autonomy. "Only scholars," wrote Kant, "can judge scholars as such."[117] Without the cultivation of expertise and the continued differentiation of knowledge through the division of intellectual labor or specialization, the university's claim to autonomy would be undercut by claims external to the university.

Such specialization, however, was not simply a matter of efficiency. It was a deeper, ethical claim about the nature of knowledge and its progressive, historical development over time. As Kant had explained when he called "young men" to work on the *Critique of Pure Reason*, to engage in specialized scholarly work was to participate in reason itself. These scholars working assiduously on Kant's book had been called to participate in something larger than themselves. They were playing a part in a whole called "science." But a group of Kantians was not enough to sustain this broader whole that Kant identified as science. What was needed was a stable institution to coordinate their efforts, a community of scholars devoted to science.

Echoing the cultural critiques of Heinzmann, Bergk, Herder, and Novalis, Kant changed the terms of the broader debate about information overload by framing the anxieties about print as an ethical problem in need of a solution that transformed people. The unregulated ubiquity of print threatened rigorous thought, so he called for a critical thinking that could be disciplined and safeguarded in smaller circles of scholars who could be formed to think well, which, for Kant, meant systematically or, more simply, scientifically. And he ultimately turned to the university as the only institution that could sustain such a practice and develop such people. In an age of media surplus, the university could be a bulwark of epistemic authority and the epitome of a critical science.

CHAPTER SIX

# The Enlightenment University and Too Many Books

In the *Conflict of the Faculties*, Kant described an ideal university whose unity was grounded not in the medieval guild of the *universitas magistrorum et scholarium* or in the political ends of the state but in the constant feud among the university's faculties, which a critical philosophy would ensure never ended. With its constant and rigorous questioning of the grounds of knowledge, the philosophy faculty would guarantee the university its legitimacy and authority. It would distinguish university-based knowledge from all other types of knowledge and, by tying the university to the authority of science and its critical capacities to distinguish good from bad, real knowledge from mere information. For Kant, such a knowledge stood in for a particular type of person and way of being in the world. The university was not simply an extension of the church or the state. It was, as Kant envisioned it, a bulwark for science and a type of thinking which had become endangered in a world awash with print.

A little more than a decade after Kant made his appeal for a university devoted to the critical work of science, the founders of the University of Berlin were preparing to dedicate their own institution on October 15, 1810. In anticipation of the event, the *Berlin Evening News* published a poem written for the occasion. Consisting of fourteen quatrains alternating between the voice of a Berlin "native" and a "chorus of arriving students," the poem celebrated the transformative promise of the new institution:

> Here you will find to science
> A castle of heroes dedicated,
> That grants you the courage, the power
> With which you will make yourself anew.[1]

In the previous day's paper, an advertisement for Clemons Brentano's cantata "Universitati Litterariae" written for the planned dedication had appeared. It too celebrated the university as an institution dedicated to science and the unity of knowledge:

> To wholeness, allness, unity,
> To universality,
> Scholarly wisdom,
> Freedom of knowledge
> Belongs this royal house!
> So I lay out the golden words to you: Universitati Litterariae.[2]

A few days later on October 20, however, another newspaper reported that the ceremony had never actually taken place: "Our university was not, as had been previously promised, dedicated on October 15th . . . because there was not enough time to take care of the necessary arrangements."[3] Although some lecture courses had unofficially begun that summer and the first faculty meeting had already taken place early that month, the official dedication had been indefinitely delayed because of absent faculty members, lack of students, and poor building conditions.[4] The "wholeness, allness, unity" that the university was to embody would have to wait for another day.

In many ways that day never arrived. The printed promises made on behalf of the university echoed those made on behalf of eighteenth-century print technologies. Both the university and its print predecessors stood in for a certain order of desire for the control and mastery of knowledge. But the poetic celebrations published in the *Berlin Evening News* marked a shift in how scholars and writers imagined such a unity and order might be achieved. They signified a shift in the reigning metaphors for what counted as authoritative knowledge. In the decade following Kant's turn to the institution of the university, the debate about the future of knowledge increasingly centered on the promises of one technology in particular: the university. If "the book," "the empire of erudition," "the encyclopedia," or even "the Enlightenment" were the operative metaphors over the course of the eighteenth century, "the university" came to stand in for a new order of knowledge around 1800.

Unlike previous technologies, however, "the university" was also an institution. It was not simply a new type of book or encyclopedia that could transmit knowledge more efficiently. It was not just another content delivery device. It was constituted not just of bricks and mortar but also of norms, practices, and people. By the end of the first decade of the nineteenth century, many German scholars and writers embraced the university as *the* technology, as *the* institution of modern science. As Wilhelm von Humboldt put it, however, the university was the pinnacle not just of scholarly labor but of the "moral culture of the nation."[5] As the poetic promises of transformation and freedom suggested, the university represented a distinct community; it was the bearer of a particular practice and culture that embodied a distinct and authoritative form of knowledge. For the idealists and romantic thinkers who rushed to embrace the idea of a university, the solution to Enlightenment information overload and anxieties about the authority of knowledge was not a more expansive encyclopedia or a more complete *historia literaria*, but an institution that organized the objects of knowledge by forming subjects of knowledge. A reimagined university, they claimed, would unite objective forms with subjective practices. It would be, as Friedrich Schlegel wrote, a "living encyclopedia"[6] organized around the character and habitus of the disciplinary self. After the failure of Enlightenment technologies to achieve universal knowledge or material completeness, the university, claimed Schleiermacher, would take their place and "emerg[e] as a whole" that would organize knowledge.[7] The university was a community of people devoted to science which endowed students with the "power" and "courage" to transform themselves in the image of science.

But as the cancellation of the events of October 15 portended, these celebrations of a finally realized—now institutional—universal knowledge were from the beginning elegiac. What was celebrated in those poems, philosophical tracts, and bureaucratic memoranda was not the actual institution but the idea of an institution that represented a new order of knowledge. They registered a longing for an order of knowledge grounded in the integrity of the disciplinary self as the figure of what counted as real, authoritative knowledge. In this sense, the desires printed on October 15, 1810, were neither a poetic flourish nor a simple ideology. They were expressions of an ethical imagination, a vision of the kinds of people who inhabited the research university. And it was this ethical imagination that became the central feature of the institution.

## The Ethical Sources of the University: From Paris to Göttingen

The university, of course, was not invented in early nineteenth-century Germany. As an institution, however, it had never been organized primarily around modern notions of science and scholarship more generally, just as science had never been tied so closely to one institution. In the sixteenth and seventeenth centuries in the wake of the so-called scientific revolution, scholars of the "new science" largely rejected the medieval university's scholastic organization of knowledge and established "extra-university" networks of communication and correspondence academies that flourished in the seventeenth and eighteenth centuries. These were epitomized by the Royal Society in London, established in 1662, and the *Académie royale des sciences* in Paris, established in 1666. Between the late seventeenth and eighteenth centuries, these scientific societies "dominated organized and institutionalized science."[8] Similarly, although universities had always been devoted in different ways to the life of the mind and the pursuit of knowledge, historically their ethical resources, the basis of their underlying norms and authority, had been tied either to the church or to the state. Grounding the university in a distinct culture of science was an innovation of the German research university.

Medieval universities were not simply extensions of the church or the state per se, but they were supported both financially and politically by one or the other, and in most cases both.[9] The privileges and regulations issued by the church to universities in Paris and Oxford in the thirteenth century, for example, "would never have been put into effect but for the collaboration of secular authorities."[10] But as a unitary corporation of students and masters, the medieval university was bound together by Christian values and scholastic practices, like the lecture and the disputation.[11] Even if they were "secularly endowed"—that is, granted their charter or funded by the state—universities had what could be described as a "religious mission."[12]

Initially without buildings or infrastructure, the first universities in Paris and Oxford appealed to Christian theology as the supreme science to ensure the university's universal authority. The medieval university was a *studium generale*, an institution of higher education whose status was confirmed by a universal authority like the pope, thus welcoming students from and granting them rights to teach anywhere in Christendom.[13] It consistently laid claim to an authority that transcended local interests and divisions. Medieval universities were supra-national institutions devoted to a pan-European learning.[14] Undergirding this claim to authority was the legal and financial support of the church, a universal curriculum and language (Latin), and ethical forms and virtues taken from Christian traditions.

For centuries, the university as a corporate body was organized according to faculties or bodies of masters and students devoted to a particular type of instruction. Generally, there were four faculties: theology, law, medicine, and arts and sciences (later philosophy). This internal faculty structure dates back to the University of Paris, which first began as an association of masters and students around 1220 and officially recognized distinct faculties by the end of the century. The term *facultas* referred not to a particular science (*ars*, *doctrina*, *scientia*), but rather, as a rector at the University of Paris put it, to "a body, an association of members devoted to special studies."[15] Each faculty issued and enforced their own ordinances, organized their own lectures, and constituted their own assemblies. They operated as semiautonomous bodies within the broader university.[16] In this sense, the faculty organization of the medieval university embodied less a particular order of knowledge than distinct juridical bodies. As an assembly of faculties, the university's primary purpose was to protect the privileges and rights of a particular class of people, namely, the masters and students of its different faculties.

Unlike the modern research university, where success in national and international disciplinary communities determines prestige and thus tenure and promotion, allegiances in the medieval and early modern university were primarily to a local body. Upon graduation, degree candidates had to take oaths swearing their allegiance not to an abstract notion of science or an external community of scholars, but to fellow members of the local university guild.[17] The structure of authority was based not on individual achievement but on an internal hierarchy of faculties, with theology on top and the arts and philosophy on bottom, and academic degrees, with doctors of theology on top and bachelors of arts and philosophy on bottom. One's place in this academic hierarchy and one's seniority were more important than any "system of knowledge."[18]

These corporate communities were bound together internally and externally by a pre-Reformation canon of texts and associated practices like the lecture and disputation, as well as a basic set of Christian and classical values and virtues legitimated in generally religious terms. Both provided for the general uniformity of medieval university faculties and programs of study and pedagogy. The shared values entailed Christian beliefs, for example, in a cosmological order accessible to human reason, man's fallen nature, the value of speculative or theoretical knowledge, and the authority of tradition. Although particular universities adopted, adapted, and even resisted specific Christian theological claims, they were generally grounded in a Christian theological tradition and dedicated to defending canonical knowledge. The medieval university scholar was characterized by a particular ethos, and the university embodied a desire to understand the rational order of God's creation and "general ethical values like modesty, reverence, and self-criticism."[19] Although they were not determined by either the church or the local state, universities drew much of their ethical orientation from the church and Christian theology more broadly. Through at least the fourteenth century, universities were basically a "unit" of medieval Christendom.[20] This was confirmed by the fact that the theology faculty functioned as the university's model science and general "censor."[21] Even in post-Reformation Europe, theology continued to dominate universities, where institutional confessions of orthodoxy continued to be the norm through at least the early eighteenth century.[22]

While the struggle between church and secular states over control of universities had been a constant since at least 1378 and the Great Schism, a renewed battle over the purpose and authority of the university began in the late seventeenth and early eighteenth centuries. Critics both inside and outside the university, especially in the German-speaking lands, began to argue that universities had changed little since they were first established in the thirteenth century. They were merely extensions of Christendom and inculcated nothing but blind submission to canonical authority and antiquated practices like the disputation. The confessional, medieval university, critics claimed, obscured the true ends of knowledge and science—not contemplation of the divine, but rather the practical needs of a broader public and state.[23] Over the course of the eighteenth century, a number of new German universities—such as Halle (1694), Göttingen (1734), and Breslau (1702)—sought to redefine themselves as institutions dedicated to a knowledge that was practical and socially useful. In particular, they sought to mitigate conflicts arising from religious orthodoxy.

These new universities tended to embrace a public or state good, the pursuit of a better future, and a progress judged by material well-being. They were guided by different ends than those of the medieval university, whose structure was actually predominant throughout Europe until at least the end of the eighteenth century. The Enlightenment university's manifold purpose was to produce state revenues (through student fees), more efficient members of the burgeoning state bureaucracy, technical solutions to particular problems, and citizens of the modern state.

The prototypical Enlightenment university in Germany was the University of Göttingen, established by King George II of England and Elector of Hannover and funded primarily by noble-dominated estates near Hannover.[24] The educational reforms that it introduced—state authority over faculty appointments, higher salaries and fees for professors, the expansion of the philosophy faculty and thus of course offerings, attempts to avoid theological divisions by not hiring doctrinaire theology professors—laid the groundwork for the rationalization and modernization of German universities. Göttingen's rise paralleled the gradual collapse of the medieval university's authority structure and foreshadowed the end of the university as an "autonomous corporation in a hierarchical, semi-feudal society of privilege and tradition."[25] It portended the eventual triumph of the rational modern university over the medieval university and the emergence at the beginning of the nineteenth century of the research university.[26] And, finally, the atrophy of religious orthodoxy and emphasis on "practical studies" oriented it toward social needs.[27] Göttingen was the first university to do away with institutionally mandated commitments of orthodoxy.[28] But it lacked the normative structure—the ethical structure of science as a practice that formed particular types of people—that would prove so crucial to the modern research university. Göttingen represented a new modern university that was co-emergent with the bureaucratic state and bourgeois society and disentangled from religious orthodoxy.

Surveying the sorry state of German universities at the end of the eighteenth century, Christoph Meiners, a professor of philosophy at Göttingen, lamented the "sad fact" that almost all German universities had "not merely declined," but done so in such a sudden and "strik[ing]" fashion.[29] Addressing his Prussian counterparts who, as we shall discuss below, were debating the future of the university, Meiners offered his "own" university as a model, whose success lay in the clarity of its purpose and the ground of its authority. The University of Göttingen found both in the state. A university, he contended, should train the public figures necessary to manage the health, wealth, and general well-being of the citizenry; it should "provide the state with able and diligent doctors, teachers, as well as all other possible civil servants."[30] This was not to suggest that Göttingen did not produce or transmit new knowledge. It was home to some of the most important scholars of the German Enlightenment and what could be considered the first research library. Its primary purpose, however, was to serve the state—be it through the production of knowledge, the education of civil servants, or the collection of student fees. Göttingen may have extricated itself from religious orthodoxy and dogma, but it tied itself to the modern state.

Göttingen was designed according to a revised and more administrative form of mercantilism, or what in eighteenth-century Germany was called "police science" [*Policeywissenschaft*].[31] Whereas previous forms of mercantilism had focused on the fiscal activities of the state and the exploitation of its properties, police science focused administrative efforts on local institutions.[32] A former Göttingen student, Friedrich Böll, likened his alma mater to a factory owned by the king: "You, Mr. Curator, are the factory director; the teachers at the university are the workers; the young people studying and their parents . . . are the customers; the sciences taught at the universities are the wares. Your king is the master and owner of his scientific factory."[33] Unlike Kant's use of the factory metaphor more than a decade later in the *Conflict of the Faculties*, which argued that the division of labor in the university grounded its autonomy, Böll highlights the state's fiscal interests in every aspect of the university. Scholars, students, and even knowledge itself had to be appropriately managed in accord with state interests.

The intimate relationship between Göttingen and the state was overseen by Privy Councilor Gerlach Adolph Freiherr von Münchhausen, who directed the daily affairs of the university for almost four decades as the electorate's leading minister and president of the Hannover treasury. For Münchhausen, the university's chief resource was the prestige of its faculties, whose selection he personally oversaw. If every faculty were filled with "famous and excellent [men]," he argued, then "wealthy people" would flock to Göttingen.[34] And once enrolled at the university, these wealthy foreigners and nobles would attract more wealthy foreigners and nobles in a virtuous mercantilist cycle.[35] In contrast to the forms of authority and prestige which would define the research university, however, Münchhausen valued the acclaim not of specialized academic audiences but of a broad reading public. As Meiners put it, the primary task of the university was not to extend the boundaries of knowledge, but rather to present the sciences as "completely and clearly" as possible, so that any student "of average talent and knowledge" could follow the teacher with relative ease.[36] Students should not be pushed too hard, lest they decide to enroll elsewhere.

Johann Justi, professor of "police science" at Göttingen, embraced the mercantilist vision and argued that the foundation of true cameral science was "police": "The police authority must sow if the cameralist is to reap. . . . Police is concerned with maintaining the total wealth and substance of the state's internal structure and increasing it."[37] In an effort to institutionalize such policies at Göttingen, Justi proposed a new faculty of economic sciences to challenge the medieval university's faculty structure and "eradicate pedantry."[38] "Scholars," he wrote, "once seemed to constitute a kind of special kingdom separated from civil society. They worked for themselves. They shared their thoughts and discoveries amongst themselves, and because they worried more about fame and the esteem of their fellows than about the benefit of the people . . . they wrote in a language wholly unfamiliar to civic life."[39] For Justi and like-minded reformers in Göttingen, the medieval, faculty-based university structure illegitimately separated the university from the needs of the state and society. The modern mercantilist university, in contrast, as Justi put it, should serve the "common good."[40]

The efficient production of civil servants and knowledge directly useful to society required an "academic police," or a mercantilist model applied directly to the university. Its primary goal was to enhance the "good reputation" of the university through an intricate policing of public lectures, printed texts, and student behavior. By acceding to such a policing, argued Meiners, Göttingen was able to maintain itself as a "privileged corporation of teachers and pupils who through the grace of their founding benefactor" were granted special prerogatives and a certain degree of autonomy. Like medieval universities, Göttingen's faculties and students were exempt from civil jurisdiction and accorded the privilege of "testing" candidates and granting academic titles, that is, reproducing themselves.[41]

Despite its tremendous successes and its historical importance as one of the prototypes of the modern university, however, the Göttingen model eventually proved to be a liability to the institution of the university in the context of an Enlightenment media ecology. By linking itself so closely to mercantilist policies and thus to the administrative interests of the state, the Göttingen model made it increasingly difficult for universities to differentiate themselves from other credentializing and specialized training institutions.[42] By positioning itself primarily as a credentializing institution and source of state prestige, the university secured its material base for decades, but it made itself vulnerable to "shocks in material support." The mercantilist conception of the university undermined its particularity and prestige as a distinct institution.[43] It risked becoming just another element of the state. Because Göttingen was already well established, financially supported, and had a renowned faculty, it did not suffer like other universities that attempted to replicate its model;[44] in fact, it flourished. But the model, as we shall see, was not easily transferable and did not withstand late eighteenth-century critiques. The state ultimately organized the knowledge of the university and determined what counted as real or useful.

While some late eighteenth-century critics of the university were hesitant to use the term "university," others saw in the institution and the tradition of the medieval university not merely a guild structure but also a model of distinction that had been devalued by the Enlightenment's basic suspicion of scholarly expertise and the more state-oriented universities. The emergence of science as a distinct cultural realm and the cultivation of a disciplinary self depended on the survival of key elements of the medieval university, namely, the structures of distinction which it supported, especially between a more narrowly defined group of scholars and the broader literate culture. The university provided a model, however antiquated and in need of reform, for self-governance, autonomy of method, and social distinction.

But a modern university also had to articulate its own ground of distinction, legitimacy, and authority, one not directly dependent on either the church or the state. This was particularly important for the more general scholar, that is, the scholar who had not studied in one of the three higher faculties of law, theology, or medicine. Without a clear and distinct university structure, and given the expansion of print and literacy, the role of the scholar threatened to dissolve into the "lay conception of culture."[45] Universities had to distinguish themselves from an increasingly literate public, the broader medium of print, and the proliferation of professionalizing and credentializing institutions; otherwise, they and the scholars they brought together risked irrelevance. In the new media environment, universities had to account for how the knowledge they produced was distinct and more valuable than that which circulated in the print market. What was their role in organizing knowledge in a modern age of increasingly accessible information?

By the late eighteenth century, the most common arguments for university reform centered on the claim that other institutions—from secondary schools to professional schools—could more efficiently credentialize and produce well-trained servants of the state and, more generally, better serve the public than could universities whose reform efforts were hampered by their medieval origins and guild-like structure. By tying the university closely to the immediate needs of the state, reformers like those at Göttingen—intentionally or not—forced the question of the university's purpose. Calls for a more immediately practical university that directly embraced the needs of the state became the object of intense criticism, especially in Prussia around 1800, and led to a second reform effort that focused on the need for a normative resource that could guide the university in a late Enlightenment media ecology. The most important effect of this debate was not on the institution of the university itself—Göttingen had already put in place many of the suggested reforms—but rather on the history of knowledge and the constitution of the modern media ecology, that is, the way the debate contributed to a new order of knowledge, a new way of ordering the desire to control and master knowledge.

## The Critique of the Enlightenment University and the Modern Media Environment

Between 1790 and 1810, writers of all sorts carried out a wide-ranging debate about the future of the university in periodicals, monographs, and lectures in Prussia. The chatter about universities became so widespread that one author collected the "results" in aphoristic form, while another claimed that so much had been printed about the question of the university that the very question had become "a science of its own."[46] Although touching on a range of topics—the relationship between the university and the state, the place of religion in the university, professionalization versus "true learning"—almost all of the contributors framed the question of the university in terms of its institutional role in producing, organizing, and communicating knowledge for a modern age. Commenting on the "worth" of universities, one anonymous writer suggested that the university be thought of as the most important "technology" [*Mittel*] for the "diffusion of learning."[47] He meant this both historically and prescriptively. The university's purpose, he wrote, had always been "to communicate . . . and disseminate" the sciences, but the decline of German universities and the proliferation of print over the course of the eighteenth century now threatened its capacity to do so.[48]

This second debate about the future of the university took place at the height of a century-long decline in enrollments at German universities. In 1720, 4,400 students were enrolled in German universities, but by 1790 that number had fallen to 3,400, and again to only 2,900 by 1800.[49] The decline in enrollment was exacerbated by the uneven distribution of students across Germany's thirty-nine universities. At any given time over the course of the eighteenth century, many universities had fewer than a hundred students. Any enrollment growth was primarily limited to Göttingen and Halle, while most other universities languished.[50] During the Napoleonic era, twenty-two universities, over half of all German universities, collapsed. Some simply closed (Rinteln, Dillingen, Helmstedt), some were temporarily closed by

Napoleon (Cologne, Mainz, Trier, and eventually Halle in 1806), and some were subsumed into other universities (Altdorf into Erlangen, Wittenberg with Halle, Frankfurt an der Ode with Breslau).[51]

Declining enrollments were especially devastating to the philosophy or arts faculties, which for centuries had struggled with lower pay and prestige compared to the higher faculties of law, medicine, and theology. Over the course of the eighteenth century, students began to enroll directly into the higher, more professionally oriented faculties and to forego the philosophy faculty altogether.[52] For the university's critics already seeking radical solutions, the decline of the philosophy faculty only reinforced the view that more specialized professional schools would be able to train students more efficiently.

Since the seventeenth century, universities' problems had been exacerbated by the success of extra-university institutions such as the new academies and learned societies, which often outproduced universities in terms of scholarship, especially in the natural sciences. In the seventeenth and eighteenth centuries, universities were not the only or even the most important institutions for the production and transmission of knowledge. Scientific societies came to fruition following the rejection of university-bound, scholastic scholarship in the seventeenth century and before the rise of a more specialized science in the nineteenth.[53] The academies founded in Paris (1666), London (1662), and Berlin (1702), scientific societies that blossomed across the continent, and more specialized schools (such as mining academies and veterinary and medical schools) were distinct institutions for the production of knowledge.[54] It was not until the early nineteenth century, in Germany at least, that universities began to monopolize the production, organization, and dissemination of knowledge. The university gradually absorbed various institutions—libraries, museums, surgery clinics, botanical gardens, and natural history collections—and developed a distinct logic and ethic that helped organize the production and transmission of knowledge. By the end of the eighteenth century, universities were forced to defend their relevance in the face of an array of challenges, not the least of which was technological.

A debate in 1795 among members of the Berlin Wednesday Society, which included the editors of the *Berlinische Monatsschrift*, officials from the Prussian bureaucracy, publishers, and, as an honorary member, Moses Mendelssohn, encapsulated the basic issues.[55] The debate was particularly notable because its initial question concerned not whether universities should be reformed, but whether they should be dissolved entirely. That a group of

Prussian cultural elites would entertain the possibility of the university's dissolution was indicative of the level of concerns about its future.

The debate began when J. G. Gebhard, a reformed Prussian pastor, delivered a lecture for his fellow society members arguing that universities should be abolished. Along with his colleagues, Wilhelm Teller, a Berlin theologian and pastor, responded with his own written comments, to which Gebhard, in turn, wrote a response.[56] Gebhard claimed that "in our age" universities had become "dispensable" because their "purpose" could be achieved by other means.[57] Although Gebhard never clarified what this "purpose" was, he implied that it was to be found in the needs of institutions and constituencies external to the university, that is, in society more broadly. Echoing Justi's call for a better "policed" university, he wrote that the university's "scholastic" organization had granted individual faculties a "monopoly" over university governance, which isolated the university from broader society. If the university were to remain influential in the modern age, he continued, faculties would have to give up their near-absolute jurisdiction over university affairs. Still guild-like juridical corporations, universities were interested not in serving the needs of the state, the general public, or even knowledge, but rather in protecting the privileges of its members.

Gebhard's criticisms echoed a growing chorus of narratives in the 1790s which portrayed the university as ungovernable, antiquated, and irrelevant. In most university towns, wrote Meiners, with the exception of Göttingen of course, townspeople saw "instead of books nothing but disputes; instead of notebooks, daggers; instead of quills, swords ——; instead of learned discussions, bloody battles; instead of industrious work, endless boozing and bluster; instead of studies and libraries, bars and whore houses."[58] These complaints were manifestations of an underlying disquiet about the perceived gap between the university and broader civic order. Johann H. Campe, the Enlightenment pedagogue and tutor to the young Humboldt brothers, argued that universities had been so corrupted that they should simply be done away with: "The ill cannot be healed. It lies in the essential form of the universities, which can only be abolished with universities. All attempted cures past and future are only palliatives that just conceal the damage from us."[59]

Beyond the boozing and cavorting, the real threat to the university, however, as Campe understood it, lay in modern forms of communication, especially print technology. Modern learning, he claimed, did not need the "organized framework" of the university. The print networks of journals, encyclopedias, and newspapers—the bibliographic order described in the first half of this book—constituted a form of exchange and communication which was not only more efficient but, given the pitiful state of universities, more productive and accessible. For critics like Gebhard and Campe, the university was an ill-suited institution for the modern media environment. If universities were to be as "indispensable as schools," concluded Campe, "then whatever is learned there must be learned only there and nowhere else."[60] In the face of the proliferation of print, if universities did not differentiate themselves, they would—and for critics such as Campe and Gebhard should—fall victim to an Enlightenment form of technological disruption. If the university persisted in its closed-off, medieval form of transmitting knowledge, then the expansion and intensification of a bibliographic order, with its lower costs and efficiencies, would replace it.

It is in this context that Gebhard argued that universities should simply be abolished and replaced with specialized schools. At the very least, he suggested, they should abandon their corporate structure and better integrate themselves into society.

In his response, Wilhelm Teller echoed these common concerns but took an altogether different approach to the question of the university. For him, the crisis of the university was less about its failure to address societal needs and more about its failure to encourage and support "true erudition."[61] Over the course of the eighteenth century, the university had ceded its position as the central institution of knowledge to the state and, increasingly, the print market. In alternating tones of nostalgia for a lost institution and prophesy of one to come, Teller warned that if scholars lived scattered about, working in their isolated studies, they would never be able to accrue "respect and prestige."[62] They could distinguish themselves and, most importantly, their "true erudition" from other social groups like "the tradesmen" and their wares only if they congregated in one institution. And the only real option was the university. The continued congregation of scholars in universities organized on a medieval model may have been a deficient and at times stultifying structure, but it was a bulwark for the authority of a university-based knowledge. The university was needed to distinguish "true erudition" from everything else circulating in the modern print market.

The difference between the medieval university and the university of the future—and the problem of imagining a new university—was that the authority of the medieval university originated from the church and the authority of the Enlightenment university rested on the state. By orienting the university to a church-based canon, advocates of a medieval university model tethered the advancement of knowledge to theological commitments. Both state rulers and church authorities protected theologians from challenges to orthodoxy and helped create the conditions in which theologians were not forced to reconcile the claims of Christianity with new empirical knowledge.[63] Similarly, by tying the ends of the university to the needs of the state, advocates of a more utility-oriented, Enlightenment model not only undermined the university's particularity but surrendered the responsibility to organize knowledge to the state and a reading public whose capacity to evaluate knowledge many scholars were beginning to doubt. The university's primary purpose, concluded Teller, was to authorize knowledge, but it needed to articulate its own ground of authority. In particular, it would have to distinguish itself from the Enlightenment's dominant technologies for organizing knowledge. It would have to differentiate, as Teller put it, "true erudition" from mere Enlightenment. Whereas, for Teller, the latter referred simply to "writing" and everything that had been printed, the former referred to the distinct, authoritative realm of the scholar.[64] Eighteenth-century scholars and writers of the empire of erudition had conflated the two, as had figures of the popular Enlightenment with its insistence that knowledge should be communicable to everyone.

In contrast to the writer and the general reader, the true scholar was defined, as Teller put it, by his singular commitment to "studying only the history of his science. . . . The more one-sidedly he thinks [for his science] that much more useful he will be to it."[65] Anticipating arguments about specialized science which would characterize the university debate in the following decade, Teller described universities not only as bulwarks against the "fragmentation" of science but also as institutions that selected, organized, and distinguished knowledge from the superficial types of knowledge which circulated in the general reading public. Universities could sustain true erudition and science only by encouraging the production of deeper, more specialized forms of knowledge. Their medieval structure helped shield individual scholars from demands to produce only knowledge that would be more immediately useful to society.

Teller concluded that the university's faculty arrangement of knowledge had to be reformed. Whereas Campe and Gebhard had criticized this arrangement because it cordoned the university off from society, Teller criticized it on epistemic grounds. A traditional, fourfold faculty arrangement of knowledge no longer accorded with a modern knowledge that in the modern age of print had fragmented into myriad sciences. The university should maintain its overall faculty arrangement, but only as the most general organizing structure. Internally, it would have to differentiate in order to account for the expansion and specialization of knowledge. Such an internal differentiation would allow for greater specialization through a range of regularizing practices that would deal primarily with the nature of research. Professors, argued Teller, should be required to focus their efforts on particular sciences. All universities should require every professor to lecture "in his public courses," those for which he was paid, on that science for which the state hired him to lecture.[66] Although this may sound obvious to contemporary readers, Teller was actually condemning a common practice that allowed, even encouraged, professors to lecture on a range of subjects and move freely among the faculties. Kant, for example, lectured on everything from metaphysics and logic to anthropology, geography, and natural science.

Not all of the Wednesday Society's members agreed with Teller, however. Echoing more established arguments like Campe's, Friedrich Nicolai, the famed Berlin author and book dealer whom Kant had mocked, disputed Teller's claim that the university should be the center of "progress of the sciences."[67] For Nicolai, an "erudite world" already existed outside the university. The production and dissemination of knowledge were not and should not in the future be limited to the university and its "systems of the sciences." The homogenous empire of erudition to which Nicolai was alluding was, as we saw in chapter 2, gradually fragmenting into more specialized worlds of particular sciences. Teller saw the university as a new technology through which the proliferation of the material forms of the sciences could be organized, managed, and unified, but Nicolai defended the bibliographic order and encouraged confidence in the enlightening and unifying potential of print. A new university-based disciplinary order of knowledge, one that emphasized the particularity, distinction, and authority of a university-based form of knowledge, presented a fundamental challenge to Enlightenment commitments to distribute knowledge broadly and make it widely accessible in print and to the principle that print could enlighten. It would fundamentally alter the basic Enlightenment assumption that knowledge could be simultaneously practical, widely communicable, and embodied primarily in the circulation of printed objects.

The disagreement between Teller and Nicolai exemplified a broader inflection point in the history of knowledge: between the ascendant bibliographic order of the empire of erudition and a university-based disciplinary order of knowledge envisaged in the debates leading up to the founding of

Berlin. Teller countered Nicolai's and Gebhard's call for the abolition of the university with a normative vision for a university organized according to a different order of knowledge, one grounded in the logic and ethos of science: specialization, borders, scientific literature, prestige, and incommensurability. Teller also provided a clear account of why such a logic and ethos were required in the first place: the university had to give an account of itself in the face of technological changes that had undermined notions of epistemic authority. Print technologies could not replicate the forms of communication and their practices.

As the anonymous writer quoted above had put it, the university was not simply a medieval corporation of masters and scholars. Similar to an Enlightenment encyclopedia or scholarly journal, it was a "technology" for organizing knowledge.[68] Any account of the university had to understand it as part of an array of Enlightenment knowledge technologies. The university stood at the intersection of various historical attempts to organize knowledge. Conceiving of the university in these terms allowed scholars to focus on what distinguished it from other technologies and social institutions and thus on the different ways in which the university produced and transmitted knowledge.

Before "the art of printing books" interwove scholars from across space and time together, continued the anonymous author, the first universities in Paris and Bologna emerged out of a "need for oral instruction."[69] For centuries, universities had assembled scholars so that they could share their truths orally in the tradition of ancient Greece. Whereas scholars bound by print disseminated their "truths" indiscriminately, without attention to their audiences, these older, university-based scholars reserved their most esoteric truths for a small group that was present—an oral, dialogical exchange protected by the university. As print technologies became more widespread, however, even the most distant scholars had been brought "closer together." Print technology had opened "the empire of truth" to those in the "most far-flung corner of the world."[70] As printed books became more affordable and more broadly available, however, the value and worth of "intellectual treasures," the archive of knowledge transmitted over time, had declined. Although the unfettered circulation of knowledge afforded by print was to be celebrated, it rendered questions of value, quality, and authority of secondary import. What mattered above all else was just getting something into print. The book market could not be relied on to filter itself and make judgments of worth.

Faced with these changes in the broader media environment, the university should stand in not just for a repository of knowledge but rather for a community of knowledge that provided authority structures and models for how to think. The university, contended the anonymous writer, distinguished between "merely having science" and "being formed" in accord with science.[71] The true scholar must not merely "store science in his memory" as though he were a walking encyclopedia. He must also be a "thinker in his science" because, as a scholar, he was also a teacher whose task was to form young men into "strictly scientific thinking." The true scholar not only possessed a bibliographic erudition but also embodied a particular way of thinking, and printed objects and the bibliographic sciences were insufficient tools for making this possible. To transform students into true scholars, what was needed was a return to the ancient Greek oral dialogue tradition that brought pupil and teacher into close, repeated contact. "*Viva Vox!*" declared the anonymous writer. This exhortation for the university to embrace dialogue assumed that oral speech afforded a certain immediacy to thought or an original idea, as though knowledge were some *thing* to be transmitted. Oral transmission was a more immediate and controlled channel than print. It allowed for a less mediated encounter with "the thoughts themselves." The voice of a Socratic teacher promised to arrest the unfettered flow of ideas and ground knowledge in the person of the teacher, whereas "writing" or "text" [*Schrift*] signified words that in turn signified thoughts. These multiple layers of mediation, suggests the author, divided the reader's attention: he "must pay attention to the sign of words, the words and finally the thoughts themselves."[72] Oral transmission—as a "single medium from thought itself"—in contrast, guarantees an "ease of comprehension" and a "clarity" of content.

This naïve embrace of orality's purity and immediacy, however, highlighted a longing for epistemic authority and deep anxieties about the sources and forms of real, authoritative knowledge. Thus, while these claims about the "clarity" of oral speech were exaggerated, they demonstrate what was at stake in these debates on the future of the university: a desire for an authoritative knowledge in a world thought to be fragmented and saturated. What was the future of the university in this new media environment? Could it escape the changes in technology? Could it adapt? Was there still a place, as the author put it, for traditional "oral instruction"? Should it be "privileged over instruction from books?"[73] Implicit in these questions was the suggestion that knowledge and science could not be reduced to the bibliographic order that had emerged over the course of the eighteenth century. The particular role and function of the university in the modern ecology of knowledge, however, remained unclear.

## Schelling on the History of Knowledge and the University

The late eighteenth-century crisis of the university was not, then, just about the university; it was another manifestation of the broader ethical crisis brought on by the proliferation of print and experiences of Enlightenment information overload. It was about the intersection of various knowledge technologies and what counted as real knowledge. Just as erudition, encyclopedias, and books had stood in for particular knowledge orders, ways of organizing desires to control and manage knowledge, so now did the university.

One of the first thinkers to synthesize these disparate anxieties and the history of knowledge technologies into a clear argument about the role of the university was the German idealist philosopher F. W. J. Schelling. In a series of lectures on the *Method of University Study* at the University of Jena in 1803, Schelling continued in the vein of the Jena encyclopedia lectures that had begun in the 1790s, but with a more explicit focus on what he considered to be the university's primary purpose: forming students into a culture of science.[74] He also embraced Kant's revaluation of the philosophy faculty, long the lowest and least regarded, to the center of the university, and he sought to integrate his own philosophical system with the lectures.[75] What Schelling termed his identity philosophy proceeded from the premise that all knowledge begins with the insight that "knower and the known are the same."[76] As we shall see, this assumption that the knowing subject, and his character, could not be detached from what was known, that ethics could not be completely separated from epistemology, was essential to his conception of what a university should be.

Schelling opened his lectures not by delving directly into the intricacies of his philosophical system, but with a description of a young man, just having arrived at the university. Confronted with the dizzying array of material on offer in lectures and textbooks, he finds himself at sea "without compass or lodestar," lost in the "world of the sciences."[77] For a new student, the university is a "chaos, a vast ocean in which he can distinguish nothing." But it is also a microcosm of the broader media environment; it offers the student no more clarity or meaning than does the market of mass print.

Disoriented and distraught, the student either stumbles into a particular science with a "mechanical industriousness and mechanical comprehension" in hopes of securing some future, post-university job—that is, he deals with the complexity of information by attempting to professionalize—or simply wanders among the various sciences without ever getting to the "seed" of any particular science.[78] In either case, he leaves the university still confused and unable to make sense of the seemingly endless amounts of information that confront him every day. Without any authoritative structures to filter the good from the bad, the important from the insignificant, he feels completely overwhelmed.

For Schelling, the eighteenth-century university reproduced the effects of information overload in institutional and pedagogical form. It not only hindered the advancement of knowledge but also threatened the integrity of the individual by producing distracted, unreflective young men. The university, especially the Enlightenment university that valued utility above all else, had been complicit in fomenting this epistemological and ethical crisis, and it was incumbent upon a vanguard of thinkers to reimagine the university as not simply an efficient institution, but rather the institutional embodiment of a distinct practice, namely, science. Only science as a practice, as a source of internal goods and virtues, not better textbooks or more complex encyclopedias, could address the epistemic and ethical effects of information overload. The task of the university was to form subjects of knowledge capable of navigating the oceans of print. It was to transform a student's vision of the world and shape their character, to fuse epistemology with ethics.

To accomplish this, universities needed to provide explicit instruction, as Schelling put it, on the "purpose [and] method of academic study."[79] Unlike the specialized encyclopedia lectures of the eighteenth century designed to introduce students to individual sciences, Schelling's lectures guided students through a series of mental steps intended to lead them to a particular insight that would in turn help them avoid what he elsewhere termed the "fundamental error in all knowledge"—the distinction between subject and object.[80] First, Schelling provided a diagnosis of the contemporary age, which was designed to cultivate in students the sense that the plethora of sciences and knowledge threatened to overwhelm them on a subjective level. Second, he offered a solution to this problem: a particular mode of thought which should henceforth characterize all university education. The lectures were an exercise designed to guide students to a particular mental attitude.

Just as Kant had offered his critical philosophy as a solution for dealing with excess, Schelling offered a method designed to lead students to a particular insight. Although a discussion of Schelling's identity philosophy exceeds our present aims, it is important to at least state the insight toward which his lectures were designed to lead: the perceived fragmentation of knowledge should be seen as just that—a *perceived* fragmentation of what was ultimately a unity. "In our first reflection on knowledge," as Schelling put it, "we believe to have distinguished in it a subject of knowledge (or knowledge when conceived of as an act) and the object of knowledge [i.e.,] that which is known. I purposely say: we believe to have discriminated for precisely the reality of this distinction is at issue here."[81] In the Jena lectures, Schelling described a particular way of thinking which would challenge this initial belief and conceive of a possible unity of knowledge. He wanted to lead students to recognize that all thinking was grounded in a prior unity, an absolute reality that preceded any differentiation or fragmentation; or, as he put it elsewhere, he wanted to help students apprehend the first presupposition of all knowledge: "that one and the same knows and is known."[82] Students would proceed through a series of steps and ultimately assent to the basic insight of Schelling's system concerning "the self-sameness of the subject and the object."[83] Such an absolute unity, as Andrew Bowie describes it, was not "the result of the overcoming of the difference between thought and being, subject and object" through a more adequate representation of the subject and object.[84] Schelling's point was that one cannot begin with difference or fragmentation and then proceed reflectively toward identity, unless what is perceived as distinct "is already the same." Without a "ground of identity . . . difference ceases even to be recognizable."[85]

Bracketing the particulars of Schelling's early systematic philosophy, we should note what is crucial for our story about the university. Schelling's goal was not simply to deliver his system in the form of transmittable propositions that students could copy down and review at their leisure or provide them with a complete compendium. His goal was to transform them and lead them to a basic recognition of the prior unity of knowledge. Students had to accede to this presupposition in order for their studies to cohere. The success of their studies depended on a change in consciousness. The university was the institution where this transformative process should be undertaken and where such thinking should be cultivated. Over the course of the eighteenth century, the empire of erudition cultivated such unity through the "tools of erudition," but its gradual dissolution had diffused its authority. The historical task now facing the university was to imagine a different basis for communication and unity, a different form of epistemic authority. And, for Schelling, this authority lay in the transformation of the character of the student.

Schelling's solution to the problem of excess, then, was to challenge the widespread perception of fragmentation and inculcate in students a different mental attitude toward knowledge. The perceived differentiation and fragmentation of knowledge could only be comprehended on the basis of a prior unity. Particular bits of knowledge or facts have no meaning apart from their relationship to other bits and facts. Meaning is a function of what binds them together, of how they relate to one another. The capacity to perceive these connections and relationships was not simply theoretical; it was practical. Schelling saw it as a practical activity, something that had to be undertaken. The recognition of the unity of knowledge toward which Schelling beckoned his students was not something to be found ready-made in the world; it was, in part, to be created through their own self-transformation into practitioners of science. The method of academic study that Schelling performed, then, both proceeded from the assumption of a unity of knowledge and worked toward its creation.[86] Science required scientists.

In their indefatigable pursuit of professional success as doctors or lawyers, however, students risked forgetting, warned Schelling, the "far higher vocation of the scholar as such—the mind ennobled through science."[87] The university provided space and time for young students to realize that whatever their future professional duties entailed they were always called to something beyond themselves and the constraints of convention. The university entrusted them with the "preservation and development" of a new world of knowledge that would change them forever. It initiated them into the life of the true scholar and its particular norms and practices. The activity of studying at the university was a "rebirth" in which students were transformed through experiences in which theoretical knowledge, the material forms of knowledge, and method were joined to subjective practices. By transforming themselves, students were able to access the ultimate grounds of all knowledge.[88]

Schelling's lectures provided a portrait of the kind of person who would emerge from such a transformation. He would be characterized by a lasting commitment to a particular science, industriousness, an ability to make judgments, obedience to standards of excellence, and a capacity to view all particular types of knowledge in relation to a broader whole. All of these capacities and virtues would be sustained by an unwavering commitment to science. Echoing Kant's definition of Enlightenment as the emergence from a "self-incurred immaturity," Schelling described such a transformative process that science represented as an "entrance into maturity."[89]

The biggest threat to such transformation, claimed Schelling, was a "certain tone of popularity" which permeated Enlightenment culture. Echoing his idealist and romantic contemporaries, Schelling warned his students that the predominant assumption of writers and thinkers beholden to Enlightenment conceptions of knowledge was that the sciences should be immediately useful and broadly communicable.[90] Knowledge had value only to the extent that it was a means to an immediate, obvious, and measurable end. Such a "utility gospel," worried Schelling, was limited not because knowledge should not be in some sense useful but rather because it inevitably reduced knowledge to a series of loosely related, often ad hoc solutions to particular problems. Questions of how these problems were related or of how the knowledge of one science related to another were rarely raised.[91] A particular science had value only as a means to a predefined, externally given end. Philosophy was only valuable as a means to doing one's duty; geometry was for constructing houses. Science was always a means to a particular action, whose value and end had been determined beforehand. Schelling did not deny that sciences could or even should provide solutions to particular social problems, but he worried that Enlightenment thinkers had no capacity for or interest in evaluating the ends to which sciences were put. Philosophy could be used to justify war and geometry to make war. If the ends to which they were put were socially sanctioned, then neither philosophers nor geometers committed to the Enlightenment's "utility gospel" had any ground to protest.

Schelling contrasted such an Enlightenment notion of science to a notion according to which science could have different kinds of ends and value. Geometry, for example, brings value to our lives, wrote Schelling, because it helps us see the "beauty" of a complexly ordered world. And philosophy, he remarked dryly, had value because it is useful in "leading a war against the shallow minds and apostles of utility in science." For Schelling, this limited notion of the value of the sciences had its basis in a conceptual failure to "grasp the absolute unity of knowledge and action." Enlightenment thinkers had detached knowledge from action and vice versa. They had separated epistemology from ethics. This separation was based on what he considered to be the false assumption that pure theoretical reflection involved no activity and that actions involved no conception of the ends or purposes toward which they were oriented. Theoretical knowledge, therefore, made no claims about what one ought to do. The character of a person had no bearing on what one could know. On such an impoverished account, knowledge could be separated from the uses to which it is put and the personal lives, practices, and commitments of those who created and shared it. Such a misconception reduced the value and worth of knowledge to the immediate and ephemeral demands of daily life. Schelling was not arguing that theoretical knowledge, knowledge disjoined from action, was an end in itself, much less that knowledge served no particular ends; instead, he was arguing that knowledge and action were inextricably bound to one another.

Schelling tied his broad philosophical attack on the "false and superficial culture" of the Enlightenment to the sad state of German universities, which he claimed had institutionalized the separation of the theoretical from the practical, epistemology from ethics. Our "contemporary institutions," he wrote, were beset by a "dreary confusion" about the true ends of knowledge. Universities had no capacity to reflect on the purpose of knowledge that they produced and transmitted. Such distinctions rendered some knowledge useful and other knowledge useless, turning universities into instruments of society and, in particular, the state. "The usual view of universities," wrote Schelling, is that

> they should educate servants for the state, perfect instruments of its intentions. These instruments, however, should without doubt be formed through science. If this is the desired purpose of education, then science must be desired (or one must want science). But science ceases to be science as soon as it is reduced to a mere means and is not at the same time advanced for its own sake. Science for its own sake will definitely not be advanced, when, for example, ideas are rejected on the ground that they are of no use for common life, have no practical application, because no use can be made of them in experience.[92]

A particular type of knowledge may seem "useless" in light of present conditions or a particular set of interests, but it was the duty of science to provide a broader account of the manifold uses and ends of knowledge. To judge the value of knowledge simply according to its immediate use value and to limit its uses to ends defined by one set of interests—the state, for example—was to foreclose other ends and other types of value that it might bring. Such a utility logic had turned universities into instrumental extensions of the state and turned university teachers and students into dolts distinguished only by their "dread . . . of exertion," "pleasant superficiality and shallowness," and lack of "thoroughness."[93]

Schelling devoted his lectures to universities because it was here that young students were inculcated into the Enlightenment culture that he so roundly criticized, but it was also in the university that these ways of thinking and acting could be undone. Schelling's insistence on the unity of knowledge and action was not just an abstract philosophical claim; it was an exhortation to his students and to his contemporaries that the knowledge of a student, that is, a future scientist, could not be separated from his character. The pursuit of scientific knowledge and its value and worth were, in part at least, a function of the character of a particular type of self. The purpose of the university was to form this self and cultivate in students an awareness of the "true harmony" of knowledge and action.[94]

Schelling's philosophical critique of the Enlightenment began and culminated in a cultural critique about the fragmentation of knowledge in the modern age. The problem with Enlightenment conceptions of the unity of knowledge was that they assumed that this unity was something to be achieved by accumulating as much material as possible in print. The compilers and editors of Enlightenment encyclopedias and periodicals were motivated by an ideal of a material completeness according to which the unity of knowledge was the last stage in a dogged, aggregative pursuit. For Schelling, such a unity was not something to be collected; it was an insight that had to be cultivated in the social environment of the university.

## Technologies of Transmission and the University

Schelling saw the university as giving the "absolute unity" of knowledge an institutional form. It could render the relationship of various forms of knowledge on a more accessible human scale. The university stood in for a social unity that, unlike the absolute unity invoked in his systematic philosophy, endured in time and space. Schelling assumed that the social character of knowledge or science was a consequence of the fact that science was ultimately a "matter of the species."[95] The social unity of science was constituted by how it "communicates itself from individual to individual, from generation to generation. Transmission is the expression of its eternal life." Such a unity manifests itself in time through processes of human communication because reason is ultimately communicative and public. As temporal and finite beings, humans cannot grasp the absolute unity of knowledge, the unity of ideal and real, in one timeless insight but rather only as it is communicated over time. The unity of knowledge can be realized only in "the species" and thus in the historical institutions and forms of communication which bind the species together. The unity of knowledge as science requires a community that is both formative and reciprocal.

If students were to recognize this social unity, they would need to be familiar with and have a capacity for the communicative forms or media through which such knowledge is transmitted over time. The history of those communicative forms was long and continuous, beginning with symbols and the invention of writing and culminating in the modern university and science. Students needed to study this history, claimed Schelling, because it revealed the "destiny" of knowledge, that is, its unity as embodied in its historical development.[96] Such a focus on the historical development of knowledge also gave Schelling an opportunity to contrast the historically distinct shapes and social forms that knowledge had assumed and the different types of communities that it shaped. This is especially evident in the contrast he made between ancient and modern forms of knowledge. The first media for transmitting knowledge were the "actions, forms of life, customs, and symbols" that communicated the ideals and beliefs of local communities. The first religious ideas, for example, were not recorded dogmas but social customs made available to the entire community. The invention of writing provided such common ideas "security" by giving them a more lasting, material form.

But even when knowledge was given material form in writing, claimed Schelling, it was never simply a possession of the individual. Ancient knowledge "lived in the light and ether of public life."[97] In the ancient world, the authority of knowledge was a function of a shared public life. Knowledge was something held in common and accessible to all and thus capable of orienting an entire people. In the "modern era," in contrast, there was no common knowledge, "especially," lamented Schelling, with respect to science. Knowledge had been fragmented into distinct sciences inaccessible to all but small groups of specialists.

Schelling blamed this modern fragmentation on "historical erudition."[98] Modern scholars treated the past as a static deposit of timeless truths and assumed that "an unbridgeable gap" separated them from the ancients. They presumed that the two worlds were connected merely through the "external band of historical transmission," and thus failed to perceive what Schelling called the "internal band" of history's progressive development. Just as the confused student could not perceive the ultimate unity of knowledge amid the excess and fragmentation of modern sciences, modern scholars could only see the sciences as separate without a sense of their common development over time. The past was some separate object "to be admired and explained" and to which contemporary scholars bore no relation. Over time modern scholars had put these historical artifacts "in the place of knowledge itself."[99] True knowledge, science, had a teleological orientation, and it was one of the scholar's tasks to reflect on its historical development. Otherwise, history would be nothing more than the accumulation of discrete facts. It would block access to the developmental unity of science over time, which was the true source of its "authority."[100]

Contemporary universities, claimed Schelling, were the products of these fragmenting historical modes of thought which confused scholars and students alike. Like other modern institutions, universities adopted and adapted these historical forms of knowledge in an effort to manage, as Schelling put it, "the great mass" of what had to be learned in order to count as erudite.[101] For scholars, it was easier simply to aggregate disparate bits of the past than to perceive the development of knowledge over time. In so doing, however, scholars had reduced themselves to "mere institutions for the transmission of knowledge." They had become institutional bibliographies, storehouses for historical facts, and were thus in part responsible for the experience of modern information overload. The modern university had "caused the organic structure of the whole to fray into the smallest elements." Instead of embodying the social and historical unity of knowledge, it "dismantl[ed] knowledge" and made the very "technologies" of knowledge the end of knowledge. It fetishized the "aids of erudition."

If universities were to become more than aggregating technologies, they had to reinvent themselves as stable institutional homes for the transmission of knowledge, capable of sustaining the communication of knowledge through the "species." The communication of knowledge from generation to generation needed an institutional form that could link scholars and thus allow them to communicate knowledge from generation to generation. In order to sustain such historically continuous communities, universities needed a unique ethos to guide and orient scholars to a common purpose. Above all, the university had to cultivate a commitment to the historical development of knowledge in the species. It needed to form scholars dedicated to the insight that knowledge developed over time.

The university's most immediate challenge, therefore, was to find teachers who exemplified the "scientific spirit" and could awaken it in students. The true university teacher "does not provide results like the writer, but rather, in all higher sciences at least, he represents how these results were reached. In every case, he allows the whole of science to emerge right before the eyes of the student."[102] University teachers present the unity of knowledge not as an abstract philosophical principle, but rather as a personal example of that unity. The university teacher appears before his students as member of a broader community, a living part of the historical transmission of knowledge over time. He appears, that is, as a member of the species through which knowledge is developed and shared. The university teacher embodies traditions, practices, and material that organize the desire to order and master knowledge. The teacher stands in for the "species" and thus the social unity of knowledge. He initiates students into a distinct community and helps them become part of that living, progressive whole of science as it developed over time. Unless the university could harmonize the objective forms of knowledge with individuals, it would be no different from the "common handbooks or the thick compilations" that past scholars had relied on.

As part of this living whole, scholars and scholars-in-formation could not reduce scholarship to the accumulation and transmission of an existing body of knowledge. They had to contribute to its progression and advancement. This is what made science "living" knowledge. It was dynamic, not "something given" but rather "something to be invented."[103] "Learn in order to create," Schelling exhorted his students, because "only through this divine capacity of production is one truly human and without [it] a pitifully assembled machine."[104] The true scholar was an artist, who worked within the historical constraints and practices of science, which allowed him to create something new. He did not create knowledge from nothing but worked with the objective material forms of knowledge and within the practice of science to produce new knowledge. Schelling tied the productive and creative potential traditionally associated with the artist to the scholar as scientist. The true scholar was a "scientific genius," the creative hero of the university who "out of the raw mass of materials" produces something new.[105]

The "scientific genius," however, was not the solitary artistic figure of a sentimental poetry. He created and participated in the progression of knowledge through time within a community of other scientists who shared a common purpose. The genius's capacity to create required membership in a community that provided the traditions, practices, and standards that focused the scientist's attention and allowed him to create. And the university was the institutional embodiment of this community. Any scholar laboring outside this university community, wrote Schelling, was like a "sexless be[e]" that simply "re-manufacture[s]" through "inorganic" activity.[106] Outside of the hive of the living university, scholars could not create knowledge, so they simply reproduced themselves through "inorganic excretions." Without community and the purpose it provides, they were "spiritless"; they had no relationship to the whole, to the unity of knowledge over time as it was embodied in the institution of the university.

For Schelling, the dialectical relationship between the individual scientist and the university community was reproduced within the university and the relationships it sought to sustain among the various sciences. The development of science depended on both particular, specialized work and a consciousness of its place within the whole of knowledge. Specialized knowledge was a microcosm of the whole of science. Only through it could the true scholar access the broader development of science, and it was also this broader unity that oriented the individual scientist and gave his particular, specialized work meaning and purpose. The more a scholar conceives of his particular field as a "purpose in itself," a "middle point of all knowledge that he wants to extend to an all encompassing totality, in which he wants to reflect the entire universe," wrote Schelling, "that much more does he strive to express a universe of ideas in it."[107] Although the individual scientist would never gaze out on the expanse of knowledge with a synoptic view, like some encyclopedic cartographer, he would experience the joy of creation and the consolation of participating in something that exceeded him, the progression of science over time, however particular and specialized his work was.

Universities as Schelling reimagined them had an "ethical vocation." They were "universal institutions of culture" which inculcated students into the social unity of knowledge by forming them into the practice and culture of science. By sustaining the practice of science and upholding its internal goods and norms, universities were able to orient students in an "ocean" of fragmented sciences and a surfeit of information. As in any art, excellence in science, wrote Schelling, could only be achieved "through practice. All true instruction should according to its vocation concentrate more on this [form] than content, exercise the organ rather than transmit the object."[108] The disciplinary self that Schelling sketched as an ideal character could not simply be imagined. It had to be shaped, formed, and created through regular exercises and habits. It needed a stable institutional home to sustain its practice and cultivate its virtues. As the singular institution of science, the university's primary task was to form people dedicated to science. "In a scientific association," wrote Schelling, "all members have a single purpose by nature. Nothing is more valid than science. There is no other distinction than talent and culture [*Bildung*]. People who are merely there to assert themselves in some other way—through profligacy, wasting their time on mindless pleasures, in a word privileged idlers like those in civil society and it is such people who engage in the most crudeness at universities—should not be tolerated here. Whoever cannot prove his industriousness and his deliberate pursuit of science, should be expelled." "The realm of the sciences," concluded Schelling, "is not a democracy. Much less an oligarchy; it is rather an aristocracy in the most noble sense."[109] It is a system of intellectual competitive collaboration and energetic engagement in which those who are not willing to work hard in pursuit of science should be dismissed. The man of letters, whose interest in science was casual and half-hearted, could not be tolerated. The scientific scholar was serious, industrious, rigorous, and devoted to the open exchange of knowledge. This was, in part, what the institutionalization of science as a practice would entail.

Given the need for sharp distinctions among scholars and a clear institutional structure that could guarantee the stability of science, Schelling even advocated for the direct intervention of the state in running universities. While universities "can only have an absolute purpose"—that is, to serve the practice of science without any constraints—they needed structures to sustain them. They needed the state to help them enforce their internal distinctions: "In order for the state to achieve its aims it needs distinctions. But it does not need those that arise from the inequality of classes, but rather those that are much more internal, those that arise from isolation and opposition of individual talents, the suppression of so many individualities, the distinction of powers to so many different aspects in order to make them better instruments of the state." The university as the institution of science required the external pressure of the state to discipline it and, when necessary, reform it. The guild-like character of the medieval university and its devotion to the integrity of the *universitas magistrorum et scholarium* encouraged nepotism and tolerated mediocrity. In order for institutions of science to flourish, "to give them dignity internally and prestige externally," the state helped ensure that the "opposite" of science was not promoted: lack of industriousness and passion for the pursuit of science. The state helped create the conditions in which scientific abilities and capacities would be rewarded and encouraged.

A community dedicated to science did not need to be absolutely autonomous. It just could not be subordinate to the state.

Following his Jena colleagues, Schelling's solution to the experience of overload was mostly an ethically oriented one, grounded in the practice of science as embodied in the university. What was needed was not a better print encyclopedia but rather "access to all *Wissenschaftlichkeit*" or formation into excellence in scientific practice.[110] Science was not a set of specific propositions, methodological prescriptions, or static taxonomies. It was a practice that depended on the crafting of a scientific character—defined by industriousness, an openness to communication and exchange, rigor, and an unrelenting commitment to science as a practice, its unity, and moral seriousness—through an immersion in particular forms of inquiry, a habituation to certain ways of seeing the world, and the conjoining of thought and action. As an "association of science," the university organized not books but people toward "one purpose"—the unity of science.[111] Because they were devoted to their particular sciences, students and teachers needed a technology to coordinate their activities and remind them of the unity to which they were contributing. The university provided an institutional home that ensured the continuity and regularity of the practices of science and distinguished the character of the scientist from the other character types—the citizen, the religious believer, the domestic family member, the professional—that had come to populate modernity.

In the two decades leading up to the founding of the University of Berlin in 1810, Prussian intellectuals, writers, and bureaucrats debated the failures and possible futures of the university in the new media environment. Many of them contended that the deplorable state of the Enlightenment university was directly related to the proliferation of print and the challenge it posed to any form of epistemic authority. They wondered how an institution as old as the university could reinvent itself and flourish in the modern print ecology. It was in this context that Schelling delivered his lectures on the method of academic study and offered an exemplary vision of the university at a moment of crisis for epistemic authority. In a formulation that was quickly adopted by his idealist and romantic contemporaries, Schelling described the university as the institution that communicated science and knowledge from generation to generation. It made knowledge a "matter of the species." It was a "scientific association" whose primary vocation was the formation of young students into the social unity of science.

CHAPTER SEVEN

# The University in the Age of Print

CONFRONTED WITH what he saw as a modern world fragmented and fractured, overwhelmed by so much to know, Schelling cast aside eighteenth-century metaphors for the unity of knowledge, such as the book and the empire of erudition, and embraced the university as an institution that organized practices and related virtues into an overarching and relatively coherent whole. And the end or purpose of the university was the legitimation of knowledge and the formation of the types of people who could generate and transmit it. The university was not just a knowledge distribution center. It was a historical institution that sustained and supported a particular type of knowledge. For most German readers around 1800, however, the image of industrious students devoted to a community of science was not the first thing that came to mind when they thought of the university. The popular image of the university was of an antiquated and failed institution that had been overrun by unruly young boys and abused by irrelevant erudites, much like the university that the German theologian and education reformer Christian G. Salzmann chided in his epistolary novel *Carl von Carlsberg, or on Human Misery*.

In the twenty-third letter of Salzmann's popular novel, Colonel von Brau receives word from the university in Grünau that his cousin Carl von Carlsberg, the novel's protagonist and a university student, has been injured in a duel with another student. The colonel rushes to Grünau to find Carl in good spirits and improving health but the university in a "miserable state."[1] Less than a day after his arrival, he wrote to his wife, students were rioting, drinking, and singing on the market square in protest of the university rector's decision to forbid dueling after Carl's injury. And they did this, noted the colonel, without any fear of civil punishment. As Salzmann's novel suggests, the university's medieval structure ensured its self-governance, but it left local towns without recourse to deal with young men who, as the colonel put it, lacked moral principles.

It was not simply its juridical organization that made the university so antiquated, however. It was also, suggested Salzmann's characters, a matter of its failure to adapt to technological change. In a conversation with the young Carl, Deacon Rollow, who had taken Carl in after the death of his parents, derides an outdated university "liturgy" that summons dozens of twenty-year-old boys before a lecturing professor who then lulls them to sleep. Rollow holds neither the students nor the professors responsible, but rather the "form" of the universities themselves: "Our universities were established at a time in which the world lacked books and when a man who could read and write was the exception. In our day, however, universities are pitiful. They are like the fortresses constructed during the crusades, in a war where bombs and canons were used to storm the fortress."[2] Traditional universities were hapless, antiquated bastions of privilege resisting technological change. For the deacon, the proliferation of print posed a challenge to these traditional universities. Why, he asked, if a young person can generally educate himself on his own—a premise encouraged and cultivated by the popular Enlightenment and the profusion of readily available knowledge in books—would he need to attend a university? What was the purpose of the university in the age of the proliferation of print?

As I showed in the previous chapter, this question was urgently posed by a range of scholars and critics of the university. Schelling in particular embraced the university as an institution dedicated to orienting young men in an age of fragmentation and excess. But it was Johann Gottlieb Fichte and Friedrich Schleiermacher who expanded on these debates and described the university not only as a new technology for managing information overload but also as the institutional embodiment of science as a practice. They pushed earlier visions for a reformed university further and even participated in attempts to bring this vision to life with a new university in Berlin. They offered detailed plans for how a university ought to go about creating a community dedicated to the practice of science. Echoing Kant, Schelling, and a litany of voices writing about the university between 1795 and 1805, they reimagined an institution that could both organize the fragmented forms of knowledge and form young scholars capable of filtering through an excess of texts. The university could sustain science as a practice and anchor a new form of epistemic authority in an age of proliferation. The university, they contended, was not just a content delivery system. Print technologies had proven more efficient and agile in that regard. The university was a bestower of epistemic authority.

## The Promise of a New University

The criticisms of Salzmann's colonel and deacon presaged the ones voiced by Prussian writers and philosophers around the turn of the century. But it was not just intellectuals and writers who were worried about the future of universities. Prussian bureaucrats were as well,[3] and they undertook a flurry of reforms in the final decades of the eighteenth century to address these concerns. As head of the new Prussian ministry for education, Justice Minister Zedlitz had been specifically charged with addressing such criticisms. Following the model of Göttingen, he immediately set out to mitigate the influence of the church and religious orthodoxy on Prussian schools and universities.[4] He ordered the introduction of the first *Abitur* exams in 1788 and oversaw the establishment of F. A. Wolf's philology seminar in Halle that same year.[5] In 1794, the state's control over educational institutions was further advanced with the publication and promulgation of the Prussian General Law Code, which declared all schools and universities state institutions and entailed a number of further changes: all educational institutions were subject to state inspection; pupils could not be excluded based on confession; high school teachers were declared civil servants; and university students were subject to local, civil law.[6] The code formalized the decades-old trend, pioneered in Halle and Göttingen, of universities becoming training institutions for civil servants.

Most of these reforms, however, especially those established by the General Law Code, focused on bringing the university under tighter state control and discipline; thus, in many ways they institutionalized the mercantilist logic of the Enlightenment university and failed to address the broader cultural questions about the purpose of the university—questions that ultimately revolved around the authority of knowledge in a modern media ecology.

This failure only continued with the appointment of Julius Wilhelm Ernst von Massow in 1798 as justice minister and director of educational institutions housed in the religious affairs department. Following the incomplete efforts of his predecessors, Massow was also charged with reforming the entire Prussian education system.[7] His official mandate read, "Instruction and education form the human and the citizen; both are entrusted, at least in the rule, to the schools, so that their influence on the state is of the highest importance."[8] Massow's task was not simply to improve "academic discipline" but to turn an array of disparate institutions into a modern, unified state system. He was to take the medieval institution mocked by Salzmann and modernize it, that is, remake it on the model of the mercantilist University of Göttingen.

Massow, a popular philosopher sympathetic to utilitarian pedagogues such as J. H. Campe, claimed that universities should serve not "only future speculative scholars" but the state and its citizens as well.[9] Massow's reform efforts typified the Enlightenment "utility gospel" that Schelling would lambast in his Jena lectures. He proposed that universities be abolished and replaced with either civil servant or professional schools that trained doctors, lawyers, and other professionals.[10] The range of specialized academies that had emerged over the course of the century in Berlin—by 1800, there were war, mining, architecture, trade, and arts academies—could more efficiently produce the kind of knowledge immediately useful for the state. Pushing for a sharper distinction between theoretical and practical knowledge, for example, he tried to move all theoretical medical education out of Berlin to Halle and Königsberg and expand the practically oriented surgery clinic in Berlin, *Collegium medico Chirurgicum*.

Against this background of repeated yet limited attempts at reform and the broader decline of German universities, Prussian scholars and bureaucrats extended the debates of the 1790s, such as those of the Wednesday Society and Kant's and Schelling's more philosophical arguments, but reformulated the question of the future of the university in more particular terms: Should there be a university in Berlin? And, if so, what would distinguish it from other universities?

In 1802 Karl F. Beyme, head of the Civil Cabinet and advisor to King Friedrich Wilhelm III, invited a number of Prussian officials and thinkers to consider whether, as he put it, an "institution of learning" should be established in Berlin. Although expressing his desire for "a new general teaching institution," he specifically avoided using the term "university," hoping to distinguish a possible Berlin institution from that guild-like, corporate institution traditionally known as the university.[11]

Although Beyme left well before the University of Berlin was officially inaugurated in 1810, he developed in more pragmatic terms the outline of an argument—formulated, as we have seen, by a number of figures—that a new university should primarily be a technology for the production and dissemination of knowledge. Berlin, he noted, was already home to an array of scientific institutions and collections, but they were distinct, fragmented, and often in competition for resources, prestige, and people. As one Berlin physician who responded to Beyme concluded, "the [Prussian] Academy, the insane asylum, the library, the natural history and art collections, the botanical garden, the chemical laboratory, the veterinary school, the riding track, etc.—everything must become a unified whole."[12] All of these institutions and collections should be organized according to one purpose, but Beyme never clarified what this purpose was.

One of the first to respond to Beyme's request for proposals on the university question was Johann Jakob Engel, a gymnasium teacher in Berlin.[13] Like other critics of German universities, Engel considered the university crisis in the context of technological changes and the shifts in epistemic authority that they had fomented. "Certain elements of instruction," he wrote, can be "presented in books . . . but never comprehended out of mere books, never taught through mere words, that require sight, presence, demonstration."[14] But this was precisely the question: what were these "certain elements of instruction" that printed books lacked, and why was the university the only technology that could provide them?

Discussing the study of the fine arts, Engel wrote that everything printed in a text is "merely the dead letter." Engaged in a broader context of use, however, these same printed texts are suddenly endowed with what they previously seemed to lack: "I can certainly sit among four bare walls and memorize the names of authors, the titles of books, the format of editions, etc.," wrote Engel, "but it is something completely different when in a great, opulent book collection, the librarian brings the works themselves and lays them down before my eyes. What an advantage the oculis subjicere fidelibus (to convince with the eyes) has over mere demittere per aures (hear say)!"[15]

The crisis of late eighteenth-century information overload was not one of books or printed matter in themselves, but rather a crisis in the ways that readers engaged them. Earlier technologies of print, suggested Engel, had reduced books and other forms of print to inert objects, distinct and somehow separate from humans. In contrast to the fetishization of the "aids to erudition," true learning, he suggested, resulted from the interaction of the technologies of knowledge—everything from books and encyclopedias to gardens for botany, instruments for astronomy, tools for surgery—and human agents, who were capable of using them well. The printed book is given new meaning and value when engaged in the institution of the library and facilitated by a librarian. Engel did not fixate on seeing the actual paintings; he focused on the images of paintings as printed in a book. It was the institutional context of these interactions that made all the difference.

During the debates about the future of the university around 1800, scholars embraced the university as the institutional context in which interactions with technologies of knowledge could be given meaning and value. It was, as one Berlin doctor and medical scholar put it, "a great technology" that would bring the persons, practices, and materials of knowledge together.[16]

## Fichte on Books and the University

In 1807, Fichte sent Beyme a detailed plan for a new university in manuscript form.[17] Although his title, *Deduced Plan for an Institution of Higher Learning in Berlin*, would suggest otherwise, Fichte did not simply deduce his vision for a future university from abstract philosophical principles. He actually began his plan with a history of the university's relationship to technologies of communication, or what he termed the "edifice of science"—the material, objective conditions of its historical development. Universities emerged in the twelfth and thirteenth centuries well before the "scientific structure of the modern world" had been erected.[18] Given how few "books" there were in the pre-print days, written manuscripts were not the primary mode of transmission; instead, orality—what Fichte called "oral propagation," especially in the form of lectures—was. Universities were an "*Ersatz*" for the lack of books. After the invention of the printing press and the reductions in time and cost which it afforded, however, print gradually offered a more efficient mode of transmission and eventually came to predominate over oral lectures.

But the efficiencies of print did not solve the problems of communication. According to Fichte's history, the shift to printed forms of communication ushered in a different set of problems. Whereas universities and schools once struggled with a paucity of texts, they were now confronted with an "excess of books."[19] And yet, even with the increased availability of printed books "in every branch of science," complained Fichte, professors, including his contemporary colleagues at German universities, continued to recite "what lies printed before everyone's own eyes."[20] Next to the printed book, they placed another type of book—the traditional lecture. After the proliferation of print, professors kept reciting books to their students as though students could not read them on their own. Even though the technologies of knowledge had changed, the university's form of communicating it had not.

For Fichte, this refusal to adapt to technological change was not only inefficient but deleterious to thought itself. Professors and students alike assumed that the oral lecture and the printed book simply distributed content. The particular forms of knowledge were incidental. This misapprehension led not only to redundancy but also to the widespread failure of students and professors to engage any text in its entirety; instead, they settled for "torn fragments."[21] Students did not read, and faculty simply repeated what had already been printed and stored in the static "archive of the universal world of books."[22] Both relied on print technologies—encyclopedias, lexica, and periodicals—to think for them. They neither read nor listened with any attention. They interacted with their printed texts poorly.

Fichte saw in this history of the university a dual failure. On the one hand, university teachers and students had reduced science to something that could be pursued at one's own discretion and in the "most convenient manner."[23] The relatively condensed, fact-based character of printed lexical compendia made for easy and superficial reading and thinking. On the other hand, universities had turned themselves into institutions for the "repetition of . . . content that was already in a book." In order to flourish in the new media environment, universities had to differentiate themselves from other communication technologies. Scholars and professors needed to reimagine them as distinct technologies that offered students something books and related forms of print could not; they had to articulate those "certain elements" of learning and scholarship that other technologies purportedly lacked. This, insisted Fichte, was the central task of the scholar. The particularity of the university as a technology lay in the fact that it was the institution of a particular type of person—the scholar. The university had to make the case that it was more than a content delivery device. Its defenders had to explain why it could bestow an epistemic authority that other technologies could not. Fichte's condensed history of communicative technologies tried to do just that by reimagining the scholar and the university.

In the decade before he submitted his proposal to Beyme, Fichte had sketched in various lectures and publications a new image of the scholar. For much of the eighteenth century, the traditional scholar was a member of the empire of erudition, who was characterized primarily by his facility with print and its various technologies. But Fichte wrote consistently of the scholar's "vocation" [*Bestimmung*], a term that in the German connotes determination both as something and toward something through an appeal to a teleological structure. Something can be destined to become something that it not yet is. In this sense, the "vocation" of the scholar referred to his defining characteristic, that which determined his particular character. For

Fichte, this defining characteristic was something to be realized or achieved. The knowledge and skills of the scholar were inextricable from the character of the scholar.

As a young professor at the University of Jena, Fichte delivered a series of lectures in 1794 entitled *Some Lectures concerning the Scholar's Vocation*, which he also described as "Morality for Scholars."[24] In a printed announcement for the lectures, he made clear his intentions:

> The sciences, as you all undoubtedly realize, were not invented as an idle mental occupation to meet the demand for a refined type of luxury. Were they no more than this, then the scholar would belong to that class to which all those belong who are living tools of a luxury which is nothing but a luxury; indeed, he would be a contender for first place in this class. All our inquiries must aim at mankind's supreme goal, which is the improvement of the species to which we belong, and students of the sciences must, as it were, constitute that center from which humanity in the highest sense of the word radiates. Every addition to the sciences adds to the duties of its servants. It becomes increasingly necessary to bear the following questions seriously in mind: what is the scholar's vocation? What is his place in the scheme of things? What relation do scholars have to each other and to other men in general, especially various classes of men?[25]

On Friday evenings at six in the summer of 1794, Fichte tried to answer these questions and exhort his students to join a community dedicated to knowledge and science.

Speaking to his young students in a lecture hall crowded with upward of five hundred people, Fichte proclaimed that scholars constituted a distinct social class, not because they were separate from society, but because they played a central role in its historical progression. The vocation of the scholar was the "supreme supervision of the actual progress of the human race in general and the unceasing promotion of this progress."[26] As the "educator of humankind," the scholar "exists only in and for society," whose progress toward perfection he supervises and nurtures. The scholar's vocation is fundamentally social. Fichte's scholar is not the erudite committed above all to the empire of erudition. He is, rather, the "ethically best man of his time," the exemplary character of the modern age. And it was in the university that a young person first encountered this "clear concept of the essence and dignity" of his real vocation and the model of science to which he was called to "dedicate his life."[27] Fichte's exalted description of the scholar represented a fundamental reimagination of the kind of person the scholar ought to be, and it tied the value and authority of knowledge to his character.

In 1805, Fichte delivered a revised version of his Jena lectures at the University of Erlangen, where he began to teach that same year, in which he explained the vocation of the scholar in terms of the contemporary media environment. In an age of mass print, the scholar's supervision and authority were needed more than ever. The "actual heart of the epoch and the true seat of all scientific illegitimacy," he wrote, lay in the uncontrolled expansion of reading made possible by the proliferation of print, which had allowed for the "disreputable" to spread and even be encouraged:

> In place of other ways of wasting time that have fallen out of fashion, reading emerged in the last half of the previous century. This new luxury requires from time to time new, fashionable wares. It is, of course, unthinkable that one would read again what he has already read, or the same thing that our predecessors read—just as it is unseemly in society to wear the same clothes repeatedly or to dress oneself according to the customs of one's grandparents.
>
> The new need produced a new business, which through the delivery of wares sought to nourish and enrich itself—the book trade. The fortunate success that the initial entrepreneurs had excited others and so it has come about that in our own day the entire network of sustenance is extremely overburdened and far too many wares with respect to the consumer are delivered. The book publisher orders like any other publisher his wares from the manufacturer simply so that he can bring his wares to the trade fair. Sometimes he trades merely in speculation. The author who writes in order that something is merely written is the manufacturer.[28]

The circulation of books as wares created a distinct realm of print disconnected from actual intellectual labor. Echoing Kant's and Herder's concerns, Fichte also worried that books had become commodities—interchangeable with one another and subject to the fleeting interests of a market that bore no relationship to thinking. The modern publisher sold books as wares, and the modern author simply produced more writing.[29] In this new world of books, all that mattered was novelty and current fashion. The "activity" of the literary market had "obliterated" and "debased" anything of worth such that it circulated in print like a "ghost."[30] Any authority or standard against which to evaluate the quality of a particular book was fleeting and insubstantial. The modern book market had undercut all epistemic authority.

Despite his critique of the "ocean" of books and the bibliographic practices that emerged to manage them, Fichte never wanted to dispense with books. The sciences were still best maintained and advanced, as he wrote in his plan for a university in Berlin, "by means of a book."[31] What was needed, however, was a different account of books and the normative context in which they were engaged. Fichte described books less as depositories of established, authoritative knowledge and more as discrete markers of the progress of science over time. The historical development of science always exceeded any one particular book. The progress of knowledge could not be bound by the codex.

Fichte even developed an idea for a print periodical as part of a much larger plan for the internal organization of the University of Erlangen. The university, suggested Fichte, should work with—not in competition against—print. Fichte proposed that the journal be called *Yearbook of the Advances of Bibliography*, which he described as a "continuous document of the uninterrupted progress" of knowledge.[32] It would counter the commodification of print by placing it within the context of science as a cohesive body of literature, and it would demonstrate the historical development of science over time. Reviews would build off one another to create a narrative of progress. The journal would not simply be a catalogue of printed commodities. If a reviewer came across a book for which he did not possess the requisite "competence," he would then pass it on to a scholar who did. Whatever was mere repetition or simple paraphrase would be identified and excluded. Unlike *historia literaria*, the aim was not primarily to tell the story of erudition, as established and authoritative knowledge, but to narrate the development of science over time. Fichte oriented the journal not toward a past of authoritative knowledge but toward a future to be created. It would bring particular sciences into greater self-consciousness by making them aware of their own historical development.[33] Fichte's plan extended the logic of *historia literaria* into a research logic that emphasized not only the production of new knowledge but its progressive character. Although *historia literaria* afforded the material conditions for research, there was a basic difference in temporal orientation. *Historia literaria* was oriented primarily toward the past. It and other related book sciences, argued Fichte, subordinated science to the world of print. Proponents of the bibliographic order assumed that books "existed on their own." They "ground themselves on the delusion of the age that the singular, authorized treatment of the sciences is the production of books."[34] It was this progressive character—which assumed the underlying unity of a past progressing toward an unknown future made by humans—that distinguished inert and living forms of knowledge.

From their late seventeenth-century beginnings, review journals and related forms of print had been oriented to what Fichte called a "mercantilist institute," which supported booksellers in circulating "their wares," and not the university. But "science," he insisted, "is not in the first instance the book, nor does it live in the book; instead, science lives in that which is produced in actual research, in the conflict of minds and lecturing. This will then become a book."[35] A book was the product of science, not its complete embodiment. The empire of erudition had reduced knowledge to its material forms, but true knowledge or science begins not in a printed object but in the interaction of people with print and each other. Fichte sought to tie the scholarly periodical more directly to the university, so that the university could help sustain print and lend it institutional authority. The university would not replace print; instead, it would interact with and help promote print and reimagine how it could be used and conceptualized.

## The Lecture as Performance

Perhaps the medium that demonstrated best how the university could interact in new ways with print and orality was the lecture. Around 1800 the lecture underwent a series of revisionary interpretations that described it as a key art of transmission and mediation for the new media environment.[36] For Fichte and Schleiermacher in particular, the lecture was not simply an unmediated form of oral communication; instead, it was a mixture of different media that combined oral and printed forms of communication to figure a distinct type of community. The lecture was a central medium for the practice of science.

The lecture, of course, had a long history in the medieval and early modern university. As its etymology from the Latin *legere*, "to read," suggests, it traditionally referred to a "reading or dictation from an authoritative text."[37] Reading and lecturing were functional equivalents, each grounded in the authority of the particular canonical text that was read. In this context, the primary purpose of a lecture was not to present or produce new knowledge but to transmit and thus safeguard by reduplication knowledge that had already been established as authoritative. At most universities a professor could even be fined for straying too far from the texts prescribed by university statutes.[38] The medieval and the early modern lecture was primarily "a site of slow oral dictation, careful memorization," and extensive note taking, which was a resource for individual students as well as an essential method of reproducing texts.[39]

As printed texts became cheaper, they emerged as possible alternatives for the efficient transmission and distribution of knowledge in the lecture. Theoretically, at least, the lecture as a communicative technology could have become obsolete, a casualty of what some today might call technological disruption. But that did not happen. The lecture survived and even flourished as a mixed practice of reading, dictation, and note taking. With the increased availability of printed textbooks over the course of the eighteenth century, for example, most lectures were based on "ordered, but relatively condensed abstracts of available knowledge," or compendia.[40] "Although reinforced by textual scarcity, which was mitigated with the printing press, this conception of textually grounded knowledge and its enactment through dictation persisted long after the era of Gutenberg."[41] The lecture's persistence—and, in the period around 1800, its discursive resurgence—makes little sense if one makes the techno-determinist assumption that a supposedly more efficient technology (i.e., printed books) will always replace a less efficient one (i.e., the lecture). This fact was especially true in the period preceding the founding of the University of Berlin in 1810, when figures like Fichte and Schleiermacher elevated the lecture as a central technology for a future university, one necessary to counteract what they considered the deleterious effects of the proliferation of print and the associated dissolution of epistemic authority.[42]

The value of a university professor, they contended, was in large part a function of his skill in the art of lecturing. One of the Enlightenment university's basic failures was that the vast majority of professors reduced the lecture, as Schleiermacher put it, to the mere "accumulation of literature."[43] They failed to treat lecturing like the "unique exercise or test" that it could be and assumed that "one could simply say what one knows," as though thoughts, concepts, and ideas were inert objects of exchange. The true lecture was a performance. It was not a "mere medium" for the transmission of knowledge. It was a central practice or art of transmitting knowledge and forming students into a broader culture of science.

Fichte tied his notion of the lecture to his idea of the true scholar, which he divided into two possible and equally distinguished types. The scholar-as-teacher was committed to forming students who were capable of grasping the "divine idea"—the unity of all types of knowledge, the unity of real and ideal, reason and sensibility.[44] The scholar-as-teacher attended to the subjective conditions, needs, and capacities of individual students, much like Schelling's description of the university teacher's task to cultivate in students the idea of a totality of knowledge and disabuse students from the Enlightenment fallacy that a material completeness could ever be achieved. The scholar-as-writer, in contrast, attended exclusively to the divine idea, or the idea of the totality of knowledge itself, and tried to give such an idea a particular form. Whereas the scholar-as-teacher could achieve his goals best through the oral lecture, the scholar-as-writer could best achieve his goals through written texts.[45] For Fichte, the oral lecture did not take precedence over print; it was simply located in a particular institutional context, the university, whose primary purpose was the formation of particular types of people.

The lecture was tied to the teacher's particular performance. Its success or failure was a function not only of the arguments articulated but also of the context in which these arguments were enacted. The vocation of the scholar-as-teacher was not simply to transmit the divine idea, but rather, as Fichte put it, to "form, express, and clothe it in the most diverse way," so that individual students could engage it in a particular context. The professor, therefore, had to "possess the idea not merely in general but in great liveliness, dynamism and inner adroitness."[46] He had to be capable of presenting it in "infinite forms" so that he could perform and present it in a variety of situations and attend to the needs of his students.[47] The lecture was a speech act crafted for a particular time and had to be attuned to the needs, capacities, and knowledge of the given audience. In order to perform a lecture successfully, a professor had to know his students.

Despite the particularity of the as-yet-formed student, however, Fichte also insisted that the scholar-as-teacher address his students as a "priest of science" speaking to "future servants" of science; that is, the scholar-as-teacher formed all students, regardless of their differences, toward a common end, science.[48] Science was not simply a matter of status or prestige. It was a practice that had its own internal goods and virtues. In contrast to the scholar-as-teacher, contended Fichte, the scholar-as-writer had "no particular reader in mind; instead, he constructs his reader and gives him the laws of how he must be"[49] because his work was "for eternity." Whereas the scholar-as-teacher addressed the particular audience sitting before him, the scholar-as-writer addressed an abstract, homogenous public that he had to imagine.

Fichte did not, however, base his distinction between the scholar-as-teacher and the scholar-as-writer on a sharp differentiation between oral and print media. He described his lectures neither as spontaneous events nor as mere dictation. His first Jena lectures on the vocation of the scholar, he claimed, were not delivered "completely free"; rather, they were worked out "word by word" and delivered as "written."[50] In other lectures, Fichte was reported to have used sketched-out notes, which when he spoke from them gave the appearance of a freer style.[51] He intended his lectures to be a performance of the act of thought with material support. The lecture was, as Sean Franzel puts it, a "scene" in which Fichte mixed different media, thematized the interplay of oral and print forms of communication, and reflected on the conditions of knowledge.[52] He never described it as an unmediated communication.

Most importantly, as already suggested, Fichte approached the lecture as an act of formation. His emphasis on performance was tied to a deep concern with the students' apprehension of the lecture. Thus, he insisted that all lectures be supplemented by pedagogical practices to test whether the students had made the lecture their "own free possession." In his *Deduced Plan* for a new university in Berlin, Fichte outlined how the "master" teacher formed the student not just through the oral lecture but through related oral exams in which the student was to use what he had learned as a "premise" to respond to a teacher's questions. These exams, what Fichte termed *Konversatoria*, consisted of a student posing a question to which a teacher responds. This exchange continued and, in its ideal form, would become a "Socratic dialogue."[53] Alongside these oral exams would be written assignments in which the student would be asked to expand on a particular premise in the "art of the written lecture." In order to demonstrate that the student had succeeded in "the artistic formation of his self," he would submit what Fichte called a "trial essay" after the lecture course, an exercise not in the display of knowledge but in the art of its production.[54] One of the basic virtues of the newly formed student was his productivity. A well-formed student created something new and did not simply imitate what was already established.

Beyond these particular curricular elements, Fichte also proposed changes to the material conditions under which students could undertake these exercises. Students would live together in a community in which the state would provide for all of their material needs, from housing to books to pocket money.[55] They would be required to wear uniforms in order to distinguish them from the nonuniversity population. Students were to be freed from all extra-university structures and authority and stand outside civil and church law, subordinate only to the pursuit of science. For Fichte, studying at a university was a vocation, and the university with all of its infrastructure was only there to ensure the practice of this vocation. The goal of all of these exercises and practices was to bring students to a more self-conscious, more determinate relationship not only to particular claims and premises but to a particular way of life. The student was an "apprentice," who needed to be transformed into a more reflective, articulate, and self-sufficient disciplinary self.

The standard of success of such an apprenticeship was whether the "budding artist [could] craft something different out of it."[56] The disciplinary self was an artist, and this meant that knowledge was not something that could be transmitted from a repository—either personal, a professor's head, or material, a universal book—to another repository, such as a student's head waiting to be filled with information. Scientific knowledge was a practical skill that one exercises and acquires through repeated personal interactions over time. Science is like an art because it cannot be specified in detail. It does not consist only of propositions that can be transmitted according to set, fully articulable rules. It can only be communicated through the example and performance of the teacher as master scientist, and it is a performance that entails personal interaction and contact. The lecture is this public performance of a community of interactions into which students are called.

Schleiermacher similarly embraced the lecture in his contribution to Beyme as a "talent of transmission" that should play a central role in the new university.[57] Alluding to the historical functions of the lecture, Schleiermacher wrote, "A professor who repeatedly reads from a notebook that has long since been written and allows it to be copied by the students, reminds us of a time when there was still no printing press and it was still very valuable when a scholar dictated his writing to a number of people, when the oral lecture also had to serve in the place of books."[58] Like Fichte, Schleiermacher considered a professor who simply dictated—one who read an already-established authoritative text before a group—a historical anachronism. In the pre-print era, a professor could dictate a handwritten or manuscript text to students, who would then assiduously copy it down and reproduce it many times over; therefore, the lecture as an act of dictation was also, by necessity, an art of transmission. In the age of print, however, "it is inconceivable why some men should enjoy the privilege of ignoring the benefit of print." Why, continued Schleiermacher, do such professors continue to gather students before themselves in a lecture hall instead of simply selling their "wisdom, already composed in static texts, in their normal way: black on white"? If the lecture were simply a technology for disseminating knowledge, then it would be redundant, inefficient, and dispensable.

But Schleiermacher, like Fichte, described the lecture as a flexible technology that did not fall into the simple opposition between text and speech. It was a performative act consisting of two basic elements. According to its first, popular element, wrote Schleiermacher, every lecture assesses and responds to the "condition in which the listeners find themselves." The "art" of the lecture is to make the audience members aware of those aspects of their present "condition" that are "insufficient." The "popular" element is not a general appeal to a common knowledge but rather a direct address to a particular audience. The lecture makes clear what the audience lacks and what it assumes to be true. Lecturing is a dialectical art directed to an audience located in a specific time and place; it concretizes knowledge by addressing a specific audience and not some abstract reading public. According to its second element, the lecture is "productive," in that the teacher performing a lecture allows "everything that he says to emerge before the listeners. He must not tell what he knows but rather reproduce his own knowledge, the art itself, so that students do not constantly just collect facts but rather can immediately perceive and reproduce the activity of reason in the production of knowledge." The lecture is the site of the moment of higher learning, when the student does not merely "collect facts" but participates in the actual production of knowledge. It is designed around the one instance that the university was to bring to life: "one moment . . . one act, namely, that of the idea of recognition."[59] The popular and productive elements of the lecture work in tandem so that a student and teacher are both differentiated (popular) and identified (productive) with one another.

Lecturing is to be a mimetic act whose authority is grounded in the context of its performance, and what is performed is the generative capacity of reason itself. The focus of the lecture is not on the grounds of knowledge, a particular first principle, or a fixed body of information, but rather on the activity of thought that is repeated in every performance. Schleiermacher assumed that the ground of thought, a foundational principle to which all knowledge can be traced, cannot be accessed. Schleiermacher's claim that a lecture produces knowledge anew and in new forms was in part a function of his presumption that an ultimate ground of thought, a foundational principle or insight from which all knowledge could spring, cannot be represented.[60] The lecture, therefore, produces a "new combination" of thoughts with every performance.[61] It points to a moment of thought that remains inaccessible to both the teacher and the student. Thus, in "teaching [the professor] will always learn," not because he provides some God's eye view or metaphysical account of the relationship of all types of knowledge with each lecture, but rather because every lecture presents such a unity in some finite and thus particular form anew.[62] The lecture conjoins the objective and subjective in the act of thought.[63] The "moment" that the lecture performs and the unity of thought that it produced were also a particular form of communication between the teacher and the student as scholar. The lecture is not a performance in which the teacher thinks and the student passively observes. It is rather a moment of engagement, in which the teacher solicits the student to partake of a process of intersubjective communication. The art of lecturing stages "reciprocal communication between those who bear likeness to one another."[64]

For Schelling, Fichte, and Schleiermacher, the institutional context of the lecture was central. In order to succeed as a performance, the lecture had to satisfy certain norms internal to it as a practice. First, it had to be performed in a university, which Schelling and Schleiermacher referred to as an "association"—an organizational form that supported the common work of different individuals and lent it institutional authority.[65] The university was a community that gave the lecture a context of meaning and authority. Conceiving of the lecture in this way was in part a protest against the perceived abstraction of a book market that concealed the true originality and productivity of writers, scholars, and authors. Second, the lecture had to be performed by teachers who possessed a certain character: a capacity to display "lively excitement and deliberateness," a commitment to science as vocation which exceeded the individual, industriousness, a capacity to relate a particular science to the unity of all sciences, and productivity.[66] The successful lecturer at a university embodied, as jurist and future Berlin University professor Friedrich Savigny put it, particular "virtues," such as an "animated enthusiasm" combined with a "reflective deliberateness."[67] Schleiermacher even suggested that the "non-virtues" of some scholars—their lack of "true vitality," for example—would make them poor lecturers.[68] The lecture was not only a material technology, an oral encyclopedia, that "endlessly rehashed" what was already known.[69] It was an exercise that formed persons.

Although the lecture, as its romantic and idealist defenders envisioned it, did not dispense with the need for print, it did resist the saturation of print and embodied a new, institutionally supported, epistemic authority.

It was precisely because of the need for institutional grounding, however, that the lecture, according to Schleiermacher, could not on its own reproduce what he called the "sanctuary" of science, or "scientific living together" that was at the core of the university.[70] The lecture was one element, albeit a vital and exemplary one, of the institution of the university, whose purpose was to bear the practice of science and protect its goods and virtues. The lecture and the "moment" of interaction that Schleiermacher claimed it embodied figured this scholarly community. The dialogue that the lecture purports to be is actually, as Sean Franzel puts it, the "fiction of dialogue, a space where listeners can take the actual institutional environs in which they find themselves as a point of departure for imagining themselves part of the scholarly community as a whole."[71] The actual lecture and the lecture hall always represent an ideal, scholarly community, a "sanctuary" of science. The lecture stands in for a particular order of knowledge and the desire to organize and master it according to certain norms and conceptions of what constitutes authoritative knowledge. And like the encyclopedia, the book, or the empire of erudition before it, the lecture figures and points to the ends of knowledge technologies: forms of communication and social interaction.

Like Fichte, Schleiermacher included in his plans for a new university additional forms of interaction and community between professors and students outside the lecture hall,[72] including small groups, individual conversations, testing, and sessions in which students share their work, all of which could foster what was essential to the university—"actual dialogue." For Schleiermacher, the seminar organized many of these parallel practices, bringing the teacher and students into a much more intimate environment and allowing for direct mentoring. Consisting of a much smaller group of students working in close proximity with the teacher, the seminar was, thus, the real site of specialized science. It was the site where students were formed into not just science generally but specialized, more highly differentiated forms of science. Seminars, wrote Schleiermacher, "attach themselves to the disciplines, which focus more on particulars. They are that being-together of the teacher and the student" in which the student performs for the first time his role as "productive" and the teacher "directs, supports, and judges" this production.[73] The seminar was in a dialectical relationship with the lecture.

More recently scholars have dismissed or challenged such accounts of the lecture as simply privileging the living voice over the "dead" letter, invoking a sense of immediate presence by masking its own rhetorical structure, or prescribing an ideal moment of prelinguistic communication in which a meaning particular to the consciousness or the spirit of the speaker is somehow recovered.[74] Many of these accounts of the lecture are part of a broader critique of hermeneutics as a reversal of the externalization of thought as speech. On this account, the lecture, as described by Fichte and Schleiermacher, was an effort to return to inner speech and establish its priority over textual, especially printed, forms. The lecture attempts to reassert the subject's control over the written or printed work. It is simply an attempt to control communication.

While powerful, these critiques overlook not only the way in which the lecture recombined various media and practices but also its iterative character. The lecture, in the highly metaphorical way that Fichte and Schleiermacher described it, reproduced not a stable meaning but an act of reason that was performative, "a speech act," as Schleiermacher put it.[75] It was more a performance than a recovery of meaning. The subject continued to be a source of meaning but in a much more contingent way than more canonical accounts of hermeneutics might have had it. Thought, for Schleiermacher, was not simply reducible to inner speech. On a purely hermeneutic account, the authority of knowledge can be safeguarded through a recovery or return to the mind of a speaker; however, Schleiermacher harbored deep reservations that such a recovery was possible for humans, who are defined by their finitude.

In either reading, however, the underlying anxiety was the same: an anxiety about the authority of knowledge in an age of print. Fichte and Schleiermacher did not consider the lecture as a recovery of meaning, but as the formation of persons. It was meant to transform students into disciplinary subjects who could judge and filter the excess of print. The filtering effect worked on the student and not simply the material of knowledge.

These arguments for the continued relevance of the lecture were also arguments for the relevance of the university in a new media ecology, because the lecture required an institutional structure. The ability to lecture distinguished teachers in the future university from those of the past and from the print market. If one considered the "ease," wrote Schelling, with which professors could assume their position and carry out their lecturing duties at certain universities, then no job could be considered easier than that of a university professor, because most universities had devolved into mere means for the transmission of knowledge.[76] The lecture stood in for what distinguished the university from the "handbooks and thick compilations" of the empire of erudition and its bibliographic order of knowledge—its focus on forming the subjects of knowledge.

## Science as a Practice

These efforts by Fichte, Schelling, and Schleiermacher to distinguish the university as a particular social institution and community devoted to the cultivation of a uniquely scholarly ethos could be described simply in terms of cultural prestige. On this account, specialized science was ultimately a romantic ideology.[77] Fichte, Schelling, and Schleiermacher sought to define a scholarly elite in opposition to an expanding bourgeoisie reading public. But to what end? Explanations based purely on ideology and social prestige are insufficient. They cannot fully account for the emergence of the modern research university because they misconstrue or simply disregard the underlying anxieties about epistemic authority which abounded in the age of the proliferation of print; therefore, they also underestimate the ethical aims of the arguments. Fichte and Schleiermacher were not simply seeking to safeguard their academic sinecures or subject unwitting students to a romantic ideology. They were trying to come to terms with what they and a broader swath of late eighteenth-century writers and intellectuals saw as a crisis in epistemic authority. The "mass of information," as Schleiermacher put it, could only be unified and science could only be "propagated" and "perfected" in the university.[78] The worry about the commodification of print was a worry about the growing inability not only to manage the overabundance of knowledge and secure social distinctions but also, and more importantly, to discern meaning and value amid an apparent excess of knowledge.

In this context, figures like Fichte and Schleiermacher turned to the university as the only technology capable of stepping into what they saw as a vacuum of authoritative sources for knowledge. They imagined that a newly conceived university could organize the fragmented forms of knowledge by forming young students into mature, disciplined scholars who could then filter through an excess of texts and distinguish real knowledge from the appearance of knowledge. In the context of the university, the purpose of such technologies, whether printed books or performed lectures, was not just to distribute knowledge but to cultivate a particular "art" of interacting with it and its technologies. If the university were to flourish, it would be as an institution that sustained science as a practice and cultivated particular virtues. The idealists and romantics who described this ideal university saw it not as a withdrawal from society but as the best way to deal with the ramifications of information overload and the attendant lack of epistemic authority.

In order for the university to flourish, scholars also had to reconceive of what it meant to know and what it meant to be a scholar. Erudition had to be reimagined not just as a distinct class consciousness—a renewed awareness of one's membership in the empire of erudition while also being a doctor, minister, or lawyer—but as a unique practice that demanded a near-total commitment. Science had to be recognized, as Henrik Steffens, a professor of natural philosophy, put it in 1809, as something that "seizes the entirety of one's being and provides a determinate direction." Fichte and company attempted to facilitate recognition by redefining science as an ethics for dealing with the fragmented experience of both individual subjectivity and the proliferation of objective forms. Science as a practice was a grounding orientation that provided confused and overwhelmed students meaning, purpose, and an account of virtue.[79] It enabled them to discern what was worth knowing and what was not.

When I describe science as a practice, I am referring to a "coherent and complex form of socially established cooperative human activity," extrapolating from Fichte's and his contemporaries' use of the term in order to move beyond the repetitive descriptions of *Wissenschaft* as simply an ideology.[80] Conceiving of specialized science as a practice will put the claims made on its behalf into a cultural and historical context that will help us make better sense of it. As practice, specialized science had its own internal standards of excellence, internal goods, virtues, and rules. These are precisely the elements that Fichte, Schleiermacher, and Schelling were trying to articulate when they insisted that science be pursued for its own sake. The failure of Enlightenment notions of science, as A. W. Schlegel put it, lay in the fact that they only ever sought "utility and applicability."[81] On the romantic and idealist account, there were, of course, goods external to science, such as prestige, status, or money, but these could have been achieved by other means. To describe, as did Fichte and Schleiermacher, science as a practice was to claim that there were some goods that could be achieved only by practicing science, that only made sense in terms of science. These goods internal to science included the excellence of what one produces as a scientific artist, the sense of community afforded by science, and the insight into the relationship of all types of knowledge and sciences that are bound together by a common practice. And ultimately, the "apprentice," as Fichte referred to the young student, gradually learns to embody science as a way of life. He comes to live his life as a participant in the practice of science.

To participate in such a life and practice required living and laboring according to standards of excellence and rules. It required the acceptance of a tradition of authority, which provided the standards of excellence of what counted as real, scientific knowledge. This was why none of the idealist or romantic advocates of the new concept of science eschewed the bibliographic practices of *historia literaria*. These were crucial to maintaining the traditions and the sources of authority for science as a practice, but they were also limited. The print objects accumulated and organized by the empire of erudition were the material foundation, but this material basis needed to be organized toward a common purpose, and, for these late eighteenth-century figures, that purpose was the practice of science. Such a notion of science, with its emphasis on standards, virtues, and formation, brought one into a community of other practitioners who shared the same goods and desires. And it was the community that provided a new form of epistemic authority.

The suggestion that science should be pursued as a distinct practice went against predominant Enlightenment ideas that science should be directed simply toward external ends and goods. Echoing Kant's and Schelling's critiques of Enlightenment conceptions of science, Schleiermacher sought to elevate conceptions of knowledge and science as valuable only as means to a practice with its own internal goods and virtues. So conceived, science was not just a storehouse of information but a source of authority that oriented a person amid the surfeit of objective forms that characterized modernity. Science, wrote Fichte, "perfects the human. . . . It is a complete life." It unites "brilliance, virtue, religion."[82] Science was a comprehensive practice, replete with its own internal goods, regulating norms, and virtues. It organized individuals, not just objects.

Just as importantly, Fichte and Schleiermacher tied science as a practice directly to the institution of the university, which, they claimed, was the only institution that could unify the various technologies of science from lectures to print and embody "an entirely new life process" in which one could become "ethical and noble."[83] Thus, while the university was, as we saw in the previous chapter, described as a technology, it was also increasingly viewed as the institutional embodiment of science as a practice. It was, as the poem celebrating the University of Berlin's official opening put it, the "house of science."

This was why Fichte considered the greatest threat to academic freedom to be not interference from the state but rather the failure of students to commit themselves fully to their practice. As Fichte put it in a fiery inaugural lecture as the rector of the University of Berlin in 1811, the university was a community in which the "learning subject" emerges and is continually transformed.[84] "One studies at a university," he continued, "not in order to reproduce in words what has already been learned for exams for the rest of his life, but rather to implement what has been learned in whatever cases might come up in life and thus to convert it into works; not just merely repeat it but rather to make something else from out of and with it. The ultimate purpose is thus in no way knowledge but more the art of using knowledge."[85] Although echoing Enlightenment discourses surrounding putting knowledge to use, the end toward which knowledge was to be put was science. "For the scholar, science is not a means for any purpose"; it is rather, or should become, "a purpose itself."[86] Science would orient the student not just during his few years at the university, however. It would be the "root of his life" regardless of what he chose to do in the future.[87]

The value of participating in science as a way of life lay in excelling at what it demanded as a practice. Science was not simply a tool to facilitate the more efficient accumulation or production of knowledge. It was a practice that oriented the subject toward the ends to which such knowledge should be put. It had its own constitutive norms that defined what it meant to successfully, and thus also unsuccessfully, perform science. These norms entailed ideals of success and failure, correctness and incorrectness. They guided a person on how to relate to himself and to others. Science oriented a person toward a particular idea of human excellence and provided distinct conceptions of ends and goods.

Against the broader critique of Enlightenment print technologies and what Fichte saw as the failure of universities to adapt, Fichte's ethical argument for the scholar and the university can be seen for what it was: an attempt to distinguish the scholar and the university from the endless circulation of print and the leveling impulses of the market. And yet, by describing science as a practice that could provide meaning and orientation, Fichte also insisted that the scholar's particular vocation was only conceivable within a broader society. The scholar, he wrote, "is destined for society."[88] The class of scholars was the only hope for arresting information overload and, as Fichte put it, "advanc[ing]" the progress of all social classes and society as a whole.[89] The difference between the true scholar and the general reader was an ethical one; it was a matter of the person and how he related to himself, others, and the ends of his scholarship. Fichte's lectures on the vocation of the scholar could have also been entitled "lectures on the moral or ethics of the scholar," because they dealt not only with rules and regulations but with "the character" of the scholar.[90] Confronting the proliferation of print and the epistemic challenges it entailed was above all a practical matter, not a theoretical one. Such an ethical framing of the problem of information overload rendered the conundrums of the empire of erudition and its pursuit of material completeness beside the point. The fullness of knowledge was characterized not by a bibliographic perfection but by the perfection of the self within a community of fellow scholars. Fichte defined the scholar in the first instance not by a corporate identity that ensured juridical privileges—membership in the *universitas magistrorum et scholarium*—but by a particular vocation, a normative character or persona into which a student could be transformed.

For figures like Fichte and Schleiermacher, the end goal of science as a practice was not merely a transformation of the self or the development of the "individual student's interests, capacities, and personality as the subjective realization of his unfettered freedom and autonomy," however.[91] A successful transformation extended the "apprentice" beyond himself. In order to flourish, the disciplinary self had to turn toward a universal conception of science. Fichte's description of science was most clearly marked by such a universal orientation. He grounded his cultural critique and vision of a scholarly ethics in his broader philosophical project, especially his attempts to account for the unity of knowledge in his doctrine of science [*Wissenschaftslehre*].

In his lectures *On the Essence of the Scholar*, he argued that all appearances have their ground in something higher and hidden, which he called the "divine idea" or simply the higher ground of all phenomena. Elements of this idea were only accessible by a certain form of reflection developed through "the free deed" of the human. And this development was only possible through a certain type of education or "learned formation," which only the reformed university could provide.[92] The scholar was thus not merely a man of erudition but also a participant in the divine. Science provided the ground for the incommensurability of the university and its members from that undifferentiated "mass of business-oriented or dull pleasure-seeking bourgeoisie."[93] It provided its practitioners the sense of belonging to something beyond themselves, something that transcended their local circumstances, a community dedicated to science. By reimaging the scholar and the university in such ethical terms, Fichte drew a stark contrast between the scholar as an erudite—a member of a social class with a university degree and Latin proficiency—and the scholar as a participant in the divine idea.[94] One was merely an adept handler of texts, while the other was a "medium" of truth.[95] The true scholar "communicate[d]" knowledge from "generation to generation."[96]

Fichte's description of scholars transmitting "the divine idea" from "generation to generation" was not simply a metaphysical claim about the absolute unity of knowledge, the perfect coincidence of real and ideal, however. It was also a claim about the social nature of knowledge and the ways in which it was created and communicated over time. The unity of knowledge was not only a metaphysical concept but a social practice. And the university embodied this social unity of knowledge because it provided a "communication" and sociability that distinguished the scholar from the mere erudite.

For Fichte, the social unity of knowledge was particularly evident in the communication between teacher and student. "Not only the teacher but also the student," he wrote, "must constantly express himself and communicate, so that their mutual learning relationship is an enduring conversation, in which each word of the teacher is a response to a preceding question that the student just posed and the posing of a new question by the teacher, which the student answers in his next comment."[97] In this constant conversation, the teacher gradually comes to know the student in an intimate and highly personal way. "Scientific instruction" is thus gradually transformed into a "dialogical form . . . in the sense of the Socratic school." The normative image of a university is that of a collaborative group engaged in reciprocal exchange.

In his proposal to Beyme, Schleiermacher envisioned a similarly social order of knowledge, one organized around the transformative potential of science as a practice. Even more so than Fichte, however, Schleiermacher described such a unity in institutional terms. A consciousness of the relationship of all forms of knowledge, of all sciences, could not be attained in isolation, in quiet meditation on some metaphysical idea. It was, rather, the product of a practical activity sustained in a community dedicated to the same end and engaged in the same practice. Membership in the "scientific community," insisted Schleiermacher, was a prerequisite for scientific insight and the generation of a uniquely scientific knowledge.[98]

Whereas Fichte tended to describe a gradual dissolution of the self into a harmonization of all individuals into the "divine idea," Schleiermacher described a gradual harmonization that would never be complete. The part and whole remained, both descriptively and normatively, inextricably linked and yet always distinct. The part and the whole in this instance were individual scientific efforts and the totality of such efforts. Science, wrote Schleiermacher, is the totality not of the objective material of science but of the diversity of "human activity."[99] The scientific activities of individuals are oriented toward a common end, but they never are subsumed by it. No other human activity, he wrote, exhibits such a unity, an "uninterrupted progressive transmission from its first beginnings," as well as such manifold diversity. Stretching through time, the unity of science as a practice compels the "scientific human" toward community and, above all, "communication" and "transmission."[100]

Science was precisely this sociality, the finite and temporal form of the absolute unity of knowledge which remains out of reach for any one individual.[101] The metaphysical unity of all knowledge had its analog in the social unity sustained by the practice of science. The university institutionalized the norms and practices that preserved "scientific exchange," the conversations and communication central to scientific knowledge.[102] As the bearer of science, it safeguarded the rules of social exchange and conduct. As Schleiermacher imagined it, it was the scientific equivalent of Henriette Herz's Berlin literary salon, where, beginning with his arrival in Berlin in 1796, Schleiermacher conversed regularly on literature, science, and the arts with contemporaries such as Wilhelm and Alexander von Humboldt and Friedrich Schlegel.[103] The salon provided a unique space between domestic and public life in which a "free sociality" devoted to the mutual cultivation of interlocutors could be practiced.[104] The rules of exchange included an openness to the views and needs of others, a commitment to "reciprocal action" (constant and equal exchange among all present), an adroit and capacious conversational ability, and, above all, a knowledge of one's interlocutors. These were also the underlying norms and virtues that Schleiermacher claimed should distinguish the university as a community of science.

When Schleiermacher called on professors and students to "behold everything particular not in itself, but in its closest scientific connections, to insert it into a great context in constant relation to the unity and allness of knowledge," he was not simply encouraging them to assume a metaphysical perspective or a timeless overview of the relationship of the sciences.[105] He was exhorting them to engage in the practical activities and exchange that constituted science as a social practice. Science, he wrote, "is not a matter of the individual. It is not something that can be brought to perfection and possessed by one person. It must be a communal undertaking to which everyone contributes his share, thus, everyone is dependent on all the rest with respect to their contribution. And alone he can only possess a fragmented part and even that, imperfectly."[106] Science was "communal work to which everyone contributes his part," and the university, as the institution of the "scientific life," sustained this ongoing collaboration by giving it an institutional form.[107]

For Schleiermacher and Fichte, the practice of science stood in for a particular order of knowledge, a particular way of managing the desire to know. Every scientific activity, every exchange, every transmission, every new creation participated in this whole of science because every science was an element of science as a practice. Following Schelling, Fichte and Schleiermacher combined two conceptions of unity: an "encyclopedic perspective"—the ideal, God-like perspective of eighteenth-century encyclopedic projects—and a "social unity" in which the practitioners of science interacted with each other in harmony.[108] And it was the latter notion of unity that Schleiermacher in particular emphasized. The university mediated the universal and particular through the institutionalized practice of science.

In order to excel in the "idea of science," a scholar must excel in "the area of knowledge to which one wants to devote himself in particular."[109] On a theoretical level, Schleiermacher's claim echoed Schelling's argument that the totality of sciences could not be known independently of its constitutive parts. On a practical level, his claim was that a student could not be formed directly into the idea of the totality of science; instead, this idea had to be developed through training in a particular science. This duel account was crucial if the integrity of science as a practice was to be maintained; otherwise, particular sciences would simply differentiate and become increasingly specialized and distinct objectivations of culture, each defined by what Max Weber a century later termed their own "inherent laws."[110] The key to maintaining the unifying force of science as a practice and thus containing the epistemological anxieties of overload was to sustain science as a practice and cultivate its virtues and internal ends. Over the course of the nineteenth century, German scholars struggled to balance increasingly specialized sciences and an ever-greater and increasingly elegiac emphasis on the unity that science as a practice was thought to afford.

As early nineteenth-century German natural philosopher Henrik Steffens put it, what distinguished the university was not "what is taught here but the manner in which it is taught."[111] The purpose of the university in a modern media age was to sustain the practice of science in order to guarantee the authority of a knowledge distinct from a simply popular knowledge that circulated in commodified form in the print market. The importance of the lecture, for example, lay not in the transmission of particular content, but in clarifying the "meaning of all scientific effort"[112]—that is, the proper ends toward which scientific labor should be organized. The university as idealists and romantics like Fichte and Schleiermacher envisioned it would embody that reason through a second nature, a set of practical capacities.[113] This second nature should not be understood as a set of "habitual responses," but as a practice into which one grows and becomes more reflective over time.[114] What Fichte and Schleiermacher offered was an alternative, more social account of the unity of knowledge. In describing science as a social practice, they sketched a new technology for confronting media surplus—an ethics and a vision of a disciplinary self. And in so doing, they made a clear distinction between the university as a community that embodied a unique epistemic authority and the university as just another content delivery device. The university was not, as Schleiermacher put it, just another "talking book."

CHAPTER EIGHT

# Berlin, Humboldt, and the Research University

IN THEIR PROPOSALS for a university in Berlin, Fichte and Schleiermacher offered a new metaphor for the organization of knowledge. They replaced the encyclopedia, the book, and the empire of erudition with the practice of science and the community of the university. The social unity of knowledge embodied in the university stood in for the desire to manage and manipulate a surfeit of knowledge, and the disciplinary self, the subject formed and cultivated in a community of like-minded individuals, provided a new figure of epistemic authority. They tied the legitimacy of a uniquely *scientific* knowledge to the character and virtue of the scholar, whose distinguishing feature was a capacity to discern unity and make connections in a modern media environment that seemed fractured and fragmented.

If his contemporaries described an ideal university, Wilhelm von Humboldt sought to give that ideal an institutional form in a particular place. "Only in Berlin," he wrote, could a university for the modern age be built, because it was only there that the necessary institutions, collections, and people were present and could be organized into an "organic whole."[1] He envisioned Berlin and its new university as a microcosm of the Enlightenment's technologies and sciences and the center of scientific communication and life.[2] Humboldt wanted to build what his idealist and romantic contemporaries, anxious about the effects of modern print but hopeful that science could find an institution to sustain, had only intimated.

When Humboldt returned from Rome to Germany in 1808, after serving as the Prussian envoy to the Holy See, he soon took over as head of the newly established "Section for Culture and Public Education," which had been charged with modernizing the Prussian school system.[3] Although in office for less than two years, Humboldt transformed a fledgling reform effort into a coherent vision for a modern system of knowledge. In a flurry of memoranda, letters, and bureaucratic maneuvering, Humboldt consolidated over half a century's worth of educational reforms and recent debates about the future of knowledge into a clear and, as it would turn out, highly portable argument for the purpose of the university in the modern media ecology. In a modernity defined by professional specialization, social fragmentation, and the collapse of epistemic authority, the university was the only institution that could both produce and translate knowledge for the good of science and society. He refined an argument for the social value of an institution dedicated to determining what counted as real knowledge in an era in which information seemed to circulate without constraint.

## The Myth of Humboldt

As we have seen in previous chapters, Humboldt was not the first to formulate these ideas, and the University of Berlin never fully realized them. Since at least the 1790s, scholars and writers had invoked the university as an institutional solution to the perceived crisis in knowledge. Most of the supposed innovations of the new university in Berlin—from the concept of research to the specialized seminar—had their precedent in Göttingen. In its first few decades, the Royal Friedrich-Wilhelms-University in Berlin struggled not only to fulfill the ideals of Humboldt and his contemporaries but simply to survive. It struggled to maintain faculty, attract students, and withstand the political turmoil of post-Napoleonic Germany.[4] And the university itself was never really Humboldt's. Humboldt resigned from his position in June of 1810, several months before the university officially opened, and was replaced by an authoritarian Prussian bureaucrat, Friedrich von Schuckmann, who immediately disbanded the university's faculty steering committee and rescinded the financial arrangement with the king, both of which Humboldt had organized.[5] Only four months after his inaugural address as the university's first rector, Fichte resigned after a very public disagreement with Schleiermacher over the question of whether attacks carried out by a few students against a Jewish colleague should be met simply with a faculty-sanctioned expulsion or with a student-led process of adjudication.[6] Like its eighteenth-century print predecessors, the university established in Berlin in 1810 never fulfilled the ideals and the longing for a unity of knowledge which preceded it. Like the "book," "the encyclopedia," or even "the Enlightenment" before it, the Friedrich-Wilhelms-University in Berlin stood in for the desire for a particular order of knowledge. The cancelation of the university's official inauguration in October of 1810 was, in this sense, fitting. The idealistic invocations of a new "house of science" were a prelude to the Humboldt myth, the appeal to Humboldt and the university he helped establish as an institutional and structural model for all universities past and present.

The real legacy of the late Enlightenment debates on the future of knowledge and the university, however, should not be reduced to the failures or successes of one institution. Humboldt and his contemporaries articulated a normative vision that has proven more lasting and influential than any one university.[7] The debates in Prussia about the future of knowledge and the university preceded Humboldt's arrival in Berlin, and Humboldt's own work on establishing a university in Berlin should be read as a contribution to this much broader debate.[8] Humboldt is a central figure in the story of the research university because he helped produce a new metaphor for authoritative knowledge and was one of the first to try to give it an institutional form, even if the modern German university never simply reproduced his plans.[9]

The gaps between reality and normative vision, the institutional and ideational elements of the Humboldtian model, were at the center of twentieth-century debates about German university reform.[10] As cultural historian Peter Hohendahl explains, "While some reformers were convinced that Humboldt's idea of the university should be abandoned to modernize the German university, others insisted on the continued importance of Humboldt's core ideas."[11] Conservatives invoked Humboldt to defend existing institutions, while reformers appealed to him to protect the university from an expanding industrial society. This debate was revisited more recently in anticipation of the bicentennial of the University of Berlin in 2010.[12] Some commentators contended that appeals to a specifically Humboldtian ideal conjure up a mythical university that never had a clear foundation in historical fact.[13] As one historian put it, no one in nineteenth-century Germany appealed to an explicitly Humboldtian model, nor did anyone speak of the University of Berlin as the *Humboldt* university.[14] For these critics, Humboldt's neohumanist university was a myth crafted by twentieth-century German scholars and their hagiographic accounts of a Prussian bureaucrat.[15] According to these older accounts that emphasized the "essence" of a particularly *German* university, Humboldt was a singular genius who "declared war" against an "encyclopedic education" and antiquated forms of scholarship and envisioned a completely new institution.[16]

It would be an exaggeration to suggest that Humboldt represented a radical disruption in the history of the university. One of the key elements of my account is that Humboldt's concepts and ideas about knowledge and the university were not all that new. He expanded on debates that had been ongoing in Prussia since the 1780s and 1790s. But it would be an exaggeration to dismiss Humboldt as either irrelevant or simply part of a conservative ideology. Humboldt's early twentieth-century hagiographers mimic eighteenth-century polemical distinctions—between passive, encyclopedic knowledge and spontaneous, true scientific knowledge—but more contemporary critics, in their fervor to deconstruct the Humboldt myth, have obscured vital elements of the modern research university's genealogy and, more importantly for our present concerns, its place in the history of the organization of knowledge.

Eighteenth-century technologies of print from *historia literaria* to encyclopedias were key components in the emergence of both the modern research university and science as a distinct cultural practice. Humboldt cast his proposal for a new university as both a bulwark against mere pedantry and utility and a technology for incorporating Enlightenment technologies into a stable institutional structure organized around an explicit ethics of knowledge, that is, around conceptions of what constituted authoritative knowledge and the kinds of persons who could legitimate it. Traditional accounts conflate Humboldt's idea of the university with neohumanism, which emphasized personal development and a recovery of classical virtues, and thereby obscure Humboldt's conception of the university as an institutional technology.[17] He saw the university as an institution that could relate and adapt the technologies of knowledge and become a center for the communicative functions of modern science. Furthermore, Humboldt's, not to mention Fichte's and Schleiermacher's, ethically oriented conception of science was not simply a romantic ideology of self-realization and "unfettered freedom and autonomy."[18] Idealist and romantic conceptions of science as community-based practice offered an innovative solution for coping with cultural and epistemological anxieties about information overload. Like many of his contemporaries, Humboldt bracketed the pursuit of material completeness without denying its realization in some possible future. The subjective, ethical conditions of knowledge were, he argued, just as important as the material, objective forms of knowledge. Like the other figures discussed in previous chapters, Humboldt was enmeshed in debates about both the university and the modern media ecology, and his attempts to found a new university in Berlin should be understood as a particular solution to a particular problem in the history of knowledge.

## The Harmony of the Subjective and the Objective

The most condensed statement of Humboldt's plan for a new university can be found in a memorandum that he wrote in 1809, "On the Internal and External Organization of Berlin's Institutions of Higher Knowledge." Bespeaking its complex history of appropriation, the text itself was not discovered until 1896, when the German historian Bruno Gebhardt found it in an archive and then published excerpts as part of his own book on Humboldt.[19] Gebhardt celebrated the text as the embodiment of the German university ideal, emphasizing Humboldt's demands for the "complete freedom and autonomy" of science and the independence of the university. It was a manifesto for the modern research university.[20] Gebhardt and generations of scholars following him extolled what they considered Humboldt's articulation of the research university's underlying principles: the integration of teaching and research, academic freedom, and the endless nature of academic inquiry or research.[21] Over the course of the twentieth century, American university presidents and reformers increasingly appealed to their own experiences as students in German universities and to these ideals to shape their own institutions and craft the modern research university.[22]

For our purposes, Humboldt's text offers a terse, highly allusive argument that consolidated the philosophically oriented claims of Schelling and Fichte, the more wide-ranging and pragmatic suggestions of Schleiermacher, and the debates about science and the future of knowledge into a perspicuous and highly translatable vision for a modern German educational system—its goals, its structure, and the place of the university within it.

For Humboldt, the university should not be an institution unto itself; instead, it should be the center of a unified system of education and knowledge production and transmission in which all institutions of knowledge—from elementary and secondary schools to libraries, museums, and academies—functioned as a whole. Humboldt envisioned the university as the center of an ever-expanding media and knowledge ecology. It was both related to and distinct from an array of Enlightenment knowledge technologies. Its particularity lay in the unique form of life that it embodied and cultivated, one defined by a devotion to the practice of science. In the burgeoning ecology of knowledge in which print remained ascendant but interacted with a host of other media and institutions, the university was to sustain the practice central to them all—science.

"The purpose" of the university, wrote Humboldt, was twofold. Internally, it was "to connect . . . objective science with subjective development."[23] The university was the only modern institution that could reconnect an objective, material world of knowledge—the bibliographic order that so many late eighteenth-century intellectuals feared had begun to exceed rational control—with the subjective formation of individuals. Humboldt's claim was an implicit critique of a culture that had been fragmented and overwhelmed by its own technologies. The university should embody and institutionalize an ethics of knowledge characterized by a subjective commitment to science and objective programs of discipline which formed and sustained such an ethics. For Humboldt, it was not enough for intellectuals and writers to stoke the longing for science in print. Sustainable and repeatable exercises and structures were needed that could keep that passion going and direct it to the goals and ends of science as a practice. If the claims of science's authority made by philosophers like Fichte and Schelling were to be realized, then the practice of science needed an institutional structure. The desire to order knowledge according to science had to be institutionalized.

Humboldt's insistence on the unity of the subjective and the objective was a direct reproach to the Enlightenment dualisms that he, like Schelling, argued had produced the modern experience of fragmentation and excess. The rapid expansion of the objective forms of science—all those printed texts—had overwhelmed subjective capacities to engage them. Condensing an idea that had been percolating since at least the 1790s, Humboldt contended that the way to manage the surfeit of information was to form people capable of navigating their way through it. What was needed was a change in subjective experiences of knowledge.

Echoing the calls of Fichte and Schleiermacher before him, Humboldt claimed that the university should lead this change as the pinnacle of a "nation's moral culture, [because] only science can change character."[24] Humboldt grounded the university's epistemic authority in the practice of science and the character of those who practiced it. The university was the only modern institution that could not only determine what counted as real knowledge and what did not but, just as importantly, form the types of ethical agents capable of doing so. This entailed instituting norms and practices for interacting with print technologies that had expanded so rapidly over the course of the eighteenth century. The university needed to form persons who could interact better with the excess of information circulating as print.

Humboldt described this new subject, this disciplinary self, in terms similar to those voiced by Schelling, Fichte, and Schleiermacher. The ideal product of the university would be characterized by his rigor and intellectual capaciousness, devoted to "uninterrupted collaboration," committed to method, productive, and capable of recognizing the value of pursuing science for its own sake, that is, capable of recognizing the internal goods of science.[25]

The internal task of the university, then, was to connect science with formation of the individual.[26] In formulating the university's purpose in these terms, Humboldt tied the emergent, institutional-dependent notion of science to the complex of concepts and connotations associated with the German term *Bildung*. Often translated into English as "self-formation," "cultivation," or even "education," *Bildung* began to acquire its modern meaning in mid- to late eighteenth-century Germany. Etymologically, it was associated with an early modern iconographic tradition, in which *Bildung* was related to the German *Bild* (imago) and the verb *bilden* (to form). In these iconographic traditions and well into the eighteenth century, *Bild* and *bilden* denoted the production or formation of some object according to an example. It was a mimetic formation. Throughout the eighteenth century, however, *Bildung* was increasingly used to refer not just to the development or formation of a material object but to the formation of a person, his character or self. In the late eighteenth century, figures like Humboldt used the term as the privileged antonym of *Erziehung* (vocational or more practical education). As opposed to mere vocational training, *Bildung* connoted a moral and spiritual depth. Kant, for example, used the term to describe an epigenetic force—that is, a purposive causality within an individual. In contrast to the mechanical, more practical oriented goals of *Erziehung*, *Bildung* connoted an epigenetic, self-organizing formation of the self. The *Bildung-Erziehung* pair became the conceptual analogue of organism-machine, life-death. *Bildung* became the activity of self-production and the imparting of purpose onto an individual life.

Humboldt, the figure most often associated with *Bildung*, used the term in this way but also stressed that *Bildung* as a process of self-organization always involved an engagement with the external world. In his use, *Bildung* denoted a relationship between the self and the world: the subject and the world existed in a prior practical relation in which the self was also formed by nature, authority structures, and persons external to himself. The subject could only organize itself in and be organized through a world of which it was already a part.[27] Thus, for Humboldt *Bildung* was not a narcissistic inwardness; rather, the very possibility of selfhood rested, in part, on an outside: "Just as sheer power needs an object on which to exert itself, and sheer form, or pure thought, material in which it can express and maintain itself, so too does man need a world outside himself. From this springs his constant attempt to expand the sphere of his knowledge and his effectivity."[28] The fundamental energies of an individual required interaction with something other, a not-I in order to take form. *Bildung* then "is the connection of our I with the world in the most general, most lively and freest interaction." Humboldt's notion of *Bildung* was mimetic, not in the sense of pure imitation but in the sense of approximating the self to things, people, concepts, images, or institutional structures in the world. The self was constituted through a mimetic engagement with the plurality of the world. The process of formation always risked "alienation," in which the individual subject may lose himself in the world.[29] Self-organization as self-formation was a recursive process in which the self constantly engages that which it is not. Formation was fundamentally social in nature.

When Humboldt described the university as an institution that formed persons in accord with science, then, he imbued both science and the university with all the ethical seriousness that *Bildung* connoted for early nineteenth-century German intellectuals. A scientific formation introduced students to science not only as a set of propositions or bibliographic material but as a calling to which they could devote their every waking moment. It endowed their scholarly pursuits in the university with the promises of a total transformation and participation in a community that provided meaning, purpose, and clear authority structures.

The university's commitment to forming students into practitioners of science distinguished it, claimed Humboldt, from vocational academies or schools, which merely trained students in specific skills for certain professions, whose goods were always external: wealth, social status, prestige. The purpose of the university was to sustain a practice in which young men could spend a few years "devoted exclusively to reflection in one place."[30] It was a community devoted to developing and inculcating the practices and commitments of science: "Insight into pure science, what only the human through and in himself can find, is reserved for the university. For this act of the self in its most proper understanding, freedom is necessary and solitude helpful. . . . Attending classes is only a minor matter; what is essential is that one is in close community with like-minded people of a similar age."[31] The "freedom" and "solitude" the university afforded students and professors distinguished it from all other institutions, gathering them into a community that provided the cultural constraints and visions that would make them productive scholars. In this sense, Humboldt, Schleiermacher, Fichte, and Schelling were not the first to offer a clear concept of science, but they were the first to tie science as a particular way of life to a particular institution.

Externally, the purpose of the university was to integrate the disparate institutions of learning and knowledge—elementary and secondary schools, scientific academies and societies, material collections, print, and people—into a unified system of science. "Freedom and solitude" were the internal conditions, contended Humboldt, that allowed the university to sustain and support the practice of science and the advancement of knowledge on behalf of the state, the church, and society. The university's internal organization made its service to the broader demands of knowledge possible. For Humboldt, the university was not to be a radically self-enclosed institution. It was a system that was open and intersected with other elements. Humboldt described it as more akin to a living, dynamic system, as the middle point of a media and knowledge ecology that extended well beyond the historical boundaries of the medieval university. If the king would allow it, wrote Humboldt, a new university in Berlin would "connect everything in Germany related to education [*Bildung*] and enlightenment."[32]

Humboldt suggested that the term "university," derived from the Latin *universitas* (totality, the whole), stood for the fact that no science, institution, or other form of knowledge would be excluded.[33] Just as "encyclopedia," "book," and even "Enlightenment" stood in for a particular order of knowledge, for what constituted epistemic authority, so too did Humboldt's use of "university." Humboldt led efforts to reform the entire Prussian education structure, from primary schools to scientific academies, and he conceived of them all as part of a comprehensive, tiered, and integrated system that revolved around the university. Even the scientific academy existed within the sphere of the university, because the university, not the academy, formed scientists. The academy merely housed them.

This is why Humboldt tied his argument for the future of the university so closely to Berlin as a particular place.[34] Drawing on the arguments of Beyme, his predecessor, Humboldt explained that the decision to establish a university in Berlin was based on the fact that there were "now already in Berlin," in addition to the academy and various schools, a range of institutional collections, anatomical theaters, art and botanical collections, and medical and veterinary clinics. But they had not yet been brought together "into an organic whole."[35] They had not, that is, been organized according to a common purpose. The materials of the Enlightenment—libraries, gardens, clinics, and print collections—would remain mere "fragments" unless they were tied to a "complete scientific instruction." Beyond the various collections and institutions, the university would also integrate various other intellectual activities in Berlin. Well before 1810, both public and private lectures had been held at various institutions and specialized academies, as well as in private residences. The famous philologist F. A. Wolf, for instance, had already been leading a course on ancient Greek literature; A. W. Schlegel lectured on art and literature, as well as encyclopedias; and Fichte had given a series of lectures, including his "Speeches to a German Nation." Berlin, wrote Humboldt, was not lacking in the "desire for serious and learned studies."[36] The publisher Friedrich Nicolai had even included Berlin's lecture scene in his travel guide to Berlin.[37]

The idea of the university as an ecosystem of technologies was even codified in the university statutes of 1816.[38] The last section of the university's lecture catalogues, found under the heading "Public Scholarly Institutions," listed opening hours and lending guidelines for the Royal Library and the various collections and theaters that were available for students to visit.[39] In addition, the Prussian Academy of Sciences was reorganized in 1812 and brought into closer relationship with the university. The university not only facilitated access to these various institutions but also sought to reorient them toward the ends of science, especially research.

These various elements of science required, claimed Humboldt, an institution to connect them.[40] As the center of an ecological system of knowledge, the university would absorb and repurpose previous knowledge technologies for the ends of science and organize them according to a disciplinary order. It was a "living whole," a living system that could adapt and accommodate the institutions, technologies, and people that it found itself in the midst of. It was adaptive. But it also had its own constitutive norms, traditions, and ideas that it developed. Anything outside the university's ecosystem would lack the stamp of scientific authority. It would be, as Fichte put it, "not-science" [*Nicht-Wissenschaft*].[41]

The idea of the university as a technology, then, was projected onto an already-existent but decentralized environment of research that had its own burgeoning but fragmented infrastructure. The "organic" unifying structure that Humboldt invoked referred not to a spontaneous process but rather to a particular relationship between part and whole and the constantly changing ways in which they interacted. On Humboldt's account, the university was not to be fully closed in on itself; it was the technology in and through which a range of institutions, practices, and material forms could remain distinct and, simultaneously, related to a whole. The "unity of science" was in this way a material, institutional project, but what bound these distinct parts together was an ethic that ensured that the "accumulation of inert collections" would not be mistaken for science itself.[42] On their own, merely material, objective collections, be they natural history cabinets or books, would "dull" the mind. The richest academies and universities were not always the places, commented Humboldt, where the "sciences enjoy the deepest and most mindful treatment."[43] The university was not reducible to its material elements, yet it emerged out of them. Its task, both internally and externally, was to relate, translate, and mediate different institutions, practices, and forms of knowledge. Humboldt envisioned the university as the center of an ecology that would supersede the print ecology that had defined the eighteenth century. And it would be bound together by the practice of science.

## The Ethics of Research

One element in particular was central to the university as a community of scientists: research. Drawing on broader shifts in the concept of science, Humboldt described research as a type of knowledge that has "never been nor will ever be fully discovered."[44] From this perspective, knowledge was not something that existed out in the world and that dutiful scholars simply had to collect and display in printed tomes of erudition or curiosity cabinets. It was "a problem that can never fully be solved."[45] The empire of erudition had been guided by an ideal of completeness, the hope and expectation that progress, however far off in the future, would one day be unnecessary. For many eighteenth-century scholars, the endless production of print made this ideal seem increasingly naïve, if not impossible. In the face of the proliferation of print and the collapse in confidence in universities, intellectuals and writers like Humboldt appealed to research as a reconceived notion of science which considered legitimate, authoritative knowledge to consist of an endless transmission of knowledge from generation to generation. Individual scholars would never be able to gain a synoptic view of the whole of knowledge in some printed encyclopedia or book, but they could contribute to knowledge as it developed over time and participate in a community of fellow researchers. Research represented a distinct orientation to knowledge; it stood in for the pursuit, not the attainment, of knowledge.

The concept of research also provided the scholar the sense that he was contributing to the historical progression of science through time. It promised a form of transcendence through his particular scholarly labor. The exhaustive accumulation of knowledge—completeness—was bracketed from the outset in exchange for an asymptotic movement oriented not to a fixed order of knowledge but rather toward a process of knowledge creation over time. Science was a task to be "ceaselessly" pursued. The key virtues of the scientist as researcher, then, were industriousness and attention to detail—the dogged pursuit of the whole of knowledge through a focus on the particular. According to such a model, there was no one external referent to discern absolute progress. The markers of progress were internal and procedural. Internally, progress could be measured primarily "in skill, method, and technique in the formulation of problems," the production of markers of such industriousness, publication and writing, and, most basically, the character of the scientist himself.[46]

The concept of research, if not the actual term, and notions of scholarly production predated Humboldt. Notions of what would come to be referred to as research guided some of the reforms introduced at the University of Göttingen during the second half of the eighteenth century.[47] University-based scholars throughout the eighteenth century published written work, so the distinction between research and other scholarly models of knowledge, like erudition, concerned not just how much these scholars wrote in comparison to later scholars, but rather the purposes of scholarly works, their form, and their intended audience.[48]

In the second half of the eighteenth century, scholarly publication by university professors was typically valued for the recognition that it brought their university. "Publishing," wrote the scholar and, briefly, professor of philosophy Christoph Wieland, "belongs in and for itself among the activities of the scholar, and it is so much the more suitable to the professor because through it he has the opportunity to make himself known abroad and to promote the honor of the university."[49] But the types of writing that universities valued, as Johann D. Michaelis, dean of the philosophy faculty at Göttingen, put it, were not "narrowly focused."[50] In Göttingen, fame (broad recognition) was valued over prestige (specialized recognition).[51] In 1768, the Prussian council overseeing universities declared that "professors should write but in accord with the prevailing fashion and taste, with grace and beauty," not according to the demands of specialized science.[52] Eighteenth-century universities encouraged professors to publish a broad range of literature that would be widely read—sermons, encyclopedias, disputations, and more generally popular literary work—and not the more specialized work that would later be characterized as research. Within the university, publication was not the primary factor for hiring decisions. A "superior ability to lecture," explained Michaelis, was oftentimes considered more valuable than the ability to publish more specialized work.[53]

Over the course of the eighteenth century, the kinds of writing that university-based scholars were expected to produce gradually changed. By the end of the century, the Göttingen history professor Ernst Brandes even worried that the general expectations for university professors were becoming unrealistic. "Thorough erudition" is a necessary requirement for the appointment of a professor, he wrote, but some "thoughtful men set too high a criterion in judging the worth of an academic; they demand that all professors be among the foremost intellects, that all be distinctive geniuses and discoverers. Aside from the impossibility of this demand, a discoverer in individual branches of learning can be a bad professor."[54] Professors, he continued, should "obtain, distribute, and, when possible, augment the treasures of human knowledge," but should all of them be required to produce new knowledge? Could one person both advance and distribute knowledge? Brandes's doubts were in part a reaction to Göttingen's academic mercantilism, which increasingly encouraged and rewarded the production of university-oriented, more specialized publication.[55] Göttingen's coupling of scholarly production and academic advancement was historically new. Before the late eighteenth century, success for a professor had not been, primarily at least, a function of scholarly publication, but rather a combination of teaching, collegiality, seniority, popularity, and, just as often, family ties.[56]

Even with its increased emphasis on university-oriented scholarship, however, Göttingen, like most other German universities of the eighteenth century, assumed that a professor's identity as a scholar should be subordinated to his identity as a teacher.[57] "An orderly, upright man with a well-ordered erudition and a gift for communicating it," wrote one commentator in response to the increasing demands for professors to publish, "is more suitable to become a professor than a scholarly monster who labors only for himself and the world or who does little for his students, or a genius who has offensive morals and who does not think it worth the labor to employ diligence on lectures for his students."[58] This comment is evidence both of a university system in which professors were not necessarily researchers and of a changing university system. Göttingen, for example, continued to define itself primarily through its efforts to distinguish its pedagogy as "more systematic and rigorous" than that of other schools and specialized academies.[59]

Over the last two decades of the eighteenth century, resistance to attempts to unify research and teaching in the person of the professor became a rallying cry for many critics of the university. "The discovery of new thoughts," wrote Michaelis, "depends so very much on chance that one can scarcely establish it as someone's duty. Most discoveries are made not because one seeks them, but rather because chance brings them into our hands."[60] University professors should not be pressured to produce something they could not control. Could "scientific genius" be rationalized and turned into a method? For critics of the emerging research ideal and advocates of older notions of scholarship, the norms of university-based writing should regulate not the production of new knowledge but the distribution of already-established knowledge.

There was, then, a clear continuity between many of the reforms suggested by Humboldt and his Prussian contemporaries, such as Beyme, and the reforms that Göttingen had implemented over the course of the eighteenth century.[61] Humboldt did not invent the structures that would come to define the modern research university. He and his contemporaries drew on innovations of their eighteenth-century predecessors. But whereas in Göttingen the emergence of a proto-research imperative was motivated primarily by an academic mercantilism, in Berlin the attempt to articulate a concept of research and science was motivated by a need for a new source of epistemic authority. Research and science were not simply organizational structures. They were ethical and epistemological concepts that grounded the very purpose and authority of the university. Research challenged the imperatives of the eighteenth-century university, in which the production of new knowledge was often tangential. It was only with the revision of the very concepts of science and scholarship that the university was organized around the production, discovery, and dissemination of new knowledge.

As the theoretically endless discovery of the new, science as research, wrote Humboldt, was "indefinite." Erudition, in contrast, was the collection of already-existent knowledge and thus, theoretically at least, could be completed. Research was the pursuit of the new, whose limits and boundaries were unknown and infinitely expandable. Humboldt's embrace of research meant that the university would have to institutionalize and regularize the production of the new and thus constrain the element of chance. It would have to institutionalize "scientific genius." Part of this process of regularization entailed how research related to other works of scholarship.[62] According to the idea of research, products of the process of research use other research to produce more research. Every contribution to science supplements and builds something new from work already done. Erudition, in contrast, primarily exhibited what was already in print on the model of the curiosity cabinet. Whereas Enlightenment technologies of print displayed or represented what was already known, research, as described by Humboldt, was a praxis, a form of productive labor that, like an art, created something new. Humboldt's concept of research was an attempt to regularize the ends and norms of knowledge production and discovery, with its focus combining objective techniques with subjective experiences. Research methods and practices could not be transmitted in the form of clear rules. They had to be passed on through example and made personal. Research was a type of formation.[63] The reformulations of the idea of the university around 1800 normalized this concept of scholarship as research.

Humboldt's concept of research had direct implications for how a university should be organized. It helped scholars better imagine nonlocal, specialized communities of science. For Humboldt, the basis of the university lay not in a local corporation of scholars and students, who were bound to each other simply by their proximity. Science was a distinct cultural practice that had a reality that exceeded the constraints of time and space. Whereas the filiations, loyalties, and alliances of the medieval and early modern university were primarily local, the communities of the research university were to extend out into the nonlocal. The modern, "dualistic" notion of the university as organized around an internal college community and an external disciplinary community has its conceptual and institutional roots in this early concept of research.[64] As the institution of science, the university mediated the local and the universal communities by sustaining the structures, infrastructure, and virtues of the practice of science, which allowed scholars to participate in the "universal" experience of science.

The concept of research, as articulated by Humboldt, also entailed a unity of teaching and research in the person of the university professor. It was a response to critics, like Brandes, who doubted that all professors should be required both to advance knowledge and to distribute it. The unity of teaching and research was not simply, as many contemporary articulations suggest, a claim that a university teacher should also produce research as an add-on activity. For Humboldt, the authority of the university rested on the assumption that professors both disseminated and produced knowledge, and as the producers of it they authorized it, that is, lent it authority. This was crucial in an age of proliferation and overload, because traditional sources of epistemic authority, the church or state, had dissolved. Research grounded that authority in a university-based activity and in the disciplinary self. Universities could only differentiate themselves from other Enlightenment technologies if they could claim not merely to distribute but to produce knowledge. In a modern print environment consumed by anxieties about the provenance and legitimacy of knowledge, Humboldt held up the university and the people it formed as a clear and authoritative source of knowledge. True knowledge was not a product of the endless circulation of disembodied, scattered voices in the print market; it was the product of an institutionally sanctioned and cultivated practice called science. Science was the mutual enterprise that bound teacher to student, research to teaching.

If teachers simply distributed knowledge that they had not produced, as Fichte had suggested, they could have easily been replaced with cheap printed books. As exemplars of a particular practice and the authors of new knowledge, they could not be so easily replaced. They apprenticed their students into a distinct practice with its own internal standards and traditions of excellence. This relationship between teacher and student was at the center of the university:

> The teacher isn't present for the sake of the student alone. Rather, students and teacher come together for the sake of knowledge. The teacher's work depends on the presence of students, without whom it would be difficult for the teacher to excel; and so if students didn't gather around him, he would have to seek them out. Indeed, the pursuit of the teacher's own goal, knowledge, is best achieved through a linking of his strengths and those of his students. The teacher is more practiced, but for that reason, he also tends to be narrower in his views, and to have a less lively imagination. The students are less experienced, to be sure; but with greater openness, they strike out intrepidly in all directions.[65]

This relationship between the university teacher and student was the culmination of an entire educational system. At every level of development, however, the final end of education—the cultivation of a "scientific insight"—remained central. The purpose of elementary education was to train the student in basic abilities and knowledge.[66] The purpose of secondary education was to enable the student to learn for himself. Both of these culminated in the university, whose purpose was to form a young scientist who could relate to his teacher as a colleague. "The university teacher is no longer a teacher, the university student is no longer the one who is learning; instead, the student himself researches, and the professor guides the student in his own research. University instruction puts one in the position to conceive of and produce the unity of the sciences."[67] The professor was no longer the singular access point to a canon of established, authoritative knowledge. His main task was to initiate students into scientific research, not as an impersonal method that could be transmitted in written or printed texts, but as a method and practice that could be shared primarily through personal interaction. The student learns by example, and the teacher adapts and alters his own knowledge in response to this interaction. In guiding the student, the professor not only informs but also forms.[68]

For this reason Humboldt tied scholarship and learning as research directly to the university and to the person and authority of the professor. According to a cabinet order of October 4, 1774, gymnasium teachers had been accorded the title of "professor." In writing about the new university as a distinct institution that protected the authority of scientific knowledge, however, Humboldt insisted that the "title of professor" should be used sparingly.[69] He based his argument on an order of November 26, 1803, which limited the title to "academic teachers," defined as those who gave public lectures and had publicly demonstrated "a unique erudition"—that is, to those who taught at a university.

The concept of research, as Humboldt articulated it, entailed a distinct notion of academic freedom. In contrast to earlier notions of academic freedom which emphasized the university's juridical autonomy, Humboldt, following Schleiermacher, argued for a more pragmatic middle ground between an academic mercantilism and the more radical autonomy that Fichte sometimes advocated for. "If the state hoped to one day make practical use of research," reasoned Humboldt, it should support and protect science both financially and bureaucratically. The state had to recognize that "on its own, it won't and can't achieve the scientific knowledge it wants. Indeed the state should always recognize that it is never anything but an impediment as soon as it meddles directly in the production of knowledge—it must recognize that knowledge production would go infinitely better without such involvement."[70] Humboldt invoked the promised goods of academic mercantilism, useful knowledge for the state, to leverage academic freedom. If the state were to reap the benefits of scientific research, it should grant universities and professors broad latitude to research and teach what they wanted.

Humboldt's underlying assumption about the external relationship of the university to the state was that the ends of the latter would, over time, inevitably coincide with those of the former. The state, he wrote, "should demand nothing from either the university or the academy that is purely a function of its immediate interests. . . . Rather, the state should foster the inner conviction that when these other institutions achieve their own ends, they also achieve the ends of the state, but from a much higher perspective. These institutions provide the state with a perspective from which much more can be comprehended, one from which the entirety of other forces and mechanisms can be mustered, including ones that the state could never come to make use of on its own."[71] Science offered the state a form of knowledge unencumbered by the vicissitudes of politics and immediate utility. To benefit from this perspective, however, it had to protect science as a distinct kind of knowledge and its institutional home, the university.

Echoing late eighteenth-century critiques of the Enlightenment university, however, Humboldt also contended that the state should protect the university from its worst guild-like tendencies. It should assume, for example, responsibility for the appointment of faculty in order to avoid what was widely perceived to be nepotistic and corporate hiring practices of universities. Criticizing the practices at his own university, Wieland had railed against a philosophy faculty that instead of focusing on the "best possible choice [had] notoriously concerned itself more with its relatives and personal friends, more with religious, fraternal, or collegiate relationships and the like in the selection of its new members than with true learned capability."[72] Given what Humboldt and critics such as Wieland saw as the faculty's inability to govern itself—Humboldt once compared leading a faculty to organizing a traveling group of comedians—a stronger state role in hiring practices was necessary. Although the state was not the end of science, it should provide "external structures and means" for science to flourish.[73] Both Humboldt and Schleiermacher acknowledged that the state would always be "self-serving" and pursue knowledge for its own short-term ends.[74] The university could be a medium between the more utilitarian demands of the state and the more internal demands of science as a practice.

## The University as Bearer of Science

Humboldt did not just articulate an ethic for the modern university, however; he also helped institutionalize specific institutional norms that he thought would ensure the continuity of practice—that is, unlike Fichte or Schleiermacher, Humboldt implemented specific bureaucratic structures designed to protect the practice of science, such as the design of university oversight committees and hiring practices.[75] Most of these administrative forms were not radically new. The university, as Humboldt imagined it, would have four faculties, both chaired and unchaired professors [*Ordinarien* and *Extraordinarien*] as well as private instructors, exam regulations, and faculty forms of government.[76] Humboldt took most of these mechanisms from other German university models, but what he did differently was frame them in terms of science and suggest that the practice of science could only survive if protected by the bureaucratic apparatus of the university. The university was organized not only around administrative structures but also around norms that emanated from science. For Humboldt, this meant that the state and the university were not opposed institutions; they depended on each other. The state guaranteed the protection of science as a practice, while the university promised to translate and dispense the products of research to the state and broader society. The university professor was, for Humboldt at least, a bureaucrat, although a highly productive and creative one.[77]

In October 1809, Humboldt wrote a detailed plan for the "Scientific Deputation," a committee within the Culture Ministry's Section for Public Instruction. According to an order from November 24, 1808, the deputation had already been tasked with producing a general plan for the reform of the entire Prussian education system, from elementary schools to academies and universities,[78] which were to function as an integrated system with one consistent logic. From its inception, the committee engaged in a detailed reform of secondary education down to how many hours of Greek per week and even what Greek texts should be read at what level.[79] Humboldt framed the deputation's purpose, however, in terms of broader university reform. The university was to mediate between the interests of the state, as represented in the bureaucracy, and the interests of science, as embodied in the university faculties. The committee, he wrote, "maintains the universal scientific principles out of which the particular governing maxims emerge."[80] It should ensure that the Culture Ministry "appropriately values" these principles and does not fall prey to the "distractions" of its daily business. The purpose of the committee was to resist the external goods, prestige and power, that the state and by extension the university would always be primarily concerned with.

Underlying Humboldt's vision for the deputation were two basic assumptions. First, there were, as he put it, "universal scientific principles," and second, bureaucratic governing procedures should be judged according to them. The state needed to be checked by a body dedicated more immediately to the practice of science and the goods and standards internal to it. The broader goal of such a committee was to find the point at which scientific principles and governing maxims could be brought into general agreement. Humboldt organized the committee around the inherent tension between the more internal interests of the deputation, oriented by "a freer scientific muse," and the more external interests of the broader bureaucracy, oriented toward the needs of the state. Neither sphere was entirely autonomous, but neither were they reducible to one another. Humboldt called on the deputation to exercise a critical role and constantly compare the normative demands of science with "what actually happens" in the daily bureaucratic operations.[81] To do so, it could not occupy a fully distinct realm; instead, it would occupy a middle space where it could test the normative against the real and vice versa.

Ultimately, however, Humboldt called on the deputation to commit itself to "the advancement of a universal education [*Bildung*]." The deputation was to ensure that "scientific education [*Bildung*] does not fragment" under the pressure to pursue external ends, but rather concentrates itself on the "attainment of the highest most universally human in one focal point." Its task was to safeguard the continuity and institutional integrity of the practice of science. One of the basic internal values was a commitment to the idea that only scholars should judge another scholar's work. Particular questions would be divided by the director among the committee members according to their particular "expertise." Even in these more specialized judgments, however, members should refrain, wrote Humboldt, from "pedantry." They should consider the issue at hand both from the perspective of a particular science and from the perspective of the broader practice of science. Did a particular question or matter satisfy the demands of the particular science—its particular historical notions of excellence, the current state of scholarship—while also manifesting the virtues and standards of excellence of the broader practice of science? Was it rigorous? Did it attend to detail? Did it engage other points of view? Did it relate itself to the traditions of science? Did it relate itself to the whole of science as a practice? To this end, Humboldt insisted that members of the deputation come only from the lower faculty, because it was there—in the faculty that embodied the imperatives of research and were not tied to a particular profession—that "individual facts could be elevated to science [and made] more productive for the mind."[82]

The hiring practices that Humboldt helped introduce were another example of how he sought to institutionalize the practice of science in Berlin.[83] As noted above, throughout the eighteenth century, a broad range of abilities was considered a sufficient qualification for a faculty position: eloquence in lecturing, publication of popular compendia, or even the inheritance of a faculty position in what some contemporary scholars have referred to as the "family university."[84] The scientific deputation sought to change these traditional hiring practices by identifying candidates who were already established in particular sciences as evidenced by their publications, with whom they had studied, and the evaluations of others from within their particular science. In particular, it tried to dispense with older practices that privileged family relationships or other local, corporate considerations. In the annual report of the Ministry of Culture, Humboldt proudly announced to the king that F. A. Wolf, renowned philologist from Halle, had been offered a position at the new university, "because with respect to philological erudition no one can measure up to him."[85] Similarly "capable" men, he reported, would be sought for theology, jurisprudence, and medicine. In subsequent reports, Humboldt clearly subordinated collegiate and corporate virtues—"of effective teaching, versatility, social and intellectual acceptability, and family ties"—to more strictly scientific virtues.[86] Berlin was one of the first university institutions to emphasize not merely "erudition and eloquence," but a potential faculty member's contribution to a particular science through research. Recommending the jurist Karl Friedrich von Savigny for an appointment as professor, Humboldt wrote that Savigny was among the "best of all living German jurists" and noted in particular his "philosophical treatment of his science," that is, his capacity to relate to it as a particular form of the broader practice of science.[87]

The shift toward more specialized academic qualifications in hiring was soon echoed in other elements of the university after Humboldt's departure. Before the university in Berlin was established, the *venia legendi* [authorization to lecture at a university] had been a prerequisite for any professorship. Traditionally, it was awarded to anyone with a "Promotion" (the German equivalent of a doctorate), but the university's ordinances of 1816 distinguished the "Promotion" from the "Habilitation" (a second doctorate, more akin to an additional academic credential) and granted the *venia legendi* only to the latter. More importantly perhaps, the *venia legendi* no longer automatically permitted one to lecture in any academic field, or *Fach*. Full and associate professors with a Habilitation were allowed to lecture in any science within the faculty to which they were appointed. *Privatdozenten* [private lecturers], who had completed a doctorate, however, were limited to particular academic fields within the broader faculty. According to the university statutes adopted in 1816, private lecturers had to "habilitate in that faculty in which they want to lecture. Thus, when registering for the habilitation they indicate which academic field they would like to lecture on. Only then will they receive permission to lecture."[88] This new arrangement institutionalized Fichte's suggestion that professors should only be able to lecture on the science in which they were specifically trained. The Habilitation prerequisite was accompanied by stricter policies on who could lecture in general. These changes affected full professors differently, however. They continued to maintain a great deal of discretion as to what they wanted to lecture on. For example, Schleiermacher lectured broadly over his career in Berlin on subjects ranging from "Encyclopedia of the Theological Sciences" and "Dogmatic Theology" to "History of Greek Philosophy" and "Dialectics, Principles of the Art of Philosophizing."[89]

These changes had long-term effects. First, they established the private lecturer as a professional, more institutionalized position, whereas the position had previously only been a category for a range of institutionally undefined positions. Second, for the first time at any German university, it limited the authority to lecture to a particular subfield. These changes played a significant role in expanding academic specialization as they spread to most German universities over the course of the nineteenth century. The Habilitation, in turn, became an even more specialized degree and took longer than it used to require to obtain.

Humboldt did not fully institutionalize the pursuit of specialized science, and the university he helped establish never fully realized it either. In the decade after it was founded, the new university in Berlin was a center of political unrest and unrealized promises. Although the logic and ethics of specialized science were first articulated in early nineteenth-century Prussia, what we today call "disciplinarity" did not fully take form until universities were more fully incorporated into state bureaucracies and enrollments and funding grew. The disciplinary system only gradually emerged over the course of the nineteenth century in Germany and then assumed its more contemporary form in American research universities of the late nineteenth and early twentieth centuries, with the development of the departmental structure and the expansion of the professional, discipline-based societies.[90]

What Humboldt did provide, however, was an account of not only how specialized science functioned but why individual scholars might choose to devote themselves to science as a practice, why they would embrace scientific specialization and research in the first place.[91] He offered a vision of the disciplinary self and the virtues, practices, and institution that could cultivate it. Specialized science, *Wissenschaft*, was not simply an ideology imposed by the vague imperatives of a rationalizing modernity. It was a normative and ethical framework that outlined particular internal goods (production of research and an ability to perceive the relationship of various sciences and forms of knowledge) and inculcated particular virtues (rigor, collaboration, intellectual capaciousness, a commitment to method). It was a way to make meaning; it was a source of passion. Specialized science formed people; it did not just organize objects of print. Humboldt's notion of research, in particular, was an ethical ideal that oriented the university. It justified and affirmed the goals and methods of increasingly specialized scholarly work.[92] The ideal of research stood in for a particular way of organizing and managing the desire to know, and this is what made specialized science such an efficient technology for managing the perceived fragmentation of knowledge. It was a cultural technology that formed people.

The disciplinary order of the university was also a bureaucratic structure, and it was one that Humboldt, as well as his idealist and romantic contemporaries from Kant to Schleiermacher, advocated for. In order for science to survive as a practice, they reasoned, it had to be institutionalized. Otherwise, it would not survive the pressures of print and the diffusion of epistemic authority in the modern age. Their arguments and visions for a reformed university may have been oriented to metaphysical notions of an absolute unity of knowledge, but the social unity of knowledge that they promulgated just as forcefully required a flourishing university. And that required institutional structures and the protection of a bureaucracy.

Critics of the modern research university, since its emergence in early nineteenth-century Germany, have often contended that the logic and ethos of specialization ultimately undercut the ideals of *Bildung*. The ascendance of the research imperative and the research university more generally, as many have suggested, threatened the purportedly loftier ideals of a Humboldtian *Bildung*.[93] According to this narrative, the ideal of the research university gradually succumbed to the "inexorable forces of academic specialization, differentiation, and professionalization," and scholars abandoned the romantic and idealist commitment to an "organic unitary ideal of *Wissenschaft*."[94] The story of the research university is another story of modernity as decline.

The research university may well have become increasingly specialized and professionalized, but it is important to understand that the logic and ethical claims of research and specialization were with it from its beginnings.[95] Accounts of the university that emphasize professionalization over all other factors overlook some of the basic particularities of the modern research university. Universities supported careers, but they also supported and sustained a unique order of knowledge and cultivated a particular culture that could not be reduced to career structures. As the research university and the ideal of specialized science were reformulated, expanded, and institutionalized over the course of the nineteenth century, these reformulations were "nonetheless consistent with its original meaning" and continued to appeal to the normative claims and aspirations of a science as first articulated around 1800.[96] The unity of science that Humboldt and his contemporaries invoked was a unity defined as progressive, developing, and never complete. It was embodied not in a particular researcher or scholar but rather in the historical community of scholars. Specialized scholarship promised scholars participation in this infinitely expanding unity of knowledge and justified their decisions to focus their scholarship in ever more detailed ways. Specialized science was a solution to a prior age of media surplus, not a problem to be solved.

CHAPTER NINE

# The Disciplinary Self and the Virtues of the Philologist

THE DISTINGUISHING FEATURE of the early German research university, as envisioned by its romantic and idealist champions, was that it was to make science into a practice with its own goods and virtues.[1] When Humboldt wrote of a research ethic and referred to the university as a "moral" institution, he envisioned an institution devoted to science and the cultivation of the disciplinary self. In a culture awash with print and unsure about what counted as authoritative knowledge, he imagined the university as the center of scientific communication and exchange, as a community where the objects of science—books, laboratories, and collections of all sorts—could be harmonized with the subjects of knowledge, the people of science. Over the course of the nineteenth century, German universities adopted and adapted various elements of Humboldt's idealist vision and developed complex structures and methods to form students into people of science.

But one discipline in particular exemplified these efforts: classical philology. Not physics, chemistry, or biology but philology was the consummate discipline for most nineteenth-century German scholars. For generations of scholars in every field, philology exemplified the virtues of modern science: "industriousness, attention to the most minute of details, devotion to method, an ethic of responsibility, exactitude, as well as a commitment and facility to open discussion" and, above all, a critical attitude.[2] In a letter recommending a German physician for a position at Heidelberg University in 1864, the famed anthropologist Rudolf Virchow praised the candidate's research as displaying "nearly philological exactness."[3] That Virchow, one of the founders of modern pathology and ethnology, would recommend a physician by comparing him to a philologist attests to the exemplary status that philology enjoyed in nineteenth-century Germany. More than any other science, philology gradually came to embody the claim that science as a practice could resolve the epistemological anxieties that had plagued late eighteenth-century German culture. It represented the unity and authority of a uniquely *scientific* knowledge that depended on the character of a particular type of person, the disciplinary self. Over the course of the nineteenth century, the ethos of the philologist as the epitome of the research scholar, and not the encyclopedia, the book, or the Enlightenment, stood in for a new order of knowledge and what counted as authoritative knowledge.

There is, of course, more than a little irony in talking about the virtues of research or the ethical claims of science, since science, committed to the "creed" of the fact-value distinction, has for well over a century regarded an impersonal objectivity and the exclusion of anything tainted by the personal or ethical as the ground of its authority.[4] But science has also laid claim to a "moral dignity of the highest order," to a particular vision of what is true, worthy, and beautiful. And it is this normative vision that has motivated and oriented research universities from Berlin to Baltimore to Beijing.

## From the Seminary to the Seminar

The primary site of this inculcation into specialized science as a practice was the university seminar, the precursor to modern graduate programs, especially as first modeled by the philology seminar.[5] The seminar was a form of intellectual training in which a small group of students, usually between eight and fifteen, were led by a professor, who worked intimately with each student both in the group and individually. In contrast to the traditional lecture, the seminar involved a range of exercises in which students interacted with each other and the professor, to whom students apprenticed themselves. The seminar was always a site of practical training and direct formation. It was here that the master teacher, the scientist, communicated science to his students, his apprentices, by personal example. The seminar was a site of intellectual apprenticeship.[6]

By 1820, twelve seminars and institutes had been established at the University of Berlin (seven in the medical faculty and the rest in the philosophical faculty). Over the next thirty years, nine more would be added. By the 1820s seminars were a basic feature of almost all German universities. Before the advent of the modern university laboratory and the ascendance of the natural and physical sciences over the course of the nineteenth century, the philology seminar embodied and cultivated the logic and ethic of science. It was the seminar that first disciplined the disciplines and institutionalized the logic of science as disciplinarity.

The nineteenth-century university seminar had its origins in the theological seminaries of early eighteenth-century Pietist Germany, especially Halle, where seminars began as institutions for training ministers. Over the course of the eighteenth century, they became institutions for training teachers, and by the early nineteenth century, they had changed once again to become institutions for training scientists and researchers, especially philologists.[7]

In 1707, August Hermann Francke, a Pietist and professor of ancient languages, established the first seminar in Halle (*seminarium selectum praeceptorium*) to train teachers for his *Pädagogium*, a secondary Latin school established a decade before. The seminar accepted only advanced students who, as Francke put it, already had a "solid foundation in the humanities [*studiis humanioribus*]."[8] It trained theology students from the university in ancient languages, modern philology, and biblical scholarship.[9] These students then went on to train teachers in the *Pädagogium*, where as "scholars" they guided younger students toward piety through "edifying discourse and good example."[10] This included evening and morning prayer, reading of scripture, and extensive linguistic training.

In contrast to traditional humanist learning, which focused on ancient Greek and Roman texts, Francke's seminar emphasized exercises in ancient languages (Latin, above all) and focused almost exclusively on the study of the Bible. The epitome of Halle's philological culture, however, was the *Collegium Orientale theologicum*, which was established in 1702 by Francke to train a select group of university students in ancient languages and biblical exegesis. Besides Hebrew, students were also expected to learn Chaldean, Syrian, Samaritan, Arabic, Ethiopian, and rabbinic Hebrew. Like the seminar, it was devoted almost exclusively to the scholarly study of the Bible.

Francke's Pietist seminar was particularly unique because Pietism, then as now, was usually associated with a "turn from orthodoxy and ecclesially-sanctioned belief to individual religiosity and subjective piety."[11] Pietism's emphasis on a highly personal relationship to God may not seem obviously compatible with modern scholarship, but these early seminars embraced the most modern of scholarly techniques and materials. Students were required to study six languages and master the methods of high textual criticism and emendation. The Pietist seminars in Halle combined these objective techniques with a rigorous moral education to produce particular subjective experiences.[12] As August Francke put it, a better, philologically enhanced Luther Bible would help people, scholars and non-scholars alike, experience God's word more intensely. Scholarly methods would help make better Christians.

These early seminars combined highly technical training in an intimate environment with the formation of a particular type of person. Training in objective, scholarly techniques produced particular types of ethical subjects. This was a crucial innovation for the development of disciplinarity and the research university almost a century later, and it guided the development of seminars from the early eighteenth century throughout the nineteenth century. The key difference was that the seminars that emerged later in the century replaced the Christian orientation of their Pietist predecessors, whose ultimate end was the production of a personal relationship with God, with an orientation toward science as a practice and the formation of the disciplinary self, the subject of modern specialized science.

Less than a year after the University of Göttingen was founded in 1737, a seminar (*Seminarium philologicum*) was established that from its beginnings was devoted not to Christian piety but to the study of antiquity. Under the leadership of Johann Gesner, the university librarian and a philology professor, the Göttingen seminar expanded Halle's focus on pedagogical training as a way to reform secondary school teaching. Initially, at least, the primary goal of the Göttingen seminar was the formation not of scholarly philologists but, as the grounding mandate of the school ordinance read, of "well trained school teachers."[13] Over time, the pedagogical focus of the Göttingen seminar gave way to a more explicit focus on scholarly and theoretical concerns.[14] And yet, the Göttingen seminar shared the Pietist seminar's methodological assumption that training in objective techniques could produce particular kinds of ethical subjects. But it also revised basic elements of the Halle model. Whereas the Halle seminar cultivated knowledge of ancient languages primarily so that students would become better readers of the Bible—more readily able to "find God's message to man in its starkest clarity"[15]—the Göttingen seminar sought to habituate students into "a high estimation of antiquity" such that the ancient writers would be "an eternal monument of human reason and other good virtues."[16] The student of the Göttingen seminar was to be formed according to antique habits of thought and a commitment to humanity.

Implicit in the Göttingen seminar's stated ends was a criticism of earlier Latinate forms of humanism as aspiring merely to rhetorical and linguistic imitations of ancient authors. Gesner reinvented humanism for his own age.

Traditional forms of humanistic study, he argued, conflated mere eloquence and linguistic facility with moral and civic virtue.[17] A true humanist education, he contended, was a function not of imitation but of "confrontation" between ancient and modern worlds. It recognized that the past could not be repeated in the present. Gesner institutionalized these aspirations through a revised curriculum.[18] He replaced the Latin compendia and anthologies with original texts and extended the canon of ancient authors to include more Greek authors, which he considered to be the original source of all sciences. He also required all students to demonstrate sound knowledge of Greek.

Under Gesner, the seminar shifted away from stylistic mimicry through the memorization of texts to a more complex form of mimesis of ancient aesthetic and moral concepts of virtue. Through his method Gesner sought to develop first a proper understanding of the given text, then a proper feeling for the "excellence and beauty of the language and thoughts," and finally an understanding of the relationship of the whole, by which he meant the character of the ancients.[19] The seminar's curricular practices were oriented toward the mimesis of an ancient persona—one that was crafted, however, from a more self-consciously "modern" position.

Gesner's successor, Christian G. Heyne, assumed the seminar's directorship in 1761 and led it for another half century. Like Gesner, Heyne saw the goal of philological training to be the shaping of students' character through mimesis. By reading the ancients and observing their character, he wrote, "we learn to think and express ourselves properly."[20] A well-functioning seminar should produce "subjects" dedicated to the "humanities" who would then go on to become "good and capable school teachers."[21] Heyne sharpened Gesner's critique of traditional humanist training, especially its focus on Latin, and argued that philology required more than just language skills. In order for philology to grow, it required the study of ancient literature, history, philosophy, myth, religion, and ethnography. It required a sharper concept of culture as the unifying whole.

Under Heyne, Göttingen's pedagogical seminar evolved into a more properly philological seminar—one increasingly devoted to training philologists and not secondary school teachers who taught ancient languages. These philologists, he wrote, "are obliged to attend several hours of collegia in the humanities each day. In addition to this, the Professor of Eloquence will offer, often without charge, lessons in which they will be practiced and instructed in interpretation, and in writing, speaking and disputing in Latin. To this end, each [seminarian] in turn will explicate, both grammatically and critically, an ancient author, as well as writing and defending an essay, written in good Latin, on a topic dealing with the philological sciences in the same manner."[22] The seminar's instruction was devoted to the intensive study of texts and the development of a range of skills, such as interpretation, translation, and the production of excellent Latin prose. The seminar cultivated, as William Clark put it, "philological abilities through participation," a practice that encouraged a more focused and in-depth knowledge of fewer texts under the close supervision of a professor.[23] Heyne's notion of interpretation also expanded philological study to include the interpretation of particular texts within antiquity's religious, mythical, political, and cultural settings.

Friedrich August Wolf, a former student of Heyne in Göttingen, founded a philology seminar in Halle in 1787 which emphasized the cultivation of the student not as a Christian or humanist but as a philologist. Wolf criticized the loosely humanist seminars for what he considered their lack of philological rigor. Originally appointed to lead the pedagogy seminar, he refused on the grounds, as one nineteenth-century commentator described it, that philology demanded "the entire activity of students" and should not have to contend with the nonscientific distractions of a pseudoscience like pedagogy.[24] Wolf triumphed, and Halle's pedagogy seminar was reestablished as a philology seminar that trained philologists, who might then go on to teach. Philology as science was its orienting logic and ethos.

As part of his directorship of the philology seminar, Wolf lectured on what he termed the *Encyclopedia of Philology*, which essentially provided a theoretical justification for a seminar devoted exclusively to philology. The lectures also provided one of the first and clearest descriptions of what constituted a specialized science or discipline. A scientific discipline could give a highly reflexive account of its internal standards of excellence and detail a history of itself as a particular practice. Wolf was attempting to establish an authoritative tradition unique to philology and thus define philology as a distinct, self-governing science. In this sense, the exemplarity of philology as a science was in part a function of the fact that Wolf made it one of the first sciences to be subjected to such intensive, reflexive analysis about its very status as a science. He framed the entire lecture course as a corrective to the confused and fragmented state of philology, which he considered little more than a "sum of facts."[25] Philological sciences had heretofore been unable to give an account of their own standards of excellence, unique tradition, and particular history. Wolf attempted to counter this situation by establishing an internal coherence for philology as a science. He did this by delineating its position within the broader field of science. The potential coherence of philology was a function of the extent to which Wolf could differentiate it from other potential sciences.

He began by pointing out that there was no clear term for the field: "It is called philology, sometimes classical erudition, sometimes ancient literature, sometimes humanities, sometimes even that foreign and very modern term, the beautiful sciences [*schönen Wissenschaften*]."[26] In order to become a distinct science, philology needed "sharply drawn boundaries," and this would require its practitioners to clarify its history, practices, and virtues.[27] By consolidating what had been regarded as various "philological modes of inquiry" under what he termed the "science of antiquity" [*Alterthumswissenschaft*], Wolf outlined a unique history of philology which he assumed would lend it authority. The political, scholarly, cultural, linguistic, artistic, moral, religious, and scientific facts about antiquity were merely the "remnants of ancient times."[28] The task of the science of philology was to seek the truth—what Wolf termed the "embodiment of the knowledge and information" about the ancient Greeks and Romans—that was not readily visible and that could only be revealed through the rigor of scientific work and practice.

Wolf contrasted his science of antiquity to early modern forms of philological inquiry, whose primary aim, as he put it, was simply to reproduce ancient works.[29] Such unreflective admiration had even led early modern scholars to mimic ancient sciences as though, wrote Wolf, they "could still be ours." But modern sciences had grown and advanced such that even the smallest textbooks of the moderns "contain more substantiated claims, more established truths than the great works of famous ancients."[30] Previous forms of philology failed to recognize the historical differences not just between the characters of modern and ancient persons but also between modern and ancient sciences. Sciences have their own history. As sciences change over time, they develop their own internally coherent values, standards, and authoritative traditions. As a "properly ordered philosophical-historical science," wrote Wolf, philology exemplified this developmental logic. It had become an "end in itself," an internally coherent practice with its own norms and goods.[31] As such, it should not be subordinated to the education of a moral or civil elite, the production of bureaucrats, pastors, or lawyers; instead, it should be engaged as an internally coherent and distinct practice. The distinguishing feature of a modern science was not its material completeness, its capacity to give a comprehensive account of everything in print, but rather its internal integrity, the extent to which it could account for itself as a self-enclosed process over time. A modern science was a practice.

Wolf combined this theoretical account of philology as a science with an emphasis on philological practice. His seminar focused on "methods of textual criticism and the use of scholarly apparatus and current and retrospective bibliography" developed by working through specific exercises and tasks.[32] Training in these particular skills, however, was subordinated to the cultivation of a scientific mode of thought, being, and work, one that put a premium on what Wolf referred to as "industriousness." He even produced tabular seminar reports to track students' progress. The names of individual students were listed on the left-hand side and the categories of evaluation—native abilities, industriousness, and moral character—across the top. These evaluations served as the basis for success or failure in the seminar.[33] Wolf also cited "*Unfleiss*" [lack of industriousness] as grounds for dismissal from the seminar.[34]

These institutional elements were designed to distinguish philology not only from other sciences but also from what Wolf considered a less rigorous, and thus nonscientific, form of philology directed to a broader public. Writing to Heyne from Halle, Wolf attributed the poor state of the "study of ancient literature" there to the failings of the local pedagogy seminar. Led by the theologian and vice-chancellor of Halle University, August Hermann Nieymer, the public seminar, complained Wolf, consisted of Nieymer simply reading Greek and Latin texts aloud instead of, "how it actually should be," having the students "practice" themselves.[35] Nieymer never trained his students in the rules of grammar or interpretation. Wolf attributed the failure of Nieymer's seminar to the fact that it drew in "everything that breathes" and thus drove students who considered it their "duty" to study "the humanities" to pay to attend Wolf's own private seminar. Wolf opposed the philologist-as-scientist distinguished by specialized knowledge to the unauthorized and amateur knowledge of the general public. Wolf was cultivating a new authority structure that organized knowledge by tying it to distinct persons.

Wolf's seminar developed excellence in philological methods by placing students in intimate contact with a single professor, who drilled his students in specific exercises. Before the students even attended their first meeting, they were subjected to rigorous evaluations. He admitted only students he deemed committed to teaching in the gymnasium and not in less prestigious rural schools. He required applicants to submit a piece of written work for admission. He excluded students interested in theology and pedagogy. Students of the seminar also enjoyed library loan privileges, a rarity for any eighteenth-century student. He submitted the seminar's best written work every year to the Ministry of Culture, which financed its publication.[36] Every week, two to three hours of study were devoted to bibliographic reviews, in which the "best books" on philology were presented to the students.[37] Above all, the format of the seminar, as opposed to the lecture, brought students in closer proximity to the professor.[38] For Wolf, this proximity placed a virtuous burden on the seminar director, who, as he put it, had to model a certain "conscientiousness" in devoting so much time to advising students.[39] All of these elements facilitated the formation of not only the individual student but also a distinct "society" of future philologists, who were characterized by their "exactness and thoroughness."[40] In his seminar, Wolf focused on modeling ways of engaging ancient texts and exhorting students to emulate his personal example. He enjoined them to submit to an intellectual apprenticeship.

In his seminar and lectures, Wolf had portrayed philology as both a self-organizing, internally coherent science and an authoritative tradition of skills, knowledge, and virtues for the broader culture. Philology's demands for technical mastery and ethical cultivation were mutually reinforcing. "By mastering and criticizing the variant readings and technical rules offered by the grammatical books and scholia," wrote Wolf, "we are summoned into old times, times more ancient than those of many ancient writers, and, as it were, into the company of those learned critics."[41] The assiduous and diligent study of ancient manuscripts, scholia, and commentaries, according to preestablished constraining traditions, enabled a better understanding of the ancient world, which, in turn, facilitated an encounter with the moral exemplars of antiquity. In his *Prolegomena*, he had even defended the time and effort he had spent studying and amending "so useless a farrago of glosses and readings" and engaging in "trifling hair-splitting."[42] Philological skill and the hard work that it demanded formed better people.

Wolf's *Prolegomena* exemplified this conviction, but it also presaged coming tensions. The *Odyssey*, argued Wolf, was not the result of one author, Homer, but the product of textual accretion over time—just as biblical scholars had eventually concluded about the Old Testament.[43] Similar to biblical scholars, Wolf, as historian Anthony Grafton puts it, thought that he faced a choice: he could either save Homer as creator and obliterate the text or save the text and destroy the author—a figure who had become such a model for classical education.[44] When he opted for the second choice, Wolf replaced one model of authority, Homer, with another, the modern philologist. His *Prolegomena* was as much an account of the *Odyssey* as it was an account of the modern philologist—his methods, questions, traditions, and virtues.[45] The modern philologist was thorough, industrious, precise, and, above all, critical. For Wolf, the philologist's critical capacities referred not only to an ability to determine the "authenticity" of an ancient text, its date, provenance, and authorship, but more generally to the disposition and character of the philologist.[46] The "critical" philologist could judge and distinguish what was legitimate and what was not. He could sift through the dross and find what was valuable and true.

Throughout the *Prolegomena*, Wolf displayed how philology could be a critical, scientific practice. With his extended footnotes that detailed the broad range of commentaries and scholia, his use of literary and linguistic evidence, and his detailed and careful historical account of the Homeric texts' transmission, he engaged the question of Homer's authorship by modeling and cultivating the tradition of philology. The *Prolegomena* gave philologists "the right to claim for their field a new intellectual weight and legitimacy"[47] by consolidating and displaying a philological canon of texts and methods, which established an authoritative tradition that the modern philologist should embody. It also justified itself through an appeal to science as an internally coherent practice that could give meaning and purpose to highly specialized scholarship.

Just as other scholars were starting to recognize philology as a distinct science, however, Wolf began to worry that its increasingly specialized character might obscure its original, ethical aims: the formation of students according to ancient exemplars of virtue. Philological scholarship, he lamented in 1806, was succumbing to "the endless striving . . . to collate the particular as particular" without taking any efforts to relate it to the broader practice of philology.[48] Philologists were reducing the science to mere aggregation and debate about "inconsequential" minutia. The persona of the philologist, whom Wolf had devoted his career to crafting, risked losing sight of the Greek ideal of wholeness—the unity of all things oriented toward an ultimate good. With its endlessly historical questions and attention to detail, philology threatened to obscure antiquity's normative and moral resources. As a consequence, it would soon be lost in a surfeit of specialized information.

Wolf's anxieties about the future of philology revealed a more basic concern: what kind of person was the science of philology forming? Wolf had laid the groundwork for the highly specialized work that he was now questioning. Wolf's predicament anticipated the emerging paradox of modern philology, namely, that its specialized methods of research obscured the ethical ends of its more traditional humanism and contributed to the fragmenting pressures of a modern media environment. Technical skill, virtuosity, risked replacing virtue.

In his attempts to establish a university in Berlin, however, Humboldt, as we saw in the previous chapter, had already offered a resolution to the tension between the production of ever more technical knowledge and the formation of an ethical character. Following his idealist and romantic contemporaries, he extolled the specialized, technical scholar as a model of modern virtue. He combined science [*Wissenschaft*] with formation [*Bildung*]. The disciplinary self—characterized by rigor, attention to detail, devotion to method, a commitment to collaboration, an openness to exchange, a critical disposition, and, above all, a love of science—constituted a new ethical paradigm in an age of excess and proliferation. And, perhaps most importantly, the disciplinary self embodied a new form of epistemic authority.

For Humboldt, it was not so much the object of study, antiquity, and its exemplary character, but the form of study, the *science* of antiquity [*Alterthumswissenschaft*], that produced the desired change in the student and formed the true scientist.[49] The ethos of science, and not the classical virtues of Cicero, was the end of science. Humboldt reconciled the tension between technical research and moral formation by embedding academic professionalization—imperatives to publish, to divide intellectual labor according to specialization, to focus on details—in a set of ideals and an ethical practice. Specialized science, he claimed, gave the scholar an ethical orientation, a source of meaning and authority, and made him a member of a community: a community of researchers who were contributing to a human knowledge project brick by brick, technical insight after technical insight. The scholar as researcher contributed to a monument that would allow him to live on and give meaning and value to his labor, however specialized and technical. The university was his home, church, and nation, and the seminar was his workshop, the site of the intellectual apprenticeship that he guided.

## Berlin Philology Seminar

The first seminar at the University of Berlin was the philology seminar established in 1812. August Böckh, one of Wolf's former students, directed it from its founding until shortly before his death in 1867. Böckh enjoyed almost complete control over all aspects of the seminar, which allowed him to craft a distinct school of philological practice and thought, known as the *Berliner Schule*. With over 1,600 graduates over the course of his career, the seminar exerted a tremendous influence on philology throughout the nineteenth century.[50]

According to its founding statutes, the seminar's purpose was to "educate those who were properly prepared for the 'science of antiquity' through a broad range of exercises that led into the depths of science, and through literary support of all kinds, so that through them this study can in the future be maintained, propagated and extended."[51] From its beginning, the Berlin seminar subordinated its pedagogical goals, so central to those in Halle and Göttingen, to the task of forming future philologists, and thereby the reproduction of philology as a science.[52] Expanding on Wolf's changes to the traditional seminar in Halle, Böckh gave his seminar an even more strictly philological character by emphasizing the selection of seminar students. He only admitted students who "primarily devote themselves to philology, not those who expect advancement from the exercise of another faculty science," a commitment he assessed by means of an entrance exam and an application essay. Böckh wanted to separate philologists from non-philologists before they even arrived at the seminar, so as to make the seminar an institution devoted strictly to the science of philology.

As already modeled in Göttingen and Halle, the training of distinct philologists extended well beyond the initial selection criteria and a commitment to philology. The entire course of the seminar focused on the transmission of basic methods in source criticism, genealogical construction, and paleographical methods, none of which were available to those outside the seminar, through close personal interaction. In the seminar, method was equivalent to science. Students met for at least two hours a week to train through various exercises designed to lead them "into the depths of science."[53] These exercises focused on exacting interpretations of Greek and Latin authors, the production of written compositions, and oral discussions of particular passages or other areas in the "science of antiquity." Additionally, every two weeks seminar members met under the guidance of the director to discuss pre-circulated essays. One student would present his work, and the others would evaluate it. As important as the written work was, the students' adherence to the procedures of the seminar was just as important. If a student submitted his paper late twice, he could be expelled.[54] When no essay was to be presented, students were to meet to discuss outstanding questions among themselves. Students were required to submit a final, capstone text, as the statutes put it, on a "philological object" of their choice. The text was then published with state funds. Successful completion was based almost exclusively on whether the text was a "specimen of industriousness and erudition." All of these exercises were carried out in a small group intensely overseen and directed by Böckh, who limited enrollment to eight to twelve students at a time.

These exercises habituated students into the methods, techniques, and standards particular to philology. Through their repetition students learned to identify what counted as a properly philological problem and what was beyond its purview. The seminar represented a sharp intensification of communication among philologists, as well. The goal of the seminar was not, primarily at least, to produce impersonal, published knowledge but rather the formation of a philological self, a scientific persona distinct from those outside the university and increasingly from those within other sciences. The seminar, as one commentator put it, made the student conscious of himself as "the historical product of his science."[55] The student was formed by and into the authority structures of his science—the methods, the literature, the canonical works, the seminar's exercises, the interaction with his professor—and his "love" for it.

The seminar's goal was to close the gap between external goods (broader notions of the good external to the university) and the internal goods of philology as a science.[56] The expectation was that students would grasp the goods internal to the practice of philology and excel at what it required. This is what it meant to pursue a science in its "ideality" or simply for its own sake. The life of the philologist entailed a devotion to something that exceeded the self—science. A seminarian, according to the Berlin seminar statutes, could be dismissed from the seminar for a "lack of scientific spirit and a sense of a more noble *Bildung*." The credibility of research was in part a function of the character of the philologist.

According to the seminar's logic, becoming an excellent philologist was a good in itself and thus required no further justification. By tying the logic of science to the institution of the university, science could become a sustainable practice replete with its own internal goods, the production of excellent works of scholarship. The philologist also comes to realize that a further internal good is that philology can constitute a way of life, a way to live and a source of joy and pleasure. As a sustainable practice, philology also had its own set of virtues: "industriousness, attention to small details, devotion to method and an ethic of responsibility and exactness, as well as a deep familiarity with group discussion."[57] And, like Wolf, Böckh related these virtues to the critical character of the true philologist. Critique referred to the ability to determine the relationship between two objects. It was the ability to filter and judge amid a surfeit of material, and as such it was for every science, as Böckh put it, the "scale of truth, which weighs the weight of reasons, and teaches one to distinguish the probable from the apparent, the certain from the uncertain, the nitpicky from the clear."[58] The critical scholar was the opposite of the "gullible" scholar, who simply inherited a given tradition and its assumptions without subjecting them to rigorous and critical analysis.[59]

One of the reasons the internal, self-regulation of specialized philology seminars like Böckh's was possible was because its students had a market for their skills.[60] Graduates of the university philology seminars had job opportunities in the *Gymnasien*. Their modes of highly specialized learning and knowledge were supported by a state-sponsored education system that, beginning in the nineteenth century, privileged a humanist education in ancient philology. In 1810, the Prussian Ministry of Culture and Public Instruction established a rigorous "scholarly exam." Whereas the higher faculties had always maintained a strong institutional link to the broader society through established professions, Humboldt helped to create a parallel link for the lower faculties by training and, with the exam, certifying professionals. The exam (officially, *examen pro facultate docendi*) served as a binding medium between the university and society. All those who wanted to teach at a public secondary school had to take this exam, which tested philological, historical, and mathematical knowledge.[61]

Previously, the typical Prussian gymnasium teacher had been a minister waiting for an office. His training had been theological, not philological. The new exam changed this by replacing theological with philological and historical training.[62] With essays to be written in Latin, translations from Greek to German and German to Greek, and oral interviews on classical Latin texts, the exam emphasized classical language ability.[63] The increasingly specialized philologists could be employed as gymnasium teachers and could continue to operate within a highly differentiated sphere. They would face little pressure to address a non-philology audience.[64] And gymanasium teachers could maintain their devotion to the practice of philology.

With the growth of the philology seminars and the changes in teacher requirements, the university arts and philosophy faculties, once way stations for the higher faculties, achieved a certain self-sufficiency.[65] They became centers not only of science and research but also of the production of secondary school teachers.[66] "Research" had an end, a purpose not wholly limited to its internal self-reproduction. This was crucial for the differentiation of philology and the broader philosophy faculty, because it was the first time that the philosophy faculty could train students for a specific professional position—philologist. At least through the first half of the nineteenth century, gymnasium teachers were generally thought to lead a life dedicated to science.[67] The ethos of science extended beyond the university and into the gymnasium, where the assumption was that the "advances" of science should guide pedagogical questions of method.[68]

## Böckh's Disciplinary Logic, or Philology as the Science of Sciences

In addition to directing his seminar, Böckh also lectured twenty-six times between 1809 and 1865 on what he termed an *Encyclopedia of Philology*, in which he attempted to define philology as a distinct science for both his seminar students and a broader university audience.[69] One of his central goals was to cultivate in his audience an awareness of a distinctly philological tradition of knowledge.[70] For his seminar students, the lectures offered a second-order account of what made philology a distinct science. They facilitated reflective engagement with the history of philology as a science—put more simply, they explained to students why they were doing what they were doing. Philology offered a coherent intellectual identity. For the broader audience, the lectures offered up philology as the exemplary scientific discipline. Philology stood in for an authoritative order of knowledge. Although filled with extensive sections on the public and private life of the Greeks and Romans, religion and art, classical knowledge, and the history of language, especially historical grammar, the lectures also contained detailed reflections on the very concepts of science and disciplinarity. The lecture and the seminar work in a dialectical relationship, with the lecture functioning as the more universal and the seminar as the more particular form of science.

Böckh's lectures opened with a discussion of what constituted a scientific discipline. "The concept of a science or a scientific discipline [*Disciplin*]," he explained, "is not given by enumerating piece by piece what is contained in it." A discipline consists in the relationship of its concept to the various parts. Every part of a discipline "represents the whole concept in itself only with particular modification that emerges from the differentiation." The totality of philology was constituted not by the sum of its parts but rather by a unifying concept of the whole. In contrast to *historia literaria*'s pursuit of material completeness, philology, argued Böckh, could not be reduced to the "material" of a science.[71] A discipline refers to a particular relationship between parts and whole, which are related through a unifying concept. If a discipline referred simply to objects of inquiry—philology is the study of X, biology is the study of Y—then, argued Böckh, the delimitation of the objects of study would be arbitrary: why these objects and not those? Various sciences could very well study the same material and still be distinct. Yet Böckh's rejection of a purely object-based notion of discipline did not imply a strictly functionalist focus on "mere method." Just as various sciences could share the same object, they could also share the same methods or activities. The methods of philology—source criticism, hermeneutics—did not fully capture philology as a discipline, since other disciplines could integrate similar methods and still remain distinct sciences.

A scientific discipline was not constituted by its material or a specific method, but rather by a relationship of concepts that developed over time; that is, a discipline was most basically the history of how a discipline had related to itself over time. Thus, although Wolf's own philological science required that texts be treated historically, it failed as science because it treated the borders and the internal arrangements of philology as though they were given and stable. Wolf "takes the disciplines as settled, instead of first showing, deriving, and constructing them in a common concept." If a science is a science and not "a mere aggregate," then it must "presume and produce a philosophical concept," that is, the idea of its own unity that develops over time. Whereas Kant described a science as organized around an a priori idea, Böckh insisted that a discipline's organizing concept was not "set." It existed in time and thus could change and develop. A discipline is most basically the history of how it has related to itself over time and how it has distinguished itself from other disciplines.[72]

Philology's organizing concept, claimed Böckh, was "knowledge of that which the human mind has produced, that is, of what is known. . . . Philology presumes that there is everywhere a given knowledge, which it must recognize again. The history of all sciences is thus philological."[73] The particularly *scientific* insight of philology, and what made it an essential element of all disciplines and specialized science more generally, was that any attempt to account for knowledge—to organize it, to transmit it, to produce it—beyond a concrete form was, as Böckh put it, quoting Hegel, a "false path." Philology recognized and organized the objective forms of knowledge: the material form it assumed in manuscripts, inscriptions, and printed texts. Philology was a *wiedererkennen*—a recognition of what has already been produced. It recognized and reappropriated the objective forms of knowledge as philology.

Underlying Böckh's notion of philology was the assumption that the various modern sciences and disciplines shared a common lineage despite the historical differences. This unity, argued Böckh, made the science of antiquity, philology in the broadest sense, the "basis of all disciplines," because antiquity contained "the beginnings and the roots of all disciplines, the primitive concepts of humanity."[74] The seemingly fragmented disciplines of the modern age shared a common antique source.

Whereas eighteenth-century scholars had turned to encyclopedias and periodicals of increasing complexity and Schelling and Fichte had turned to philosophical conceptions of science, Böckh embraced philology as the science of science which could organize and unify all knowledge, because it was the science of the historical transmission of knowledge from antiquity to the modern age. With its methods of critique and understanding, philology sought to clarify what made the disciplines possible in the first place: the historical transmission and communication of cultural traditions over time. It was the "methodological propaedeutic for all sciences."[75] On the one hand, philology engaged the particular with philosophical care. It sought to see the particular within the broader development of knowledge over time. It sought to discern relationships among particulars and thus dispensed, as Böckh put it, with the "raw *Polyhistorie*" and *historia literaria*, which simply sought to display particulars with no sense of historical development.[76] On the other hand, philology engaged the idea of the unity of knowledge with empirical care. It sought to see the universal within the terms of concrete, historical particulars. Philology sought to keep the particular and universal in dialectical relationship. It was, as Böckh put it, a "smart empiricism."[77]

Part of philology's exemplarity as a science lay in its ability to organize and reflect on the transmission and development of knowledge over time. Philology was the process of knowing or recognizing, and its object was that which was already known: "*das Erkannte wiedererkennen*." Its purpose was to produce knowledge of what is and has been known. Philology's predecessors, such as the *historia literaria* that Böckh disparaged, were primarily auxiliary sciences. They were always subordinate to other sciences and simply provided information, texts, literary history, and general bibliographic data to other sciences, especially theology.[78] Extending the work of Heyne and Wolf, Böckh extricated philology from these subsidiary relationships by insisting on the historical nature of all sciences and disciplines. All disciplines were reliant on processes of curation and transmission. In this sense, Böckh's conception of philology echoed *historia literaria*, but he saw these processes as developmental. Echoing Schelling, he assumed that the modern sciences were organized around the cultivation and transmission of authoritative traditions that bound a living present to a living past. Such *living* traditions could not be reduced to the material forms, however. They had to be constantly reinterpreted, submitted to criticism, and thus made new in the present. In other words, living human agents interacted with the material forms of knowledge to keep them alive and dynamic.

Unlike the Enlightenment knowledge projects that pursued a material completeness and tried to map a set of static relationships, philology sought to discern the underlying historical processes of development. Philology's task was endless not because there was a surfeit of information, but because the material itself was constantly being remade and reimagined. The philologist was the consummate "researcher," the scientist envisioned by Humboldt who endlessly pursued a problem that could never be fully solved.[79] Through his tireless pursuit, the philologist-as-researcher contributed to the universal project of science and thus participated in a community of fellow philologists. "In so far as science is actualized," wrote Böckh, "it exists . . . only in the totality of its agents, in a thousand heads, partially fragmented, broken, no doubt even eccentric."[80] Philology allowed the individual scholar to transcend his particular work and participate in something that exceeded the individual and gave him meaning, purpose, and the promise of a future of which he would be a part through his work, however insignificant and specialized it seemed when considered outside of this broader historical framework.

Like Humboldt and his idealist and romantic predecessors, Böckh tied his notion of an eternally developing, historical unity of knowledge to the institution of the university. The unity of science was actualized in the historical community of researchers who together constituted a distinct group defined by a particular practice. The university was the only technology, the only institution that could coordinate and unify these individual labors of philologists into a community of scientists who together could transcend their particular, seemingly inconsequential work.

The unifying idea of science was a consolation for the individual researcher who often labored in isolation on work that was oftentimes difficult to relate to the whole of the science. "The full reward," wrote Böckh,

> for the arduous acquisition of material that science requires is given to us through the consciousness that the building blocks that we gather will gradually join together into a unified work of art and magnificent building of human knowledge, even if its perfection lies far off in the distance because of the infinity of science. If the division of labor is a necessary law for the construction of the building and the individual of the greatest talents cannot embrace the whole in its expanse, still the purpose of science, like the purpose of the civil state, is perfected in its members, is perfected in the totality of the learned class.[81]

Philology could never be "exhausted or completed. And therefore it can never be perfected in its full expanse in one particular mind. In its full meaning, it can only be realized ideally in the totality of its confessors in innumerable, manifold minds."[82] Although the individual scholar was only part of the infinite progress of science, his specialized work found its justification in relation to this progressive whole. The practice of science sought to maintain this unity, and the university's purpose was to safeguard the practice of science. The university ensured "that scientific formation does not fragment itself according to external ends and conditions," but rather unites all scientific efforts into "one focal point for the attainment of what is the highest universal-human."[83] The institution of the university "immerses" individuals and their labor in "*Wissenschaftlichkeit*"—the practice of science.[84] The university maintained a distinct order of knowledge, whose authority and legitimacy rested on the formation of people dedicated to science and its historical continuity.

If the university's purpose were simply to disseminate knowledge, reasoned Böckh, then the unparalleled "expansion . . . of the realm of books . . . by means of printed books" would have rendered it "redundant."[85] But print had not replaced the university. And that was because "our institution," as Böckh wrote of the University of Berlin, has a higher "destiny." It was established to provide a stable, institutional home for the practice of science. The university guaranteed the continuity of knowledge by cultivating both its objects and subjects. After the failures of lexica, dictionaries, and encyclopedias, Böckh, echoing the claims of the early idealists, defended the institution of the university and its disciplinary order as the last hope for the "uninterrupted transmission [of knowledge] from generation to generation."[86] On such an account, the university was not conflated with the practice of science. Like any institution, it could pursue other goods, prestige above all, but in the modern age it was the "bearer" of science as a practice.[87] As the singular institution of "scientific formation [*Bildung*]" embodying the virtues particular to the practice of science, the university would not need the church, the state, or ancient Greece.[88] Science provided the university with its ethical resources and normative vision, and the university provided science with institutional stability.

## On the Limits of Philology as a Practice and the Crisis of the University

Within a decade of its founding, the Berlin philology seminar left broad questions about the good life and a common good aside and focused instead on historical reconstructions of particular passages, methodological and technical innovations, and debates within an increasingly restricted circle of specialists. For some nineteenth-century scholars, this apparent departure from the ideals of eighteenth-century philology was a travesty. As one German philologist put it in 1820, "We're turning out men who know everything about laying the foundations but forget to build the temple."[89]

But the philology seminar as it was conceived by Böckh and then developed over the course of the nineteenth century offered a different account of the authority of knowledge. Philology increasingly became less a history of classical texts and antique moral exemplars and more a history of itself as science.[90] German research universities increasingly produced hyperspecialized scholars who understood themselves and their work only in terms of philology as a science. Internally, the success of the philology seminar led to ever-greater fragmentation of the broader discipline as philologists focused on increasingly specific and technical questions. Externally, philology gradually detached itself from the broader, nontechnical culture. Philologists saw no need to justify their activities or commitments to a public who could not understand their work anyway. As the historian Barthold Niebuhr described the situation of philology in 1811, even before the Berlin seminar was founded, "In Germany during recent times philological studies have taken on a dynamism which the most famous philologists and schools of earlier times never experienced. Rigorous interpretations [and] precise grammatical analysis link up with exploratory research into collective scientific knowledge and opinions, as with the history and institutions of antiquity."[91]

Philology became a discipline in the modern sense of the term: self-regulating and self-justifying. Philology seminars produced not modern-day Greeks but rather philologists. The historical and methodological work of philology crowded out other more traditionally humanistic ends. Its specialized inquiries into the technical details of philology were filters that determined what questions to address, what books to read, and what to ignore. Specialized, disciplinary science became the basic technology for organizing and producing knowledge. The technological efficiency of specialized science—its capacity to keep knowledge from becoming too unwieldy and overwhelming by lending it coherence and affording practitioners a coherent intellectual tradition—helped it supersede Enlightenment technologies. It came to stand in for what counted as authoritative knowledge.

The ethics of such a notion of science was committed to specialization from the beginning. The unity of knowledge was projected onto a community in which knowledge was thought to progress over time. It was located not in a synoptic capacity of an individual scholar but in the historical community of scholars, in the practice of science. Without the practice of science, the authority and integrity of the university and scientific knowledge would not exist.

And yet like the print technologies that had preceded it, the research university, for some critics, failed to live up to these promises of community, primarily because science was not lived out as a practice. Early nineteenth-century critics condemned philology not only for abandoning its neohumanist commitments but also for forsaking its commitment to science as a practice. Philology, they worried, was moving away from the ethical and epistemological conception of science as envisioned by Fichte, Schelling, and Schleiermacher and endorsed by Humboldt. The charges of pedantry, obscurity, and irrelevance which eighteenth-century scholars had leveled against humanist scholars were increasingly leveled against scholars who now considered themselves scientists. Whereas eighteenth-century critics of *historia literaria* called for "true erudition," early nineteenth-century critics demanded, as Adolf Diesterweg, the director of Berlin's pedagogy seminar, put it in 1836, "true *Wissenschaftlichkeit*."[92] The persons of science were not living out the virtues of science as a practice. Less than thirty years after the founding of the University of Berlin and the promise of a new culture of science, universities, he claimed, were failing to form students into the culture of science. Scholars no longer sought to reintegrate their particular research into the systematic whole of science. And furthermore, most of the research that purported to be new was really no more than a "mass of unproductive historical knowledge," that is, supposed researchers repeated each others' work with only slight, inconsequential additions.[93] For many critics, philology was failing both as a technology for information and as a practice.[94] Not only had it failed to solve the problem of information overload; it was exacerbating it.

Böckh voiced similar concerns in 1820, less than a decade after his philology seminar had been established. Most philologists, he lamented, had "almost no concept of the internal relationship of the different parts of philology"; instead, they pursued their specialized scholarship with a "certain thoughtlessness, as a familiar business."[95] Philology may have achieved autonomy as a science, but it had done so by becoming a highly rationalized, bureaucratic system. Addressing the Society of German Philologists in 1850, Böckh described how philology's embrace of a strict division of labor—with each philologist focusing on one highly specialized philological problem—had produced a "mass, we could even say flood, of monographs" and thus had greatly expanded philological knowledge.[96] This division of labor also threatened the ability of individual philologists "to keep alive the idea of the unity of all knowledge"—to cultivate the micro/macrocosm relationship that Böckh, following his romantic and idealist predecessors, argued was necessary for philology to flourish. Philologists had become reticent or unable to embrace broader concepts that would relate their research to other research.[97] As thinkers from Kant to Böckh had argued, the division of labor or specialization had always assumed a whole that could be divided. Specialized scholarship could only be justified on the grounds that it was part of a larger whole. It had value and worth insofar as it contributed to the historical development of knowledge. Without this basic assumption, specialized research was fragmented and illegitimate.

In public lectures for the university and the Berlin Academy of Sciences delivered over four decades, Böckh struggled with what he came to regard as the basic tension of specialized science and disciplinarity. On the one hand, specialized or "micrological" forms of labor, he wrote, had made philology more productive and more distinct as a science. Philology should not be denied its "micrology," just as natural science could not be denied its "microscopy."[98] Specialization produced new knowledge. On the other hand, these practices and the kinds of research that they engendered threatened the integrity of philological knowledge—that is, any sense of the ends to which all this microscopic labor and hyper-production should be put. If the virtues particular to philology had no harmonizing element, what was the purpose of philology beyond the maintenance and curation of the material archive of human knowledge? By ignoring such broader questions and limiting itself solely to the pursuit of "specialized research," philology had "largely fragmented itself." Most philologists "lack universal ideas, an overview," he lamented. "Everything is dismembered in their hands. They, therefore, have neither a concept of the extent nor an intuition of the essence of philology, but only know details in which their thought collapses."

The fragmentation of philology cast the very premise of specialized science into doubt. Scholars such as Böckh had conceived of specialized science and scientific disciplines as a solution to the epistemological and cultural anxieties of the Enlightenment, but for many nineteenth-century scholars science was beginning to manifest the same problems of proliferation and fragmentation. Specialized science was now responsible for the unparalleled "expansion of the realm of books in printed texts."[99] In Germany, the differentiation of the humanistic, natural, and physical sciences over the course of the nineteenth century was undergirded by a fundamental paradox between an ever-finer differentiation of intellectual labor in the sciences and a constant effort to unify them.[100] As the physical and natural sciences began to expand and assume the preeminent role as the model sciences, especially after 1880, anxieties about specialization and fragmentation and elegiac longings for unity only intensified.[101]

The internal battle over the shape and future of philology continued throughout the nineteenth century.[102] On the one hand, philologists such as Theodor Mommsen claimed that the purpose of philology was to organize the "the archive of the past" and to "secure the source material for its own sake."[103] As archivist, the philologist should not ask whether particular materials were worth being saved or not; his task was simply to collect. The purpose of science was to maintain and expand the archive of human knowledge. Echoing Enlightenment notions of completeness, Mommsen celebrated the ever-expanding archive as the totality of science and the unity of knowledge. He shifted the focus of philology from the formation of particular types of people to the accumulation of material forms of knowledge.

On the other hand, Böckh warned that if philology conceived of its purpose only in terms of curating the material archive of knowledge, it would forsake what distinguished it from other forms of knowledge: its formative character. "All science when carried out well," he insisted, "forms human beings."[104] Philology needed to model science as a practice, as an ethical commitment to the "development of science" as a whole. Böckh was not arguing that philologists should cease their specialized research. Given the "amount of work" to do, the "division of labor" characteristic of modern disciplines was necessary. But this divided labor needed to be undertaken within the practice of philology, that is, out of a devotion to the habits and virtues that distinguished the discipline of philology as a science from the accumulating methods of *historia literaria* and the technologies of the "empire of erudition."

The increasingly divisive debates of nineteenth-century philology, observed Böckh, presented the discipline with an "ethical challenge."[105] Although he considered a commitment to method and frank, open discussion a key philological virtue, he worried that the singular focus on the material of philology—particular problems or research questions—without attention to the broader science of philology would ultimately undermine the unity and thus the authority of the discipline. If philologists failed to recognize themselves as philologists dedicated to the same practice, then philology would forfeit whatever authority it had accrued over the course of the nineteenth century.

Böckh's worries about nineteenth-century philology represented an elegiac longing not for a unity of knowledge understood as a synoptic, God-like view of the relationship of all knowledge, or a material completeness on the model of the empire of erudition; instead, they represented a longing for the unity of science as a practice. They signified a lament that philologists might soon no longer pursue science as *Bildung*, as formation, but rather as a positivist, strictly empirical project of accumulation. Philology had to recover the ethical orientation that Kant, Schelling, Fichte, Humboldt, and Schleiermacher had given it; otherwise, it risked becoming just another modern profession and the university just another modern bureaucracy.

The story of the nineteenth-century philology seminar raises broader questions about the legacy of specialized science. Beginning with figures like Böckh, scholars have long wondered whether, as Anthony Grafton puts it, Humboldt "fatefully tie[d]" research to ethical formation [*Bildung*].[106] Did *Wissenschaft* gradually dissociate itself from idealist and romantic notions of a social unity and become bound "to growing positivist assumptions, empirical research, and the seemingly inexorable forces of academic specialization and differentiation?"[107]

In an address to the Academy of Sciences in Berlin entitled "On the Relation of *Wissenschaft* to Life," Böckh claimed that science afforded a distinct type of unity and harmony, but one that was insufficient to fund a comprehensive form of life: "We are not suggesting that science should have no influence on life. Indeed, I maintain just the opposite that the most fully perfected human life is attained precisely in the unity of theory and practice, of knowledge and action, of thought and deed. But until this ideal, or more accurately this ideal-real condition, the divine in the human, has been fully achieved, any one-sidedness, as long as it is competent, should be allowed. Not everything should be expected from everyone but only that for which he is capable. Not all standpoints and domains of life should be mixed."[108] Less than forty years after the romantic and idealist invocations of science as a new order of knowledge, Böckh acknowledges the utopian character of the entire project. And he encourages his fellow academics simply to carry on in their "quiet, mostly solitary work," while they wait for their fragmented labors to find some ultimate resolution in a future they could only imagine.[109]

Specialized science or *Wissenschaft* may well have proven to be an insufficient resource to sustain the longings for unity and purpose that its idealist and romantic forbearers endowed it with. It may well have been an inadequate replacement for the manifestly religious-based practices that offered a comprehensive way of life. But these questions are of a different sort than the ones I have tried to pursue here. Doubts about the validity and sufficiency of science as a distinct form of life intersect with questions surrounding secular forms of knowledge. My goal here is to tell the story of how specialized science or disciplinarity emerged in an age of media surplus not as a problem to be solved but as a solution. I offer a genealogy of scientific specialization, not a defense of how it developed over the course of the nineteenth century and beyond. My intention is to show how scientific specialization came to stand in for a desire to control and manage knowledge, just as the "book," the "encyclopedia," or "erudition" had before it. It came to stand in for epistemic authority.

Böckh's insistence that philology be pursued as a practice that organized not just the material of knowledge but also the subjects represented a shift in what stood in for real, legitimate knowledge. Whereas the seminars of the eighteenth century entailed an external justification—philological research led students to acquire ancient habits of thought and virtues—the seminars of the nineteenth century entailed a primarily internal justification. They led students to acquire the habits of thought and virtues of science. Research was justified not because it led to a more humane person but because it led to the formation of the disciplinary self. "The humanities," the models of scholarship underlying eighteenth-century conceptions of philology, wrote Böckh, should not be conflated with science.[110] The latter may well "serve the development of the purely human," but that was not what distinguished it and lent it authority. Specialized science was a practice with its own ends, virtues, and tradition. Subordinating it to external ends risked undermining its authority and what distinguished it from other forms of producing and transmitting knowledge.

## The Virtues of Specialization

In 1918, the German sociologist Max Weber described with devastating and detached effect what had become of the modern German research university and the specialized science that it embodied in the century following the founding of Berlin's philology seminar: science had loosened itself from the ethical demands that its forefathers in Berlin had placed upon it. The integrity of the disciplinary self had fractured as the university no longer sustained and defended the practice of science and its virtues and standards of excellence. In fact, the research university had gradually separated science from ethics, so much so that modern, specialized scientists had nothing to say about questions of value, meaning, worth, and the good. Questions about how one should live, claimed Weber, were of no concern to science. For the modern specialized scientist, private and public were distinct. The logic of specialization was ineluctable, and attempts to glean ethical guidance from science were misplaced, elegiac longings for an enchanted world that had long since disappeared. Science, wrote Weber, was "meaningless."

The worries of Wolf, Böckh, and all those philologists who feared that philology would fracture and fail as a practice had, on Weber's telling, proven all too prescient. And the institution designed to embody science as a way of life had become just another modern, rationalized bureaucracy.[111] With the sundering of science from the practice of science, the archive of knowledge from the people who created, interpreted, and interacted with it, disciplinarity from the disciplinary self, the modern research university had undone the legacy of its idealist and romantic predecessors. It had detached epistemology from ethics.

Despite Weber's claims, however, which continue to haunt the research university, there was a clear ethical imperative to what has been referred to as Weber's "value neutrality" [*Wertfreiheit*]. Just listen to his exhortation for how a scholar should conduct himself: "To the person who cannot bear the fate of the times like a man, one must say, may he return silently without the usual publicity buildup into the wide and comforting arms of the church. . . . For me, [such a return] stands higher than academic prophecy [pronouncing values and ends in the classroom], which does not clearly realize that in the lecture rooms of the university no other virtue holds but intellectual virtuousness [*Rechtschaffenheit*]."[112] Despite his insistence on the non-normativity of science, Weber's exhortation to act "like a man" contains a deep appeal to virtue. The question of whether one embraces the meaninglessness of science or the comforting enchantments of the church is not simply propositional—whether the dogmas are true or false—but rather, and more importantly, a reflection of whom the scholar is as a person, a statement about the scholar's dispositions. The capacity to exclude systematically broader questions of meaning and value, to focus simply on the "needs of the day" (the university's rules, procedures, and practices)—these are the virile virtues of the modern and, for Weber, always masculine scientist.[113] There was an "intellectual virtuousness" that characterized the modern scholar of the research university, and these are also the virtues that characterized the nineteenth-century philologist: industriousness, attention to detail, rigor, adherence to method, openness to discussion, a commitment to science, and a critical attitude. This was the bedrock of the modern university's epistemic authority. Despite Weber's claims to the contrary, science did lay claim to value, meaning, worth, and, ultimately, a uniquely modern epistemic authority grounded not in divine revelation or the ends of the state but in the character of the disciplinary self. Weber's scholar of "intellectual integrity" was another formulation of the virtues of the seminar and the practice of specialized science which crafted and formed the disciplinary self.

The modern research university has always been based on normative commitments and claims about what counts as authoritative knowledge—visions and claims about what is worthy, valuable, and good. It has defended scholarly virtues; it has promoted unique goods such as the production of scholarly work that both embodies these virtues and adheres to standards of excellence that have developed over time; and, above all, it has constituted a community of people who participated in a shared practice. Advocates of the research university have surely conflated the normative and the descriptive. Their aspirations, as Böckh's elegiac critique of nineteenth-century philology intoned, were never fully fulfilled. But the research university and its underlying ethic had come to stand in for a new order of knowledge, a new way of organizing the desire to know. It had come to represent a new and distinct form of epistemic authority, and the disciplinary self had come to embody the virtues and habits of this authority.

# Afterword

## *Too Many Links*

On June 26, 2012, after two weeks of criticism and national scrutiny, the University of Virginia's governing board reinstated President Teresa Sullivan. In the months that followed, however, calls for a "deep, radical, and urgent transformation" of the research university only intensified as Charlottesville became the epicenter of the ongoing debate on higher education. In her first public initiatives after faculty had rallied for her reinstatement, President Sullivan offered substantial grants to faculty "to integrate technology" into their courses, ordered a university-wide audit on the use of digital technology, and oversaw the debut of the university's first three MOOCs (massive open online courses) with Coursera. Elsewhere, consultant reports, newspaper articles, and, increasingly, panicked university administrators warned of the traditional university's imminent collapse. Fearful that they would be left behind, universities across the country clamored to sign up with Coursera and edX, MIT and Harvard's joint online course start-up, to bolster their technological prowess. By the end of 2012, which the *New York Times* dubbed the "year of the MOOC," the language of crisis and technological disruption had become ubiquitous.[1]

In those first few tumultuous months, many MOOC advocates embraced it as though it were a singular, salvific technology. Those who touted its promised efficiencies and common-sense appeal—why teach forty students when you can teach forty thousand!—rarely mentioned its predecessors in distance learning and other online instructional technologies, much less the fact that the MOOC had been around since at least 2008.[2] The MOOC had no history, just a future full of promise. No one knows, noted William Bowen, former president of Princeton, "how effective . . . online learning has been in improving . . . learning outcomes," but such ignorance did not stop him from declaring such technologies "transformative."[3] These new technologies and the "existential" threat they posed to universities were justified by their mere existence. The twenty-first-century prophets of digital disruption fetishized MOOCs, just as eighteenth-century scholars had their printed encyclopedias and lexica.

Already by the spring of 2013, however, the prophesies of a MOOC-enabled revolution in higher education seemed to falter, especially for increasingly self-doubting university faculty. Professors at Duke and Amherst voted against joining online ventures; the San Jose State philosophy department repudiated their dean's attempts to make them middlemen for Michael Sandel's commoditized "Justice" course; Coursera abandoned its singular focus on MOOCs for a business plan that included more traditional course management platforms; and Sebastian Thrun thought better of his prediction that in fifty years there would be only ten institutions in the world "delivering education."[4] The MOOC didn't disappear; it just looked more like a new digital textbook and less like an elixir for the university's troubles.

The "MOOC moment" of 2012 and 2013 was not just about differences in "market prognostication"—could universities withstand the purported disruption of online instruction?—just as the "print moment" of 1800 was not just about whether universities could compete with affordable books.[5] The sometimes nervous, sometimes euphoric debate about MOOCs crystalized our confusions about the place of the university and exposed our anxieties about what constitutes authoritative knowledge in a digital age. As our institutions and digital technologies change so quickly, the capacity of the research university to fulfill its historical purpose—to generate and transmit authoritative knowledge by forming people in the practice of science—has been cast into doubt. Can the research university still orient us in our new media environment? What is its role in producing and disseminating knowledge today?

In this book I have tried to explain how and why the research university came to be a central institution of knowledge, organizing other institutions (libraries, museums, natural science collections), print technologies, and practices into a relatively coherent whole whose purpose was the management and legitimation of knowledge. I began with the premise that before we rush either to defend or to attack the research university, we need a better understanding of what is about to be disrupted or destroyed. What kind of technology is the research university?

Although the research university fits into a genealogy of Enlightenment technologies, it is not just another, more efficient encyclopedia. It never simply "delivered" information, as one of Coursera's cofounders described the function of his company.[6] It is not just an information hub. Had this been the case, its purported "inefficiencies" would have doomed it long ago.[7] The research university has persisted because it embodied a community that realized a range of technical as well as cultural functions. Even in its incomplete and imperfect form, the research university helped scholars and students navigate an ocean of information; it offered collective wisdom and insight; it helped students and scholars decide which books to read, which to avoid, and what sources of information to trust. It made knowledge less "abstract and overwhelming."[8] It organized what had been experienced as a "flood" of print into coherent, culturally complex intellectual identities. It was a source of authority amid moments of technological change and epistemic anxiety.

Some of the same critics who have recently urged radical and immediate changes to the research university have also suggested that its various functions be "unbundled" and served by "providers that are not universities at all."[9] The components of a university—research, granting of degrees, faculty, students, governance and administration, curriculum, teaching and learning, assessment, and educational experience—should be broken down. Some universities would then focus on undergraduate teaching; others on research; some on vocational training; some on producing curricula, others on delivering them. Universities should specialize.

While some of this is already happening, such an "unbundling" would represent a fundamental reimagination of the purpose and institutional integrity of the research university. Since its inception in early nineteenth-century Germany, the research university has understood itself as a unified whole. It was a system or, as F. W. Schlegel put it, a "*living* encyclopedia" in an age thought to be beset by fragmentation and proliferation. And it was designed to organize the institutions, materials, practices, and people of knowledge into a relationship, to order into a coherent whole whose end was science. The research university and its ethos of specialized science were a solution to a particular problem in the history of knowledge. The research university stood in for a particular way of managing and legitimating knowledge.

The contemporary research university faces a range of problems, from a collapse in public funding, to worries about the future of the liberal arts, to perpetual complaints about specialization. But even more profound than any of these challenges is that of how to orient itself in an age of technological change, in an age of anxiety about what counts as real knowledge. Our challenge is strikingly similar to the one that late eighteenth-century intellectuals faced. The research university's exceptional authority to generate and transmit knowledge is being called into question. Since its emergence in Germany in the early nineteenth century, and since its reinvention in the United States decades later, the research university has maintained a near monopoly on knowledge by managing its acquisition, conservation, distribution, and interpretation. It has been, as Johns Hopkins's first president, Daniel Coit Gilman, put it, modernity's primary "agency" for the advancement of knowledge: "The idea is often vague, sometimes perverted, commonly half developed, at times inflated; nevertheless it contains the principles of life, that in every civilized community there must be a high school, capping, crowning, binding all other institutions for the advancement of knowledge."[10] The research university melded the "agencies" and technologies of knowledge by cultivating particular skills, habits, and virtues oriented toward the ideals of the practice of science. The "unbundling" of the research university would not simply create efficiencies by outsourcing the university's various components, or so goes the logic of many a critic; it would also threaten a particular practice, a particular way of knowing and, by extension, the central technology for organizing knowledge over the past two centuries. This does not mean that the research university ought to be defended simply on conservative grounds, but we simply have to confront the questions of what we might gain and lose were the university to change radically or simply dissolve into a fragmented, "unbundled" institution.

Before considering the possible consequences of such a loss, as well as the opportunities for transformation, let me clarify what is not necessarily at risk of being destroyed or disrupted. The university as imagined by the German romantics and idealists, as well as the idea of the university adopted and adapted by American university presidents such as Charles Eliot and Gilman, should not be conflated with the conglomerate that is today's university. The unified institution they imagined was organized around the advancement of knowledge, but over the course of the twentieth century this institution, in the United States at least, has been absorbed into what Clark Kerr, former president of the University of California, termed in 1963 the "multiversity." Such an institution, he wrote, "is not one community but several."[11] The University of Virginia, for example, is an institution of knowledge, but it is also a health care provider, entertainment center (its basketball arena hosts everything from Bruce Springsteen concerts to monster truck rallies), sports business, start-up incubator, sustainability coordinator, and industrial and government research center, not to mention a mental health provider, food service, and employment agency for more than twenty thousand students.

Such an institution, as Kerr sardonically noted, has neither devotees nor defenders. It has "practitioners," primarily administrators who struggle to manage a panoply of enterprises and competing visions of the university's purpose. It functions not like an organism in which parts and whole are bound by a common purpose, but rather like a "mechanism held together by administrative rules and powered by money." The contemporary research university is but one component of a larger enterprise increasingly indistinct from any other large-scale, multinational corporation. The provost at most universities, after all, is merely the "chief academic officer," just another manager alongside a host of vice presidents for health systems, finance, and operations.

The idea of the research university should also not be conflated with the American college. Whereas Harvard, Yale, and Princeton Colleges originally formed clergy until the late nineteenth century, the research university has always formed scientists. From its beginnings in Germany, it was primarily committed to reproducing a distinct form of knowledge by means of reproducing the disciplinary self. These original aims live on in everything from international university rankings to the perennial devaluation of undergraduate education in the research university.[12] Since its inception, the research university has struggled to reconcile its ends with a collegiate ideal, from Schelling's embrace of the liberal arts tradition to Columbia University's implementation of its Contemporary Civilization program around the First World War. Some of these efforts have managed to protect a cherished college experience within a broader university, but none have successfully combined the two models. The "wall" that Charles Eliot erected between undergraduate and graduate programs at Harvard in the late nineteenth century remains.[13] Recent attempts to reform undergraduate curricula at Stanford, Harvard, and a number of public universities have continued to struggle with this basic tension. Efforts to focus research university faculty on undergraduate teaching and advising, for example, inevitably run counter to the incentives of the disciplinary order. No one gets tenure for spending time with undergraduates. The college and the research university continue to be distinct institutions.

The future of the institution—the research university, not the multiversity or college—that emerged as a solution to epistemological anxieties surrounding information overload had always been in doubt, however. Critics have long dismissed the research university as a product of a romantic ideology, or else a bureaucratic machine. Concerns about its imperative to specialize, as we saw in the case of German philology, have been around since at least the early nineteenth century. They only increased as research universities became established in the United States. In his inaugural presidential address at Princeton in 1896, Woodrow Wilson worried that academic specialization had led universities to abandon their more basic responsibility to form "conscious and broad-minded" citizens of the world.[14] What had become, he lamented, "of the general foundations of knowledge, what of the general equipment of mind, which all men must have who are to serve this busy, this sophisticated generation"?[15] But questions about the legitimacy of the research university, as Wilson's own worries make clear, have been inextricable from deeper, more pervasive epistemological anxieties about what counts as authoritative knowledge.

Misgivings about specialized science and disciplinarity have returned in recent jeremiads about the research university from within its most elite ranks. Harvard professor Louis Menand writes that the "structure of disciplinarity that has arisen with the modern research university is expensive; it is philosophically weak; and it encourages intellectual predictability and social irrelevance. It deserves to be replaced."[16] Similarly, CUNY professor Cathy Davidson has criticized the research university as an "archaic, hierarchical, silo'd apparatus of the nineteenth century."[17] Our institutions of higher learning have "managed to change far more slowly than the modes of inventive, collaborative, participatory learning offered by the Internet" and other online and digital technologies. Unlike some of the more general critiques of the university's disciplinary structure, however, Davidson's critique is more focused on what is actually at stake. Our universities are "stuck," she writes, "in an epistemological model of the past."[18] Our digital age entails not just new and better technologies but an entirely different notion of what constitutes true knowledge: how it is produced, authorized, and disseminated. The disciplinary organization of knowledge is antiquated and dispensable. The very structures and forms of knowledge are changing, and, for Davidson at least, the disciplinary research university is being left behind.

In her more recent work on the future of education, Davidson embraces the potential of digital technologies to undo the authority structure of the research university and spur "collaborative" forms of knowledge production. And yet, in what she describes as a "field guide and survival manual for the digital age," her *Now You See It: How the Brain Science of Attention Will Change the Way We Live, Work, and Think*, she relies on that same authority structure she seems eager to escape. She bases her "guide" for the digitally perplexed on what she calls "the science of attention."[19] She grounds her argument in the authority of modern, disciplinary-based science as she cites study after study, all of which are legitimated by the authority of the disciplinary order of the modern research university.

Davidson's bad faith is a testament to just how enduring a system the research university ethic is. But it has endured not because it was a rigid, hierarchical system, a Weberian iron cage, a Foucauldian panopticon, but rather because it has sustained communities of people engaged in a common pursuit. Research universities have never overcome the fragmentation of knowledge or realized anything like a universal knowledge. But what they have done is organize intellectual labor, traditions, and desires more effectively over the past two hundred years than any other technology. To dismiss the research university as an antiquated bureaucratic "apparatus" defined by constraint and enforceable standards is to overlook the ways in which its continuity and stability depended on the transformation of actual people. Disciplines at the center of this institution are not timeless "silo'd" taxonomies. They are practices guided by norms of knowledge production and transmission which have developed over time. Nineteenth-century philology, for example, depended on particular actors who developed and embodied practices and habits of labor, criteria for excellence, and systems and forms of communication, all of which were tied to the disciplinary self. The practices of the nineteenth-century philology seminar were not simply expressions of an ideology or an abstract self; they constituted the character of the philologist. Specialized science "wedded" epistemology to ethics within a stable institutional context.[20] The survival of the disciplinary system depended on the formation of persons, who could then reproduce through their own participation in practices and modeling of its virtues. There was no disciplinary order without the disciplinary self. Any account of the research university that downplays the formative element of the research university fails not only as history but as prognostication. It underestimates the capacity of the research university to reproduce itself as a distinct culture, because it underestimates its cultural functions. Academic disciplines are structures that focus our attention and craft our intellectual identities. They provide limits and frameworks within which scholars can create knowledge. Considering how scholars generate knowledge in this way helps us better understand how our production of knowledge relies on constraints and filters, and how scholars rely appropriately on authoritative communities that form them.

The arrangement of the research university has both internal and external functions. Internally, as alluded to above, it offered the university immense cultural and epistemological advantages. In this sense, scientific specialization is not a new problem to be solved, but the answer to an older problem of media surplus we still inhabit. It was a way of dealing with anxieties about the ubiquity and quality of information. And its effectiveness was tied to the assumption that it was not simply a professional structure—that is, the possibility that scientific work could support a career and provide a professional identity. It was also an ethos. It gave meaning and formed people. It cultivated dedication. It provided a way of being in the world and a resource for making sense of it when the very sources of what constituted authoritative knowledge had dissipated or become highly contested. It also kept knowledge from becoming too abstract by providing standards of excellence and authoritative traditions. But it was also highly flexible. As we saw in the case of nineteenth-century philology, the boundaries of a discipline were not fixed and externally imposed; instead, they were functions of internal distinctions and the history of the discipline itself. It is this flexibility that has undergirded the system's endurance and effectiveness.

Externally, disciplinarity provided the epistemological and ethical grounds for the university's relative autonomy with respect to other social institutions, such as the church and the state. It embodied a particular type of autonomy, one based on expertise and the power of a community of specialists. In the research university, the logic of disciplinarity is the basis for faculty claims to self-governance and academic freedom. The disciplinary order did not create a completely autonomous institution, but it did establish the university as a buffer between the internal demands of science and the external demands of society. The disciplinary order did not completely separate the research university from an extra-university public; instead, it mediated the particular, more specialized interests of scientific research with the broader interests of an extra-university public. If universities were to somehow abandon "disciplinarity" as an antiquated relic, then they would not only undermine their claims to self-governance and academic freedom but further subordinate themselves to the imperatives of markets and immediate social needs. The disciplinary order and the continuity of the practice of science and its virtues have been the ground of the university's claim to its unique authority, and a continued defense of them is crucial to the university's survival.

Some critics, however, contrast the constraints of disciplinarity and the research university with the liberating promises of digital technologies. "Neither the Internet nor the WWW," writes Cathy Davidson, "has a center, an authority, a hierarchy, or even much of a filter on the largest structural level."[21] With the advent of digitally supported learning, "conventional modes of authority break down."[22] Digital technologies will liberate us from the constraints that academic disciplines impose on us. They will deliver us into a future without filters.

Such messianic hopes mistakenly suppose that there can somehow be a vacuum of epistemic authority. But in truth forms and functions of epistemic authority, be it the disciplinary order of the research university or *Wikipedia*'s fundamental principles or "Five Pillars," are themselves filtering technologies, helping us orient ourselves amid a surfeit of information. They help us discern, attend to what is worthwhile. Google searches point us in the direction of some resources and not others. Technologies are normative, evaluative structures to make information accessible, manageable, and, ultimately, meaningful. It is not a question, then, of the presence or absence of epistemic authority; it is about better or worse forms of epistemic authority.

If the Internet and the World Wide Web lack, as Davidson puts it, a "centralized authority" and a "filter," they do so only on the most abstract level. Our daily interactions with the Web are made possible by a host of technological constraints and filters. People access and engage information through technologies that allow them to select, filter, and delimit. Web browsers, hyperlinks, blogs, online newspapers, computational algorithms, RSS feeds, Facebook, and Google help us turn all those terabytes of data into something more palatable, that is, something that can be made useful to an embodied person. These now-ubiquitous technologies help us to sort, to google a needle in the haystack; in so doing, they have become central mediums for the way we experience the world.

In this sense, technology is neither an abstract flood of data nor a simple machine-like appendage subordinate to human intentions. It is the very manner in which humans engage the world and each other. To celebrate the Web, a printed book, or the research university as inherently liberating or constraining is to ignore the human scales at which we engage them. Our access to imagined expanses of pure information, be it a flood of digital data or printed books, is made possible by technologies that are constructed, designed, and constantly tweaked by human decisions and experiences. These technologies do not exist independent of the individuals who design and use them. We constantly interact with them.

Think of it this way: the Web as a whole, in its entirety—just like those terabytes of information that we imagine weighing down upon us—is inaccessible to the ill-equipped person, just as a "comprehensive" account of everything that had been printed was always just beyond the reach of eighteenth-century German readers. *The Web* does not exist, at least for us, any more than the universal "book of books" or "science of sciences" did for eighteenth-century readers. Technologies make the Web or the totality of books accessible and real by making them much smaller and more manageable than they are. To suggest, as critics such as Davidson do, that the advent of the Internet and the Web portends the end of epistemic authority and thus the technological filters that make it possible is both naïve and false. These new technologies are designed around particular notions of what counts as real knowledge. The singular insight of Google cofounders Sergi Brin and Larry Page was to recognize that not all Web pages were equal. They modeled their PageRank algorithm on the normative value of one page over another. In designing it, they were concerned not simply with completeness or managerial efficiency but with value, as defined on massive scales of human input. Google's PageRank is one example of an algorithmic authority that is now ubiquitous.

If, then, the Web and search engines are also technologies that orient us in a world of easily accessible information, are they all that different from the research university? Is there anything unique about the way the university helps us filter and search in our contemporary media environment? I think so, but in order for the university to flourish in this new environment, university-based scholars need to continue to develop and articulate the research university's distinct role, what distinguishes it from other technologies and institutions. Although the research university is no longer, as Gilman claimed at the end of the nineteenth century, the singular "agency" of authoritative knowledge, it does have a distinct role to play in the broader media and knowledge ecology.

New information technologies have taken over some of the research university's traditional functions. For well over a century, the research university organized bibliographic information, because the university research library was the primary repository for print-based knowledge. But now Google's search engines and related online tools make possible the creation of bibliographies outside the university library; its book-scanning project provides access to many of the world's books without the mediation of a university library. MOOCs and related digital technologies serve a similar communicative function. Companies such as Coursera and Udacity organize and then redistribute university-based, disciplinary knowledge, as do blogs authored by academics. They are digital platforms for university faculty to distribute their disciplinary knowledge, their research, to a wider public.

In the case of both Google and MOOCs, however, research universities continue to produce the knowledge that these new technologies disseminate. These new digital technologies rely on the authority of disciplinary knowledge. Without its university partners, almost all research universities, Coursera would have nothing to distribute. Even commodified knowledge still has to be created somewhere.[23] Similarly, Google's book-scanning project partnered primarily with university libraries. For now at least, the disciplinary structure of the university continues to maintain its authority to generate knowledge, even if its capacity to control its dissemination is declining.

These new for-profit corporations, however, will inevitably repackage, reorganize, and redistribute "academic" work for their own purposes. While the increased access that such digital and online projects promise is to be applauded, there is also reason to be concerned about such shifts in the transmission of knowledge. Although some critics worry that professors will not get compensated for their intellectual work, I worry that the university's traditional authority to transmit and repurpose knowledge will be forfeited to institutions that do not share its norms and ethical commitments. The norms that guide how companies like Coursera and Google organize and disseminate knowledge are primarily market based and have little in common with the formative practices and intellectual virtues that constitute the core of the research university. So far, Coursera and Google Books have touted their services as "free" and available to all, but these companies are under pressure to return a profit to their investors. Commodified curricular materials have been around for centuries in the form of the textbook, but granting companies such as Coursera, Udacity, or Google the authority to distribute, even as platforms and not necessarily owners, university-produced knowledge could cede control over the dissemination and organization of knowledge to institutions primarily oriented to profit making. As with most changes in media, the central issue concerns authority—"who gets to say what to whom."[24]

These broader concerns about epistemic authority and media change were at the center of the San Jose State philosophy department's denouncement of university administrators' attempts to implement Michael Sandel's "Justice" MOOC as an off-the-shelf curricular module with little to no input from its philosophy department. Philosophy faculty members were concerned about what it might mean for their students, largely working class, to watch Sandel lecture Harvard students on social justice.[25] Their concerns were not simply the last gasp of an antiquated institution clutching onto power, however. They represented anxieties about who gets to determine what counts as "real" knowledge and how that knowledge ought to be distributed in our digital age. Their implicit assumption was that a university was a community in which students are formed into distinct intellectual traditions and virtues, not simply served prepackaged content.

Despite the overwrought promises and anxieties unleashed by the "MOOC moment" of 2012, digital technologies have been transforming our cultural traditions for almost two decades now, and the pace is only increasing. As Jerome McGann and Bethany Nowviskie have put it, we "stand before the vast, near whole-scale transformation of our various and shared cultural inheritance."[26] The digital transformation of the history of knowledge and our varied cultural legacies raise new questions and possibilities for reflection on how and what we know, questions about interpretation, method, preservation, collation, collaboration, and transmission. University-based scholars must fully engage this shift, both conceptually and practically, from print-based to digital forms of knowledge transmission and production, not least because it is already happening regardless of whether they participate or not. As our cultural traditions are digitized and reorganized, universities and scholars need to participate in these processes of remediation and help guide the gradual transition from a print to digital archive, even as we continue to argue for the importance of maintaining our print legacies. Universities are key institutions, in part, because they sustain the types of communities and practices that will help ensure the sustainability and integrity of these transformations.

As research universities engage the new media ecology, they should also recognize that they are distinct institutions within it. The "MOOC moment" brought to the surface urgent questions about what institutional and technological forms might be best suited for generating and sharing knowledge today, and it also helped clarify the distinct role that universities have had in the past and how they might adapt in a digital future.

Like their eighteenth-century print predecessors, today's new information technologies are outstripping our normative and ethical capacities to engage them. We need practices and norms to guide our interactions with new digital technologies. In an age in which our ways of knowing and the forms of knowledge we generate and transmit are changing, we need criteria for what constitutes knowledge, models of systematization, forms of communication, practices of work and interaction, and exemplars of knowledge. Pure philosophical thinking will not be enough. We do not yet know how MOOCs and other technologies will develop, although efforts by universities to take charge of their development and mold them according to their own ends are to be encouraged. Technologies are changing so quickly that our norms and practices will inevitably lag. I am not suggesting that we need to articulate abstract principles in advance and simply apply them or simply deduce what our universities should become. New norms and practices of knowledge will have to develop alongside and, most likely, in response to technological developments. Given the pace of change and the need for ethical guidance, we need structures that, though not removed from these developments, are also not fully subordinate to them.

This is where the idealized claims made on behalf of the research university by Fichte, Schleiermacher, Humboldt, and their romantic and idealist contemporaries have a vital role to play today. The research university has cultivated epistemic norms and maintained the infrastructures for sustaining them for nearly two centuries now. Despite its failures and shortcomings, the research university was always more than a mere delivery device. It has stood in for a community in which the habits of science were passed on in personal interactions between teacher and student. The research university does not just deliver content or distribute information; it bestows epistemic authority. It is a distinct technology that MOOCs, Google, *Wikipedia*, and myriad other digital technologies have yet to replicate. It is an institution that sustains practices and their standards and goods, thus grounding a particular form of epistemic authority.

If the research university is judged simply in terms of how efficiently and widely it distributes information, then it has no future. And it shouldn't. But the research university has never been simply about broadcasting knowledge. When German idealists and romantics put forth a vision for the university in the age of print, they never claimed that the university could more efficiently distribute knowledge. They argued that the university generated real knowledge by forming people into a community of science. They never imagined that the university would replace or supersede print, but they did envision an institution that would integrate old and new technologies and embed them in practices oriented toward the legitimation of knowledge, an end that took on special urgency amid the cultural anxieties of media surplus. The university they imagined formed not just the things of knowledge but the people of knowledge.

There are limits, of course, to holding onto the idea of the research university as a historically unique community of knowledge, especially given its highly rationalized and bureaucratic structure. Structural accounts of the research university such as Menand's—whose guiding metaphors could be Weber's "iron cage," Foucault's "panopticon," or Luhmann's "system"—are perhaps the inevitable consequence of the push initiated by Schelling, Fichte, Humboldt, and Schleiermacher to imagine science as a practice. Their efforts to distinguish true knowledge as science from the surfeit of print led to the emergence of an institutionally grounded, theoretically complex system of knowledge. They also led to a bureaucracy of knowledge.

Over the course of the nineteenth century, German universities embraced the logic of specialized science and integrated it into a complex, highly rationalized bureaucratic structure. "The rapid bureaucratization of German academic life in the wake of the so-called Humboldtian reforms," as William Clark writes, was "not a sign of their failure . . . [but rather] a mark of their singular success."[27] The fervor of the German idealists and romantics for science was matched only by their commitment to an institution that could sustain it. Over the course of the nineteenth century, the research university expanded and gradually attained something akin to a monopoly in authoritative knowledge. It became a bureaucratic system that justified itself, oftentimes as an afterthought, by an idealist ethos and logic. The institution of the university provided stability and continuity, but it also tied knowledge to a highly rationalized institution. It transformed the ideal of a "living" community, a scientific sanctuary, into a bureaucratic machine.

Perhaps this was inevitable, but it does not mean that we should discard the traditions and ideal of the research university because they were never perfectly realized. At this particular moment of technological and institutional change, we need motivating ideals to orient our institutions and ourselves. The idea of the research university is more than its bureaucratic structures. However haltingly, the research university embodies ideals and virtues that scholars both inside and outside the university hold dear. This is where primarily structural accounts of the research university as simply a bureaucratic system, seemingly lacking human agents who endow it with meaning and life, can offer no compelling vision for a future research university. These cool, distant accounts of the research university, so redolent of Weber's description of any other modern, rational system, see nothing at stake, just the inexorable logic of another modern bureaucracy. They overlook the persons and norms that have always been the core of the research university. Anthony Grafton describes this attitude best: the "loss of patience, or faith, or interest in specialized knowledge" is ultimately a capitulation to the absoluteness of the bureaucratic system of the contemporary research university.[28] Such an attitude belies a thoroughly structural account that omits the research university's most basic feature: its underlying ethic. These more radically functional accounts, however descriptively illuminating, can never answer a basic question: why would anyone choose to devote herself to specialized knowledge and an institution such as the research university? The research university reproduces itself by forming people into its culture. Its survival relies on the decisions of actual people, not simply on the abstract totalizing mechanisms of an institution. Advocates of the contemporary research university need to recognize and embrace its most central feature: the fact that it embodies a set of norms, practices, and virtues central to modern knowledge. Whatever its myriad failings and bureaucratic functions, the research university sustains what scholars hold in common and commit themselves to—an ethics of knowledge. I am not suggesting that we should naïvely embrace the current structure of the research university, but we do need to recover its underlying normative core and adopt, adapt, and interpret it for our current age.

There is, as William Bowen writes, "something maddeningly difficult . . . to measure" in the modern university.[29] This "something" harkens back to the stubbornly idealist and romantic attempts of Schelling, Schleiermacher, Fichte, Humboldt, and their contemporaries to reimagine the university as a distinct community bound together by a shared practice and dedicated to organizing and legitimating knowledge. The research university became a vital institution of modernity not because it had a monopoly on the materials and technologies of knowledge, but because it provided ethical and normative resources for making sense of them. This was its primary purpose in the past. And we would do well to let it guide us into the future.

# NOTES

*Introduction*

1. Brewer, "Coup That Failed." My former colleague Mitch Green made this argument on the steps of the Rotunda.

2. See e-mails of May 30, 2012, and June 3, 2012, as reported on by the *Cavalier Daily*.

3. Hennessey, quoted in Auletta, "Get Rich U."

4. Bowman, *Higher Education in the Digital Age.*

5. Davidson and Goldberg, *Future of Thinking*, 53.

6. See also Taylor, *Crisis on Campus*; Menand, *Marketplace of Ideas*; Kronman, *Education's End.*

7. On the perpetual "crisis" in universities, see Donoghue, *Last Professors.*

8. Arum and Roksa, *Academically Adrift.* See also Karabel, *Chosen*; Bowen, Chingos, and McPherson, *Crossing the Finish Line.*

9. See, for example, Delblanco, *College*; Kronman, *Education's End.*

10. Nussbaum, *Not for Profit.*

11. Christensen and Eyring, *Innovative University*, xxiii.

12. Gilman, *University Problems in the United States*, 73.

13. This percentage increase is based on the number of listed titles in the Leipzig Book Fair catalogue, which went from 755 titles in 1740 to 1,144 in 1770 and then to 2,569 in 1800. See Jentzsch, *Der deutsch-lateinische Büchermarkt.* On the limited but useful picture of the German book market afforded by these catalogues, see Wittmann, *Geschichte des deutschen Buchhandels*, 111. For a broader discussion of such numbers, see chap. 4.

14. See, for example, discussions in Wittmann, *Geschichte des deutschen Buchhandels*, 111; see also Rudolf Schenda, *Volk ohne Buch*, 173–75. For a more detailed discussion and a comparison of the production of printed books in Europe more broadly, see Buringh and van Zanden, "Charting the 'Rise of the West.'"

15. Fichte, *Deduzierter Plan einer zu Berlin zu errichtenden höheren Lehranstalt*, in Fichte, *J. G. Fichte Gesamtausgabe der Bayerischen Akademie der Wissenschaften*, II, 11:83. Cited henceforth as *Deduced Plan.*

16. Ibid.

17. Schelling, *F. W. J. Schelling, Werke*, 2:551.

18. Here and throughout this book I translate *Wissenschaft* simply as "science." As will become clear, however, "science" refers not simply to the physical or natural sciences but much more broadly to a certain type of knowledge.

19. Here and throughout this book, I use the term "practice" in the basic way that Alasdair MacIntyre defines it in *After Virtue* (187). For an extended discussion of my use, see chap. 7. Foucault's notion of "technologies of the self" offers another way of thinking about these sorts of practices; see Foucault, *Technologies of the Self*, 28. I will return to this discussion in chap. 9 in particular.

20. See, for example, Graff, *Professing Literature*; Haskell, *Emergence of Professional Social Science*; Kuklick, *History of Philosophy in America*.

21. For one more recent example, see Casper, "Teaching and Research."

22. Louis Menand has recently traced the emergence of the modern research university to late nineteenth-century America and the founding of universities like Johns Hopkins. But the disciplinary organization of the university into *the* modern institution for the production of knowledge has a much broader history. Menand's singular focus on the American research university obscures not just a historical fact. It also conceals the fact that this system of knowledge was built on an explicitly ethical foundation. See Menand, *Marketplace of Ideas*.

23. All Kant citations, with the exception of those to the *Kritik der reinen Vernunft*, refer to the *Akademie Ausgabe*: Kant, *Gesammelte Schriften*. Henceforth, all citations will include a short title followed by volume and page number. With occasional emendations, translations of the *Kritik der reinen Vernunft* follow the Cambridge University Press translation, *Critique of Pure Reason*, trans. Paul Guyer and Allen W. Wood. Unless noted, all other translations are my own. Citations to the *Kritik der reinen Vernunft* refer to the A and B editions. Here, Kant, "Was Ist Aufklärung?," 8:35.

24. Daston and Galison, *Objectivity*, 191–251; see also Hadot, *Philosophy as a Way of Life*.

25. Foucault, *Technologies of the Self*; Hadot, *Philosophy as a Way of Life*; A. Davidson, "Ethics as Ascetics," 63–80; Daston and Galison, *Objectivity*, 191–251.

26. Daston and Galison, *Objectivity*, 204.

27. See, for example, Carr, "Is Google Making Us Stupid?" See also the expansion of his argument in Carr, *Shallows*.

28. Ann Blair, "Information Overload, the Early Years," *Boston Globe*, November 28, 2010, www.boston.com/bostonglobe/ideas/articles/2010/11/28/information_overload_the_early_years/.

29. Blair, *Too Much to Know*, 58.

30. Ibid., 2.

31. Postman, *Technopoloy*, 18. Unlike Postman, however, I do not use the term as an implicit condemnation of new communications technologies on public discourse. I use the term as a heuristic, a model to describe technological change over time.

32. For a fuller account of this use, see Wellmon, "Why Google Isn't Making Us Stupid . . . or Smart."

33. Johns, *Nature of the Book*, 40–41.

34. Novalis, *Novalis Schriften*, 2:661.

35. See Jentzsch, *Der deutsch-lateinische Büchermarkt*. On the limited but useful picture of the German book market afforded by these catalogues, see Wittmann, *Geschichte des deutschen Buchhandels*, 111.

36. Kant, "Was Ist Aufklärung?," 8:35.

37. Daston and Galison, "Image of Objectivity," 83.

38. Clark, *Academic Charisma*, 442.

39. See, for example, Hansen, "New Media"; Gitelman, *Always Already New*; Koschorke, *Körperströme und Schriftverkehr*.

40. See, for example, de Ridder-Symoens and Rüegg, *History of the University in Europe*. Especially important in this respect is the work of Clark, *Academic Charisma*.

41. In this respect, see especially the more recent rich conceptual work of Piper, *Dreaming in Books*; Siskin, *Work of Writing*; Price, *How to Do Things with Books in Victorian Britain*; Lupton, *Knowing Books*; Johns, *Nature of the Book*; Chartier, *Order of Books*. See also the pioneering work of Darnton, *Business of Enlightenment*; and, more recently as an overview of the field, Darnton, "'What Is the History of Books?' Revisited"; McKitterick, *Print, Manuscript and the Search for Order*. Much of my thinking on these issues has been indelibly shaped by my participation in "Interacting with Print," an international research group of book history scholars led by Tom Mole and Andrew Piper at McGill University.

42. See, in particular, Daston and Galison, *Objectivity*; Frasca-Spada and Jardine, *Books and the Sciences in History*.

43. See esp. MacIntyre, *After Virtue*; Hadot, *Philosophy as a Way of Life*; Herdt, *Putting on Virtue*.

*Chapter One. Science as Culture*

1. I will discuss the popular image of the German university in the second half of the book. On the widespread mockery of German universities, see McClelland, *State, Society, and University in Germany*, 80–85.

2. Zimietzki, *Das akademische Leben im Geiste der Wissenschaft*, 80, preface, 89, 117.

3. Ibid., 21, 95.

4. Markus, "Changing Images of Science."

5. See, for example, Zedler, *Grosses vollständiges Universal-Lexicon*, 57:686. In tracing this history, I am deeply indebted to Markus's work in "Changing Images of Science."

6. Aristotle, *Nicomachean Ethics*, bk. 6, sec. iii, lines 1–4.

7. Lear, *Aristotle*, 6.

8. Aristotle, *Metaphysics*, 993a30.

9. Markus, "Changing Images of Science."

10. Aristotle, "On Interpretation,", I.16a4–7.

11. Aristotle, *Rhetoric*, bk. II, 1413b. See Enos, "Ancient Greek Writing Instruction and Its Oral Antecedents," 26.

12. See, for example, *Phaedrus*, 274b.

13. Peters, *Speaking into the Air*, 47.

14. Markus, "Changing Images of Science," 4.

15. McNeely and Wolverton, *Reinventing Knowledge*, 37–76.

16. Parkes, "Reading, Copying and Interpreting a Text in the Early Middle Ages," 92–96.

17. Cassiodorus, quoted in McNeely and Wolverton, *Reinventing Knowledge*, 53.

18. Aquinas, *Summa Theologiae*, Question 1, Articles 2–10. See N. Schneider, "Experientia-ars-scientia-sapientia."

19. See Schrimpf, "Disciplina."

20. "Scientia," in Glick, Livesey, and Willis, *Medieval Science, Technology, and Medicine*, 457.

21. Markus, "Changing Images of Science."

22. Mauch, *Der lateinische Begriff Disciplina.*

23. Hugh of St. Victor, quoted in ibid., 30.

24. See Heilbron, "Das Regime der Disziplinen," 28.

25. Kelly, introduction to *History and the Disciplines.*

26. *Lexicon Philosophicum* of Etienne Chauvin (1692), quoted in Kelly, *History and the Disciplines*, 16.

27. Heilbron, "Das Regime der Disziplinen."

28. Kelly has described these very well. See his introduction to *History and the Disciplines.*

29. Heilbron, "Das Regime der Disziplinen," 29. See also Schmidt-Biggemann, "New Structures of Knowledge"; and Brockliss, "Curricula."

30. Vidal, *Sciences of the Soul*, 4.

31. Markus, "Changing Images of Science."

32. Minnis, *Medieval Theory of Authorship*, 10. The term *auctor* denoted someone who was at once a writer and an authority, someone not merely to be read but to be respected and believed.

33. Markus, "Changing Images of Science," 8.

34. Blair, *Too Much to Know*, 6. See also Eisenstein, *Printing Revolution in Early Modern Europe*, 88–126.

35. Blair and Stallybras, "Mediating Information, 1450–1800," 143, 148. Mar-

shall McLuhan notes something similar: "The first two centuries of printing from movable types were motivated much more by the desire to see ancient and medieval books than by the need to read and write new ones. Until 1700 much more than 50 percent of all printed books were ancient or medieval" (*Understanding Media*, 171).

36. Quoted in Blair, *Too Much to Know*, 56. On the logic of early modern compilation, to which I will return, see Gierl, "Kompilation und die Produktion von Wissen im 18. Jahrhundert."

37. Blair, *Too Much to Know*, 6. See also Eisenstein, *Printing Revolution in Early Modern Europe*, 88–126.

38. Yeo, *Encyclopedic Visions*, 3.

39. Bacon, *New Organon*, 6, 4.

40. Ibid., 70.

41. Bacon, "Description of the Intellectual Globe," 97.

42. Bacon, *New Organon*, 2.

43. Blair and Grafton, "Reassessing Humanism and Science."

44. Bacon, *New Organon*, 290.

45. See Febvre and Martin, *Coming of the Book*, 319–32.

46. German-language texts made up only 29% of printed texts at the beginning of the century and over 62% by 1700.

47. For more on the decline of Latin pedagogy, see Stichweh, *Die Ausdifferenzierung der Wissenschaft*, 27–29; on the history and eighteenth-century decline of Latinity in schools, see La Vopa, *Grace, Talent, and Merit*, 58–70; Paulsen, *Geschichte des Gelehrten Unterrichts*, 1:345–87.

48. See Schmidt-Biggemann, *Topica universalis*, 289.

49. Kelly, *History and the Disciplines*, 14; Cassirer, *Das Erkenntnisproblem*, 1:80.

50. Cassirer, *Das Erkenntnisproblem*, 63; Grafton, "World of the Polyhistors."

51. Mencken, *Charlatanry of the Learned*, 61, 70, 68.

52. I will discuss these periodicals more fully in chap. 2. See Martens, *Die Botschaft der Tugend*.

53. Gottsched, *Die vernünftigen Tadlerinnen*, 433.

54. Ibid.

55. *Hamburger Menschenfreund* (1737), quoted in Martens, "Lasterhaftes Lesen," 103.

56. Israel, *Enlightenment Contested*, 409–21.

57. Lessing, *Der junge Gelehrte*, 1:178.

58. Martens, "Vom Thomasius bis Lichtenberg."

59. Iselin, *Ueber die Geschichte der Menschheit*, 205.

60. Eberhard, *Synonymisches Handwörterbuch der deutschen Sprache*, 85.

61. Velkley, *Freedom and the End of Reason*, 23.

62. Kant, "Bemerkungen zu den Beobachtungen über das Gefühl des Schönen und Erhabenen," 20:175:29.

63. Kant, *Kritik der reinen Vernunft*, A832/B860.

64. Ibid., A833/B861.

65. Such a systematic unity, writes Kant, is "only a projected unity, which one must regard not as a given in itself." Ibid., A647/B675.

66. Ibid.

67. "Wissenschaft," in Adelung, *Grammatisch-kritisches Wörterbuch der hochdeutschen Mundart*, 4:1582–83.

68. Zedler, *Grosses vollständiges Universal-Lexicon*, 57:678, 714.

69. Walch, *Philosophisches Lexikon*, 4th ed., 1:1578–79.

70. See Bumann, "Der Begriff der Wissenschaft im deutschen Sprach- und Denkraum," 73.

71. Markus, "Changing Images of Science."

72. Kant, *Grundlegung zur Metaphysik der Sitten*, 5:433.

73. Kant, "Was Ist Aufklärung?," 8:35.

74. Blair, *Too Much to Know*, 253.

75. Terlinder, *Versuch einer Vorbereitung zu der heutigen positiven*, 20.

76. Ibid., 19.

77. Kant, *Kritik der reinen Vernunft*, A693/B721.

78. Valenza, *Literature, Language, and the Rise of the Intellectual Disciplines*, 11.

79. Humboldt, "Über die innere und äussere Organisation," in *Wilhelm von Humboldts Gesammelte Schriften*, 10:251.

*Chapter Two. The Fractured Empire of Erudition*

1. The German term *Wissenschaftler* was first used in 1811 and predates the English "scientist," which was not commonly used until 1870. In the French context, a range of terms were used throughout the nineteenth century—*savant*, *home de lettres*, *gens de letters*—but *un scientifique* was first used in the twentieth century. See Stichweh, *Die Ausdifferenzierung der Wissenschaft*, 233n; see also Bumann, "Der Begriff der Wissenschaft im deutschen Sprach- und Denkraum"; Hahn, "Scientific Research as an Occupation in Eighteenth-Century Paris."

2. The second figure was often referred to as *der gebildete Mensch* of the *Bildungsbürgertum*, the cultured, generally educated member of the German bourgeoisie that would flourish over the course of the nineteenth century.

3. Trunz, "Der deutsche Späthumanismus um 1600 als Standeskultur."

4. Turner, "*Bildungsbürgertum* and the Learned Professions in Prussia," 108.

5. Ibid. The rest of this paragraph draws directly on Turner as well. For a broader account of the Republic of Letters, see Kühlmann, *Gelehrtenrepublik und Fürstenstaat*, 285–454; Grafton, "Sketch of a Lost Continent"; Goldgar, *Impolite Learning*.

6. Grimm, *Letternkultur*, 4–5.

7. Michaelis, *Raisonnement über die protestantischen Universitäten in Deutsch-*

*land*, 1:112. On this situation, see Zammito, *Kant, Herder, and the Birth of Anthropology*, 40–41; Schneiders, "Akademische Weltweisheit."

8. Zedler, *Grosses vollständiges Universal-Lexicon*, 57:1499.

9. Ibid., 57:1502, 1517.

10. Walch, *Philosophisches Lexikon*, 1169. The text was first published in 1726.

11. See also, for example, Brucker, *Ehrentempel der deutschen Gelehrsamkeit.*

12. Jöcher, *Compendiöses Gelehrten-Lexicon*, 1:1777.

13. Jöcher, *Compendiöses Gelehrten-Lexicon.*

14. For an extensive bibliography, see Kirchner, *Bibliographie der Zeitschriften des deutschen Sprachgebietes bis 1900.*

15. See Habel, *Gelehrte Journale und Zeitungen der Aufklärung*, 47n.

16. "An den Leser," *Nöthigen Beytrags zu den Neuen Zeitungen von gelehrten Sachen*, vol. 5 (1740), quoted in Habel, *Gelehrte Journale und Zeitungen der Aufklärung*, 152.

17. Ibid.

18. Habel, *Gelehrte Journale und Zeitungen der Aufklärung*, 74–75, 50–51. See also Dann, "Journal des Şcavans zur wissenschaftlichen Zeitung."

19. *History of the Works of the Learned* (1702), quoted in McKitterick, "Bibliography, Bibliophily and Organization of Knowledge," 202.

20. *Göttingische Zeitungen von gelehrten Sachen I* (1739), Vorrede.

21. "Vorrede zu der vorigen Auflage," in Jöcher, *Compendiöses Gelehrten-Lexicon*, 1:4.

22. Jöcher, *Compendiöses Gelehrten-Lexicon*, vol. 1.

23. Jöcher, "Vorrede," *Allgemeines Gelehrten-Lexicon*, vol. 1.

24. Jöcher, *Compendiöses Gelehrten-Lexicon*, vol. 1.

25. Kappens, "Vorrede," X.

26. See Blum, "Bibliographia," 1067–69.

27. Gierl, "Bestandsaufnahme im gelerhten Bereich." The following draws on Gierl's argument.

28. Bacon, *De Augmentis*, 300.

29. See Grunert and Vollhardt, "Einleitung," in *Historia Literaria*, vii.

30. Nelles, "Historia litteraria and Morhof," 42.

31. Bacon, *De Augmentis*, 301.

32. The full title reads, "*Polyhistor sive de notitia auctorum et rerum comentarii. Quibus praeterea varia ad omnes disciplinae consilia et subsidia proponutur.*" Only the first two books were published in his lifetime. The last two volumes were published in 1704. For a broader account of Morhof's work, see Waquet, *Mapping the World of Learning*, 91.

33. Morhof, *Polyhistor*, quoted in Blair, "Practices of Erudition according to Morhof," 61.

34. Zedler, *Grosses vollständiges Universal-Lexicon*, 10:378.

35. Reimmann, *Versuch einer Einleitung in die Historiam Literariam*; Gottlieb Stolle describes his *Historia literaria* in the same terms; see Stolle, *Anleitung zur Historie der Gelahrtheit*.

36. Reimmann, *Versuch einer Einleitung in die Historiam Literariam*, "Vorrede."

37. Ibid.

38. See, for example, Stolle, *Anleitung zur Historie der Gelahrtheit*, 1.

39. Kappens, "Vorrede."

40. Gundling, *Vollständige Historie der Gelahrtheit*, 1:5.

41. Bernhard, *Kurtzgefasste Curieuse Historie derer Gelehrten*.

42. Ibid., 859.

43. Ibid., 867.

44. Fabricius, "Vorrede," *Abriß einer allgemeinen Historie der Gelehrsamkeit*, 1:4.

45. Ibid., vol. 1.

46. Here and in what follows I am extending Martin Gierl's argument in "Historia literaria," in *Historia Literaria*, 113–27, 119, 117. See also Gierl, "Kompilation und die Produktion von Wissen im 18. Jahrhundert," 102.

47. Israel, *Enlightenment Contested*, 416.

48. See Thomasius, *Einleitung zu der Vernunft-Lehre*, 75.

49. Thomasius, *Monatsgespräche*, 28, 29.

50. Thomasius, *Monatsgespräche* II, quoted in Schmidt-Biggeman, *Topica universalis*, 273.

51. Christian Wolff, *Logica*, para. 889, quoted in Wolff, "Einführung," in *Vernünftige Gedanken* in *Gesammelte Werke*, 91. See also Wolff, *Vernünftige Gedanken* in *Gesammelte Werke*, 222.

52. See Wolff's rules for improvement of *historia literaria* in para. 784 and 785 of *Philosophia Rationalis*.

53. See Heumann, *Conspectus Reipublicae literariae*, 3.

54. Zedler, *Grosses vollständiges Universal-Lexicon*, 28:673.

55. Vierhaus, "Einleitung," 14.

56. In more recent German scholarship, the period from roughly 1750 to 1780 has been referred to as the *Hochaufklärung*, in contrast to the *Frühaufklärung*, associated with Thomasius, and the *Spätaufklärung*, generally associated with Kant. See Schneiders, "Der Philosophiebegriff des philosophischen Zeitalters," 60.

57. Nipperdey, "Verein als soziale Struktur in Deutschland," quoted in Zammito, *Kant, Herder, and the Birth of Anthropology*, 22. On this shift, see Vierhaus, *Wissenschaft im Zeitalter der Aufklärung*.

58. Ernesti, *De Philosophia Populari*, quoted in Zammito, *Kant, Herder, and the Birth of Anthropology*, 36.

59. Turner, "University Reformers," 501.

60. Zammito, *Kant, Herder and the Birth of Anthropology*, 23.

61. See van der Zande, "In the Image of Cicero"; Schneiders, "Der Philosophiebe-

griff des philosophischen Zeitalters." By the end of the century, *Popularphilosophie* was used as a polemical term, oftentimes in contrast to a critical or actual philosophy.

62. Garve, *Von der Popularität des Vortrages* [1793, 1796], in *Gesammelte Werke*, 1:4:1:348.

63. Walch, *Philosophisches Lexikon*, 1162.

64. Fabricius, *Abriß einer allgemeinen Historie der Gelehrsamkeit*, 1:12. On this passage, see also Grimm, *Letternkultur*, 159.

65. Fabricius, *Abriß einer allgemeinen Historie der Gelehrsamkeit*, 1:4.

66. Ibid., 1:10.

67. See Kocka, "Bildungsbürgertum," 22.

68. Iselin, *Ueber die Geschichte der Menschheit*, 1:319–20.

69. Ibid.

70. Schulte-Sassen, "Das Konzept bürgerlich-literarischer Öffentlichkeit," 101. On the concept of the "public sphere," see Habermas, *Structural Transformation of the Public Sphere*, esp. 51–79. For accounts of Habermas, see Calhoun, *Habermas and the Public Sphere*.

71. Turner, "Historicism, Kritik, and the Prussian Professoriate," 483.

72. Kant, *Kritik der reinen Vernunft*, BII, BV.

73. Turner, "*Bildungsbürgertum* and the Learned Professions in Prussia," 111.

74. Fabricius, *Abriß einer allgemeinen Historie der Gelehrsamkeit*, 14.

75. Ibid., 8.

76. On the topic of the German *Bildungsbürgertum*, see Kocka, *Das Bildungs bürgertum im 19. Jahrhundert*.

77. "Einige Bemerkungen, welche sich über den deutschen Meßkatalogus machen lassen," in *Deutsches Museum*, ed. Christian Heinrich Boie (Leipzing, 1780), 2:182. Reprinted in Rietzschel, *Gelehrsamkeit ein Handwerk?*

78. Jentzsch, *Der deutsch-lateinische Büchermarkt*, 147.

79. According to Jentzsch, the percentage of books listed in the book fair catalogues which he categorized under "*Allgemeine Gelehrsamkeit*" had declined from 5.298% in 1740 to 1.44% in 1800. See ibid., 186.

80. Ibid., 37. See also Beaujean, *Der Trivialroman in der zweiten Hälfte des 18. Jahrhunderts*, 183.

81. Jentzsch, *Der deutsch-lateinische Büchermarkt*, 96–98, 186.

82. Kirchner, *Die Grundlagen des deutschen Zeitschriftenwesens*.

83. For on overview of this category and the related literature, see Schenda, *Volk ohne Buch*, 22–26.

84. Martens, *Die Botschaft der Tugend*, 16–17.

85. Nicolai, *Das Leben und die Meinungen des Herrn Magister Sebaldus Nothanker*, 121.

86. On these changes in reading habits in the second half of the eighteenth century, see Engelsing, *Analphabetentum und Lektüre*, 56–68.

87. On these new social forms, see Wittmann, *Geschichte des deutschen Buchhandels*, 189–200. See also Dann, *Lesegesellschaften und bürgerliche Emanzipation*.

88. The subtitle of Bodmer's *Anklagung des verderbten Geschmack* reads "*Oder Critische Anmerkungen Ueber Hamburgischen* PATROTEN *und die Hallischen* TADLERINNEN"; Schenda, *Volk ohne Buch*, 59.

89. Frömmichen, *Einige Bemerkungen* (1780), quoted in Schenda, *Volk ohne Buch*, 59.

90. "Von der Journalsucht in Deutschland," *Haunauisches Magazin* (1783), 432–36, quoted in Raabe, "Die Zeitschrift als Medium der Aufklärung."

91. "Einige Bemerkungen, welche sich über den deutschen Meßkatalogus machen lassen," in *Deutsches Museum*, ed. Christian Heinrich Boie (Leipzig, 1780), quoted in Rietzschel, *Gelehrsamkeit ein Handwerk?*, 32–33.

92. My discussion of particular journals draws on Kirchner, *Das deutsche Zeitschriftenwesen, seine Geschichte und Seine Probleme*, 1:72–112.

93. "Vorrede," in *Göttingische Zeitungen von Gelehrten Sachen*, 1:2.

94. "Vorrede," in *Göttingische Anzeigen von Gelehrten Sachen*, 4.

95. For a more detailed description of these journals, see Lindemann, *Deutsche Presse bis 1815*, esp. 202–32. See also Fabian, "Im Mittelpunkt der Bücherwelt," 261.

96. Here are a few of the exemplary titles: *Chymischen Experimente* (1753), *Beyträge zur Natur- oder Insectengeschichte* (1761), *Der Naturforscher* (1774), and *Natur- und Kunstkabinet, oder Sammlung von physikalischen Aufsätzen* (1753).

97. "Vorrede," in *Physikalische Belustigungen* (1751), 1:6.

98. *Physikalische Belustigungen* (1751), 1:8.

99. Fabian, "Im Mittelpunkt der Bücherwelt," 255.

100. "Vorrede," *Berlinische Monatsschrift*, vol. 1 (January–June 1783).

101. "Ephemeriden," in *Deutsche Encyclopädie oder Real-Wörterbuch aller Künste und Wissenschaften*, vol. 8 (1783), quoted in Habel, *Gelehrte Journale und Zeitungen der Aufklärung*, 14.

102. *Allgemeine Deutsche Bibliothek*, 67, no. 1 (1786): 260–61. See Habel, *Gelehrte Journale und Zeitungen der Aufklärung*, 326.

103. "Vorrede," *Das Gelehrte Deutschland*, 4th ed. (1783), ix.

104. "Vorrede," *Das Gelehrte Teutschland*, vol. 1, 3rd ed. (1776), v–vi.

105. "Vorschlag und Aufforderung," 73.

106. Ibid.

107. It was actually one of four distinct publications overseen and published by the *Allgemeine Literatur Zeitung*. The *Repertorium* was published in three volumes. The first two volumes, published in Jena in 1793, contained all the *Fächer*, and the third volume contained an alphabetical index according to article title and author. The other two related publications were the *Intelligenzblätter* and the *Revisionsbände*.

108. Ersch, "Vorrede," in *Allgemeines Repertorium*, i.

109. This encyclopedic category was taken from a review of Johann Eschenburg's *Lehrbuch der Wissenschaftskunde* in the *ALZ* in 1798; see *ALZ* 125 (1798): 164.

110. *Allgemeines Repertorium der Literatur für die Jahre 1791–1795*, vol. 3, iv–vi, quoted in Habel, *Gelehrte Journale und Zeitungen der Aufklärung*, 329.

111. Beutler and Gutsmuth, *Allgemeines Sachregister ueber die wichstigsten deutschen Zeit- und Wochenschriften*, 1:iv.

112. Ibid.

113. Ibid., vii.

114. Ibid., 26.

115. Ibid., xv.

116. Humboldt, "Über die innere und äussere Organisation," in *Wilhelm von Humboldts Gesammelte Schriften*, 10:253.

*Chapter Three. Encyclopedia from Book to Practice*

1. Schlegel, *Vorlesungen über das akademische Studium*, 38.

2. On this broader history, see Henningsen, "Enzyklopädie."

3. Yeo, *Encyclopedic Visions*, 2.

4. Ibid., 7.

5. Dierse, *Encyklopädie*, 9.

6. Yeo, *Encyclopedic Visions*, 3.

7. Dierse, *Encyklopädie*, 19.

8. Alsted, quoted in Schmidt-Biggemann, *Topica universalis*.

9. Thomas Blunt, *Glossographia*, quoted in Yeo, *Encyclopedic Visions*, 7.

10. For a broader discussion of Zedler's lexicon, see U. J. Schneider, "Die Konstruktion des allgemeinen Wissens in Zedlers Universal-Lexicon."

11. "Preface," in Zedler, *Grosses vollständiges Universal-Lexicon*, 1:2.

12. Carels and Flory, "Johann H. Zedler's Universal Lexicon," 165–96.

13. "Introduction," Siskin and Warner, *This Is Enlightenment*.

14. "Preface," in Zedler, *Grosses vollständiges Universal-Lexicon*, 1:6.

15. Ibid., 15.

16. Yeo, *Encyclopedic Visions*, 225.

17. Diderot, "Encyclopedia," 291.

18. D'Alembert, *Preliminary Discourse to the Encyclopedia of Diderot*, 4.

19. Diderot, "Encyclopedia," 291.

20. Voltaire, "Men of Letters."

21. D'Alembert, *Preliminary Discourse to the Encyclopedia of Diderot*, 47–48.

22. Diderot, "Encyclopedia," 308.

23. Ibid., 307.

24. Bates, "Cartographic Aberrations," 10.

25. Cassirer, *Philosophy of the Enlightenment*, 14.

26. Ibid., 23.

27. Diderot, "Encyclopedia," 306–7.

28. Ibid., 307.

29. Ibid., 314.

30. On the particular strategies of this, see Anderson, "Encyclopedic Topographies."

31. Quoted in Rosenberg, "18th-Century Time Machine."

32. Yeo, *Encyclopedic Visions*, 30.

33. *Kurzer Begriff aller Wissenschaften und anderer Theilen der Gelehrsamkeit* appeared in six editions from 1745 to 1786. In what follows, I quote from the second edition; Sulzer, *Kurzer Begriff aller Wissenschaften.* See Maatsch, "Jenaer Vorlesungen zur Enzyklopädie und Wisschaftskunde."

34. Sulzer, *Kurzer Begriff aller Wissenschaften*, 5.

35. Ibid., 6.

36. Ibid., 7.

37. Ibid., 8.

38. Diderot, "Encyclopedia," 291.

39. See Maatsch, *Naturgeschichte der Philosopheme*, 17–33.

40. See, for example, the discussion of the two types of encyclopedias in Ersch, *Allgemeines Repetorium der Literatur für die Jahre 1791–1795*, v.

41. *Allgemeine Literatur-Zeitung*, May 22, 1790, 409.

42. Martini, *Allgemeine Geschichte der Natur in alphabetischer Ordnung*, 1:vii.

43. On this same point, see Maatsch, *Naturgeschichte der Philosopheme*, 31–32.

44. "Vorrede," in Köster, Gottfried, and Roos, *Deutsche Enzyklopädie*, vol. 1.

45. Köster, Gottfried, and Roos, *Deutsche Enzyklopädie*, 8:374.

46. Ersch and Gruber, *Allgemeine Encyclopädie*, viii.

47. Quoted in Maatsch, *Naturgeschichte der Philosopheme*, 30.

48. C. H. Schmid, "Ueber die Klassifikation."

49. Ibid., 237.

50. See Behler, "Friedrich Schegels Enzyklopädie der literarischen Wissenschaften."

51. Herder, *Briefe zur Beförderung der Humanität*, 2:92–93.

52. Lichtenberg, quoted in Puschner, "Mobil gemachte Feldbibliotheken," 63.

53. Goetschel, Macleod, and Snyder, "*Deutsche Encyclopädie*," 55.

54. Diderot, "Encyclopedia," 291.

55. *Lehrbuch der Wissenschaftskunde: Ein Grundriß encyklopädischer Vorlesungen* was published in three editions between 1792 and 1813. I cite the second edition published in Berlin by Friedrich Nicolai in 1800.

56. Ibid., 9.

57. Ibid., v and vi.

58. Ibid., vi.

59. *ALZ* 125 (April 1798): 164.

60. Burdach, *Der Organismus menschlicher Wissenschaft und Kunst*, 8.

61. Eschenburg, *Lehrbuch der Wissenschaftskunde*, 4, 5, 9.

62. Ibid., 9, 16, 7.

63. *Revision der Literatur in Ergänzungsblättern* 1, no. 1 (1801): 1. This claim was echoed in a review of Karl Heydenreich's *Encyclopädische Einleitung in das Studium der Philosophie* (1793); Papst, "Vollständigkeit und Totalität."

64. Kant, *Kritik der reinen Vernunft*, B860.

65. See Dierse, *Encyklopädie*, 103–24.

66. Krug, *Über den Zusammenhang der Wissenschaften*, 17.

67. Krug, *Versuch einer systematischen Enzyklopädie der Wissenschaften*, 1:9.

68. Ibid., 12.

69. See Maatsch, *Naturgeschichte der Philosopheme*, 38; Dierse, *Encyklopädie*, 116.

70. Maatsch, *Naturgeschichte der Philosopheme*, 39.

71. Krug, *Versuch einer neuen Eintheilung der Wissenschaften*.

72. Krug, quoted in Maatsch, *Naturgeschichte der Philosopheme*, 38.

73. Maatsch, *Naturgeschichte der Philosopheme*, 43.

74. Kant, *Kritik der reinen Vernunft*, B502.

75. Maatsch, "Jenaer Vorlesungen zur Enzyklopädie und Wisschaftskunde," 132. The following draws directly on Maatsch's historical reconstruction of these lectures. On the unique role that the University of Jena had in the development of late eighteenth century German intellectual life, see Müller, Ries, and Ziche, *Die Universität Jena*.

76. Dierse, *Encyklopädie*, 73–74.

77. Pütter, *Neuer Versuch einer Juristischen Encyclopädie*, 74.

78. Ibid., 2, 3.

79. Feder, *Grundriß der Philosophischen Wissenschaften*. On these courses see Maatsch, "Jenaer Vorlesungen zur Enzyklopädie und Wisschaftskunde," 126–29.

80. Maatsch, "Jenaer Vorlesungen zur Enzyklopädie und Wisschaftskunde," 129.

81. Quoted in ibid., 130.

82. Heusinger, *Versuch einer Encyklopädie der Philosophie*. The text was published in Weimar in 1796, the same year that he delivered his Jena lectures, in two volumes. I cite this version.

83. Ibid., vi, vii.

84. Ibid., xxiii, xxv, xxvi, xxxi.

85. Ibid., xlii, 13, xlii, xli, 5.

86. *Critik der reinen Vernunft im Grundrisse zu Vorlesungen nebst einem Wörterbuche zum leichtern Gebrauch der Kantischen Schriften* (Jena, 1786).

87. C. C. E. Schmid, *Allgemeine Encyklopädie und Methodologie der Wissenschaften*, 7, iii.

88. Ibid., 47, 26, 50, 45.

89. Krug, *Versuch einer neuen Eintheilung der Wissenschaften*, v.

90. Ibid., 6, 47, 54.

91. Schlegel, *Vorlesungen über schöne Literatur*, 484, 534, 535, 486.

92. Ibid., 481, 484.

93. Schlegel, *Vorlesungen über Encyklopädie*, 8. In his *Vorlesungen über schöne Literatur*, Schlegel offered a second solution: the concept of *Literatur* as a particular kind of writing that had been filtered and sorted from among the surfeit of all that had been printed. What was needed to remedy the sorry state of German literature and thought more generally, claimed Schlegel, was a normative, critical history that would separate the good books from the bad ones and help readers make their way through the surfeit of print.

94. Schlegel, *Vorlesungen über Encyklopädie*, 8.

95. Ibid., 3, 17.

96. Ibid., 8.

97. Ibid., 4, 10.

98. Ibid., 10.

99. Ibid., 122.

100. Friedrich Ast, *Grundriß einer Geschichte der Philosophie* (Landshut, 1807), 12, quoted in Maatsch, "Jenaer Vorlesungen zur Enzyklopädie und Wisschaftskunde," 134.

101. Schlegel, *Vorlesungen über das akademische Studium*, 38.

102. Ibid., 40.

103. Ibid.

*Chapter Four. From Bibliography to Ethics*

1. Schlegel, *Vorlesungen über schöne Literatur*, 531, 523.

2. Stichweh, *Zur Entstehung des modernen Systems wissenschaftlicher Disziplinen in Deutschlands*, esp. 40–63.

3. Lichtenberg, *Gedanken, Satiren, Fragmenten*, 92.

4. For Luhmann's work on the university, see Luhmann, *Universität als Milieu*; see also Luhmann, *Soziologische Aufklärung*; Stichweh, *Der frühmoderne Staat und die europäische Universität*; Parsons, *American University*.

5. Hohendahl, "Humboldt Revisited," 183.

6. Habermas, "Die Idee der Universität—Lernprozesse," 91. The rest of this paragraph follows Habermas's argument very closely.

7. This is the core project of Clark's *Academic Charisma* even if, as the title intimates, key elements of the modern university cannot be accounted for by a strictly administrative account.

8. See Habermas and Luhmann, *Theorie der Gesellschaft oder Sozialtechnologie*, esp. 151.

9. Schlegel, *Vorlesungen über schöne Literatur*, 484.

10. Heinzmann, *Appel an meine Nation*, 125.

11. Ibid., 413.

12. Hoche, *Vertraute Briefe über die jetzige Lesesucht*, 68. For a further survey of such texts and a discussion of *Lesesucht*, see Schenda, *Volk ohne Buch*, 57–66.

13. On similarly morally charged criticisms of print in the sixteenth century, see Giesecke, *Der Buchdruck in der frühen Neuzeit*, 169–91. Giesecke argues that these criticisms were, in most cases, a "qualification of praise, as a warning addendum to an otherwise positive valuation of the new technology."

14. The book flood metaphor has a longer history as well. See Werle, "Die Bücherflut in der frühen Neuzeit."

15. See Jentzsch, *Der deutsch-lateinische Büchermarkt*. On the limited but useful picture of the German book market afforded by these catalogues, see Wittmann, *Geschichte des deutschen Buchhandels*, 111. For a broader picture of book production, see Schenda, *Volk ohne Buch*, 173–203.

16. Wittmann, *Geschichte des deutschen Buchhandels*, 112.

17. Ibid., 142.

18. On the necessity of considering these discursive traces in book history, see Lupton, *Knowing Books*.

19. Schulte-Sassen, "Das Konzept bürgerlich-literarischer Öffentlichkeit," 103–5.

20. See Bürger, "Literarischer Markt und Öffentlichkeit am Ausgang des 18. Jahrhunderts in Deutschland."

21. Ibid., 78.

22. I cite from the third edition; Nicolai, *Das Leben und die Meinungen des Herrn Magister Sebaldus Nothanker*, 1:90.

23. Ibid.

24. Herder, *Briefe zur Beförderung der Humanität*, 2:92–93.

25. One is reminded here of Alexander Pope's *Dunciad*. See Paula McDowell's recent unpublished work on oral and print cultures presented at Indiana University Bloomington's Eighteenth-Century Studies Workshop.

26. Herder, *Briefe zur Beförderung der Humanität*, 2:195.

27. Lichtenberg, *Gedanken, Satiren, Fragmenten*, 1:78.

28. Quoted in Behler, "Friedrich Schegels Enzyklopädie der literarischen Wissenschaften," 246.

29. For the original thesis, see Engelsing, *Der Bürger als Leser*.

30. Wittmann, *Geschichte des deutschen Buchhandels*, 195; Schenda, *Volk ohne Buch*, 88.

31. See Valenza, *Literature, Language, and the Rise of the Intellectual Disciplines*, 1. Kant, "Was Ist Aufklärung?," 8:34.

32. Novalis, *Novalis Schriften*, 2:661; Wellmon, "Touching Books."

33. Ibid.

34. See Kim, *Als die Lumpen Flügel bekamen*, 42.
35. Novalis, *Novalis Schriften*, 2:662.
36. Ibid.
37. Ibid., 2:663.
38. See Knodt, *Negative Philosophie und dialogische Kritik*.
39. Bergk, *Die Kunst Bücher zu lesen*.

*Chapter Five. Kant's Critical Technology*

1. Kant, *Kritik der reinen Vernunft*, Aiii, Biii.
2. Kant, *Logic*, 9:44.
3. Kant, *Philosophie Encyclopädie*, 29:30.
4. Ibid.
5. Kant, "Was Ist Aufklärung?," 8:35.
6. Ibid.
7. On the more particular relationship of philosophy to literature, see Nancy, *Discourse of the Syncope*.
8. Kant, *Logic*, 9:47.
9. Schaper, "Taste, Sublimity, and Genius," 378.
10. The following section draws directly on Markus, "Changing Images of Science," 27.
11. Kant, *Kritik der Urteilskraft*, 5:238.
12. Ibid. Here, I use the translation in *Critique of the Power of Judgment*, trans. Paul Guyer and Eric Matthews, 122.
13. These discussions exceed our present concerns, but we should note that the transcendental possibility of this universal communicability is grounded in the "necessarily propositional structure" of everything that counts as real knowledge; Markus, "Changing Images of Science," 27. For a detailed discussion of these issues, see Guyer, *Kant and the Claims of Taste*, 252–73. Guyer, however, discusses the communicability of rational thought only in relation to aesthetics.
14. Markus, "Changing Images of Science," 27.
15. Kant, *Anthropologie in pragmatischer Hinsicht*, 7:219. For a similar discussion in the *Critique of Pure Reason*, see the distinction that Kant draws between knowing and believing and conviction and persuasion:

Persuasion is mere illusion, because the ground of the judgment, which lies solely in the subject, is regarded as objective. Such a judgment has only private validity, and the holding of it to be true does not allow of it being communicated. But truth depends upon agreement with the object, and in respect of it the judgments of each and every understanding must therefore be in agreement with each other. . . . The touchstone whereby we decide whether our holding a thing to be true is conviction or mere persuasion is therefore external, namely, the possibility of communicating it and finding it valid for all human reason. (B848)

16. Kant, *Anthropologie in pragmatischer Hinsicht*, 7:219. I should note that Kant writes here of the "subjective" and not the "objective" necessity of such a testing. We are, in Kant's terms, in the realm of the empirical, and the force is not categorical.

17. Neiman, *Unity of Reason*, 98.

18. Kant, *Anthropologie in pragmatischer Hinsicht*, 7:219.

19. Markus, "Changing Images of Science," 28.

20. Kant, "Von der Unrechtmäßigkeit des Büchernachdrucks," 8:83.

21. Jordheim, "Present of the Enlightenment."

22. See Cassirer, *Kants Leben und Lehre*, 150–52.

23. *Gothaische gelehrte Zeitungen. Gotha. Neun und fünfzigstes Stück, den fünf und zwangzigsten Julius*, 1781. S. 488; *Gothaische gelehrte Zeitungen. Acht und sechszigstes Stück, den vier und zwanzigsten*, August 1782. S. 560–63; *Altonaischer Gelehrter Mercurius*. 33. *Stück Greifswald den* 30. August 1783. S. 257–58. All of these reviews have been reprinted in Landau, *Rezensionen zur Kantischen Philosophie 1781–87*.

24. Garve to Kant, July 13, 1783, in Kant, *Prolegomena zu einer künftigen Metaphysik*, 208.

25. Ibid.

26. Petrus, "Beschrieene Dunkelheit und Seichtigkeit."

27. See Beiser, *Fate of Reason*, 165–66.

28. Garve, *Popularität des Vortrages*, in *Gesammelte Werke*, 1:4:1:353.

29. Beiser, *Fate of Reason*, 138.

30. Garve, *Von der Popularität*, in *Gesammelte Werke*, 1:4:1:348.

31. Garve, "*Die Garve-Rezension*," in Kant, *Prolegomena zu einer künftigen Metaphysik*, 219.

32. The "thread of Ariadne" is a term, based on the myth of Ariadne, used to describe a method of solving a problem that proceeds step by step.

33. Kant, *Prolegomena zu einer künftigen Metaphysik*, 220.

34. Garve, *Von der Popularität*, 346.

35. Kant, *Kritik der reinen Vernunft*, Bxliv.

36. Kant, Letter to Moses Mendelssohn, April 8, 1776, 10:70.

37. Kant, *Kritik der reinen Vernunft*, Bxv.

38. Kant, Letter to Moses Mendelssohn, April 8, 1776, 10:56–57.

39. Zammito, "Rousseau-Kant-Herder and the Problem of *Aufklärung* in the 1760s," 212.

40. See Kant, *Kritik der reinen Vernunft*, Bxxxiv.

41. Reflections on the *Kritik der reinen Vernunft*, quoted in Cassirer, *Kants Leben und Lehre*, 150.

42. Ibid., no. 9.

43. Kant, *Reflexionen zur Moralphilosophie*, 19:247.

44. Kant, *Kritik der reinen Vernunft*, Bxxxii.

45. Kant, *Logic*, 9:43.

46. D'Alembert, *Preliminary Discourse to the Encyclopedia of Diderot*, 3.

47. Kant based his lectures on Johann G. H. Feder's *Grundriß der Philosophischen Wissenschaften nebst der nöthigen Geschichte zum Gebrauche seiner Zuhörer* (1767). On the differences and similarities between Feder's text and Kant's lectures, see Stern, *Über die Beziehungen Christian Garve's zu Kant*.

48. Kant, *Philosophische Encyclopädie*, 29:5.

49. Kant, 16:189, from Kant's own marginalia to G. F. Meier's *Auszug aus der Vernunftlehre* (1752 and 1760).

50. Kant, *Philosophische Encyclopädie*, 29:5, and editor's notes in 29:672.

51. Only once in the *Critique of Judgment* did Kant speak of the encyclopedic, when he wrote that "every science must have its determinant position in the encyclopedia of the sciences" (5:416).

52. In *Opus postum*, Kant contrasted encyclopedia as a collection of empirical perceptions [*Wahrnehmungen*] to a system of pure reason. See Kant, 21:96 and 21:106, 109.

53. Kant, *Physische Geographie*, 9:158.

54. Kant, *Kritik der reinen Vernunft*, B502.

55. Ibid., A832/B860.

56. Kant, *Philosophische Encyclopädie*, 29:30. Kant made a similar distinction throughout his oeuvre. See, for example, *Logic*, 9:72. The causality of such systems is most fully articulated in the "Critique of Teleological Judgment" in the *Critique of Judgment*. See, for example, the distinction between a *nexus effectivus* and a *nexus finalis* in para. 65, 5:372–73.

57. Kant, *Kritik der reinen Vernunft*, B502.

58. Ibid., A832/B860.

59. Kant, *Logic*, 9:148.

60. Kant, *Kritik der reinen Vernunft*, A833/B861.

61. Eco, *Semiotics*, 83–84.

62. Kant, *Kritik der reinen Vernunft*, Bxxiv.

63. Ibid., Bxxxiv.

64. Ibid., Bxxxviii.

65. Ibid., Bxxxviii.

66. Kant, *Streit der Fakultäten*, 7:113. In the *Prolegomena*, Kant wrote that he had "the entire system in [his] head"; Kant, *Prolegomena zu einer künftigen Metaphysik*, 213.

67. Kant, *Streit der Fakultäten*, 7:113. See also 7:280 for a similar description.

68. Ibid., 7:113.

69. Kant, *Prolegomena zu einer künftigen Metaphysik*, 220.

70. Kant, *Kritik der reinen Vernunft*, A839/B867.

71. Kant, *Streit der Fakultäten*, 7:69.

72. Kant, *Kritik der reinen Vernunft*, A839/B867. The practical imperative of philosophy can be seen throughout Kant's writings. See, for example, Kant, 16:872.

73. For more on the fundamentally practical character of Kant's critical philosophy, see Velkley, *Freedom and the End of Reason*.

74. Kant, *Die Metaphysik der Sitten*, 6:252.

75. Kant, *Kritik der reinen Vernunft*, Bvvxvi.

76. Kant, *Logik Dohna-Wundlacken*, 24:704.

77. Ibid.

78. Kant, *Kritik der reinen Vernunft*, Bxliii.

79. Kant, Letter to Garve, 10:339; Kant, *Kritik der reinen Vernunft*, Bxliv.

80. Kant, 5:488.

81. Kant, *Reflections on the Logic*, 16:862. For a discussion of this passage to which I am indebted, see H. L. Wilson, *Kant's Pragmatic Anthropology*, 109–22.

82. Kant, *Prolegomena zu einer künftigen Metaphysik*, 4:262.

83. Ibid., 214.

84. Kant, *Die Metaphysik der Sitten*, 6:206.

85. Kant, *Kritik der reinen Vernunft*, Bxliv.

86. Kant, *Philosophische Encyclopädie*, 29:30.

87. Hunter, *Rival Enlightenments*, 280.

88. Kant, *Kritik der reinen Vernunft*, Bxxxiv.

89. Ibid., A710/B738.

90. Ibid., A711/B739.

91. Ibid., A836/B864.

92. See also ibid., A751/B779. On this process of reflective reenactment, see Meld-Shell, *Kant and the Limits of Autonomy*, 225.

93. Kant, *Kritik der reinen Vernunft*, A837/B865.

94. Kant, *Logic*, 9:13, 8.

95. Ibid., 9:47.

96. Eberhard, "*Prüfung der Frage*," 7.

97. Ibid., 7.

98. Kant, *Die Metaphysik der Sitten*, 6:223.

99. Kant, "Von der Unrechtmäßigkeit des Büchernachdrucks," 8:83. See "*Geschäft*" in Adelung, *Grammatisch-kritisches Wörterbuch*, 601–3.

100. Kant, *Über die Buchmacherei*, 8:431.

101. "It is so easy to be immature if I have a book that has understanding for me [*das für mich Verstand hat*]"; Kant, "Was Ist Aufklärung?," 8:35.

102. Kant, *Philosophische Encyclopädie*, 29:30.

103. Blair, *Too Much to Know*, 253.

104. Kant, *Streit der Fakultäten*, 7:28.

105. Ibid., 7:22.

106. Ibid.

107. Ibid., 7:19.

108. Ibid., 29.

109. Ibid., 29.

110. Ibid., 18; slightly altered translation from Kant, *Conflict of the Faculties*, 247–48.

111. Ibid., 7:34.

112. Compare Kant's distinctions to those made by Wittmann to describe the three different reading publics in eighteenth-century Germany: (1) the broad but uneducated readers of the *Lesebibliotheken*, (2) the educated of the higher classes, and (3) *die Gelehrten*, formed by professors, students, and reviewers. See Wittmann, *Geschichte des deutschen Buchhandels*.

113. Meld-Shell, *Kant and the Limits of Autonomy*, 251.

114. Kant, *Streit der Fakultäten*, 7:17.

115. Smith, *Wealth of Nations*, 18.

116. Kant, *Grundlegung zur Metaphysik der Sitten*, 4:388. Translation is taken from Kant, *Practical Philosophy*, 44.

117. Kant, *Streit der Fakultäten*, 7:17.

*Chapter Six. The Enlightenment University and Too Many Books*

1. Von Arnim, "Der Studenten erstes Lebehoch," 418–19.

2. Brentano, *Universitati Litterariae*.

3. Quoted in Steig, *Heinrich von Kleists Berliner Kämpfe*, 301.

4. See Ziolkowski, *German Romanticism and Its Institutions*.

5. Humboldt, "On the Internal and External Organization," 10:250.

6. Johann Samuel Ersch and Johann Gottfried Gruber, *Allgemeine Encyclopädie der Wissenschaften und Künste in alphabetischer Folge*, quoted in Maatsch, *Naturgeschichte der Philosopheme*, 29; Friedrich Schlegel, quoted in Polheim, "Studien zu Friedrich Schlegels poetischen Begriffen," 379.

7. Schleiermacher, "Gelegentliche Gedanken über Universitäten in deutschem Sinn," 215.

8. McClellan, *Science Reorganzied*, xix–xx.

9. Schelsky, *Einsamkeit und Freiheit*, 15.

10. Nardi, "Relations with Authority," 92.

11. See, for example, Rüegg, "Themes," 3–34.

12. Howard, *Protestant Theology*, 50.

13. Verger, "Patterns," 41–45.

14. Nardi, "Relations with Authority," 102–3.

15. Duboulay quoted in Compayré, *Abelard and the Origin and Early History of Universities*, 107.

16. See ibid., 111.

17. These oaths were not abolished until 1785 in Austria and 1804 in Bavaria, for example. See Clark, *Academic Charisma*, 202.

18. Ibid., 47.

19. Rüegg, "Themes," 32–33.

20. Verger, "Patterns," 57.

21. See Hammerstein, "Epilogue," 626.

22. See Gregory, *Unintended Reformation*, 330.

23. McClelland, *State, Society, and University in Germany*, 217–31.

24. See Hofstetter, *Romantic Idea of a University*, 7.

25. McClelland, *State, Society, and University in Germany*, 104.

26. Clark, *Academic Charisma*; McClelland, *State, Society, and University in Germany*, 34–57.

27. McClelland, *State, Society, and University in Germany*, 59.

28. See Legaspi, *Death of Scripture*, 27–52; McClelland, *State, Society and University in Germany*, 35–57; Howard, *Protestant Theology*, 104–21.

29. Meiners, Über die Verfassung, 1:v.

30. Ibid., 1:1, 42.

31. Wakefield, *Disordered Police State*, esp. 49–80.

32. Walker, *German Home Towns*, 161.

33. Böll, *Das Wesen der Universität in Briefen*, 4. For two other discussions of this passage, see Wakefield, *Disordered Police State*, 49–50; and Clark, *Academic Charisma*, 379–80.

34. Quoted in Stichweh, *Die Ausdifferenzierung der Wissenschaft*, 45.

35. Wakefield, *Disordered Police State*, 48–50.

36. Meiners, *Ueber die Verfassung*, 2:31.

37. Clark, *Academic Charisma*, 12; Wakefield, *Disordered Police State*, 70; Justi, *Grundsätze der Polizeywissenschaft* (1756), quoted in Walker, *German Home Towns*, 162.

38. Justi, *Grundsätze der Policeywissenschaft*, 255–56.

39. Quoted in Wakefield, *Disordered Police State*, 70.

40. Ibid.

41. Meiners, *Über die Verfassung*, 44, 100, 342.

42. See Hammerstein, "Epilogue," 621–40.

43. Collins, *Sociology of Philosophies*, 1007n.

44. Hammerstein, "Die Universitätsgründungen im Zeichen der Aufklärung," 280.

45. Collins, *Sociology of Philosophies*, 643.

46. Wachler, *Aphorismen über die Universitäten*, 9; Ernst Gottfried Baldinger, *Über Universitäten und Unwesen*, quoted in König, *Vom Wesen der Universität*, 47.

47. "Bemerkungen über den Werth der Akademien," 7–32, 12. Here and throughout this chapter I translate "*Mittel*" as technology. Usually the term is translated as "medium" or "instrument," but I translate it in this context as "technology" to emphasize the fact that the authors are using it to refer not simply to a mere tool and extension of the rational mind but to something that has a formative power and a complex relationship to humans.

48. Ibid., 8, 12.

49. Turner, "University Reformers," 498. McClelland offers similar if slightly different numbers: between 8,000 and 9,000 from 1700 to 1755, below 7,000 from 1791 to 1795, below 6,000 from 1795 to 1800, and 4,900 from 1811 to 1815. He argues that even accounting for the disruptions of the war, university enrollments showed a real decline and "seemed even worse because the trend ran counter to the movement in size of the general population." See McClelland, *State, Society, and University*, 63–64.

50. For more exact numbers, see Scheel, "*Die deutsche Universität.*"

51. Ziolkowski, *German Romanticism and Its Institutions*, 227. See also p. 226nn23–24.

52. Turner, "University Reformers," 499.

53. McClellan, *Science Reorganized*, xix.

54. See ibid. McClellan describes the eighteenth century as the "hey-day of the general scientific society" and argues that by the nineteenth century the learned societies "ceased to be the premiere institutions for the organization and pursuit of science" (xix).

55. For a background discussion of the society, see Birtsch, "Die Berliner Mittwochsgesellschaft."

56. In what follows, I draw on these texts as collected, edited, and excerpted in Stölzel, "Die Berliner Mittwochsgesellschaft."

57. Ibid., 204. Gebhard's lecture was actually lost. These quotations are from an extended excerpt of Gebhard's text in Teller's response, which Stölzel found and published.

58. Meiners, *Über die Verfassung*, 1:4.

59. Campe, *Allgemeine Revision des gesammten Schul- und Erziehungswesen*, 16:164.

60. Ibid., 218.

61. Stölzel, "Die Berliner Mittwochsgesellschaft," 204.

62. Ibid.

63. Gregory, *Unintended Reformation*, esp. chap. 6.

64. Stölzel, "Die Berliner Mittwochsgesellschaft," 204.

65. Ibid., 205.

66. Ibid., 207. Public lectures were generally held by *Ordinarien*, professors em-

ployed directly by the state, and required neither registration nor fees. Private lectures were often held in a professor's own lodging and tended to focus on newer material.

67. Stölzel, "Die Berliner Mittwochsgesellschaft," 212.

68. "Bemerkungen über den Werth der Akademien," 7–32, 12.

69. Ibid., 12.

70. Ibid.

71. Ibid., 15.

72. Ibid.

73. Ibid.

74. They were delivered as *General Academic Methodology* in 1802 and as *University Methodology* in 1803; first published in 1803. I cite Schelling, *F. W. J. Schelling, Werke.*

75. Especially in *Darstellung meines Systems der Philosophie.*

76. Schelling, "System of Philosophy in General," 141.

77. Schelling, *F. W. J. Schelling, Werke*, 541.

78. Ibid., 541.

79. Ibid., 542.

80. Schelling, "System of Philosophy in General," 141.

81. Ibid.

82. Quoted in Bowie, *Schelling and Modern European Philosophy*, 60.

83. Schelling, "System of Philosophy in General," 143.

84. Bowie, *Schelling and Modern European Philosophy*, 60.

85. Ibid., 61.

86. See Schelling, *F. W. J. Schelling, Werke*, 543; Zöller, "*Die 'Bestimmung alles Wissens,'*" 65.

87. Schelling, *F. W. J. Schelling, Werke*, 542.

88. Ibid., 543.

89. Ibid., 558.

90. Ibid., 570.

91. Ibid., 551.

92. Ibid.

93. Ibid.

94. Ibid., 588.

95. Ibid., 554.

96. Ibid., 547.

97. Ibid., 555.

98. Ibid.

99. Ibid., 557. Helmut Zedelmeier refers to Schelling as a "radical analytical thinker of the historicity of all existence." See Zedelmeier, "Schelling's Vorlesung *Ueber das Studium der Historie und Jurisprudenz*," 208.

100. Schelling, *F. W. J. Schelling, Werke*, 557. Schelling alluded to the Wolffian-Kantian distinction between historical knowledge as "the mere science of fact" and rational knowledge (572).

101. Ibid., 557.

102. Ibid., 562.

103. Ibid., 564.

104. Ibid., 571.

105. Ibid., 547, 571.

106. Ibid., 547.

107. Ibid., 635.

108. Ibid., 571.

109. Ibid., 560.

110. See "Einführung," in Breidbach and Ziche, *Naturwissenschaften um 1800*, 19.

111. Schelling, *F. W. J. Schelling, Werke*, 566.

*Chapter Seven. The University in the Age of Print*

1. Salzmann, *Carl von Carlsberg*, 155.

2. Ibid., 341. Salzmann was, in part, giving voice to *Philanthropismus*, especially as articulated by Johann Bernhard Basedow (1729–1790) in *Vorstellung an Menschenfreunde und vermögende Männer* and *Elementarbuch*.

3. See Ziolkowski, *German Romanticism and Its Institutions*, 226. For further examples, see König, *Vom Wesen der deutschen Universität*, 22ff.

4. For more on these reforms, see Albrecht and Hirsch, *Das niedere Schulwesen im Übergang vom 18. zum 19. Jahrhundert*. For a detailed account of the reforms undertaken by the *Oberschulkollegium*, see Schwartz, *Die Gelehrtenschulen Preussens unter dem Oberschulkollegium*. There were also many non-Prussian reforms prompted by Napoleon. See, for example, Cobb, "Forgotten Reforms."

5. Although the *Abitur* was introduced in 1788, it was initially established only as a qualifying exam for the university and not as an exit exam for secondary schools.

6. See Koselleck, *Preußen zwischen Reform und Revolution*.

7. For a full discussion of his plans, see König, *Vom Wesen der deutschen Universität*, 63–78. The call for a complete reformation of the structure of the university was published in *Jahrbüchern der preussichen Monarchie* in 1798. For an overview of his reforms, see Heubaum, "Die Reformbestrebungen unter dem preussichen Minister J. v. Massow," 87ff. For a detailed account of the reforms undertaken by the *Oberschulkollegium*, see Schwartz, *Die Gelehrtenschulen Preussens unter dem Oberschulkollegium*.

8. *Jahrbücher der preussischen Monarchie*, 7:287, quoted in König, *Vom Wesen der deutschen Universität*, 287.

9. Massow, February 16, 1801, quoted in Köpke, *Die Gründung der Königlichen Friedrich-Wilhelms-Universität*, 14. Massow and the king's cabinet criticized the academies as well for not directing their work.

10. Massow, quoted in Köpke, *Die Gründung der Königlichen Friedrich-Wilhelms-Universität*, 12. See also Massow, *Ideen zur Verbesserung des öffentlichen Schul- und Erziehungsesens . . . Annalen des Preussischen Kirchenwesens* I (1800); and Lenz, *Geschichte der Königlichen Friedrich-Wilhelms-Universität*, 37.

11. Beyme, quoted in Lenz, *Geschichte der Königlichen Friedrich-Wilhelms-Universität*, 35. See also Kluge, *Die Universitäts-Selbstverwaltung*, 78–79.

12. Justus Loder, quoted in Lenz, *Geschichte der Königlichen Friedrich-Wilhelms-Universität*, 110.

13. See Engel's letter to Beyme, March 13, 1802, in Müller, *Gelegentliche Gedanken über Universitäten*, 5–6.

14. "Denkschrift über Begründung einer großen Lehranstalt in Berlin," March 13, 1802, in Müller, *Gelegentliche Gedanken über Universitäten*, 7.

15. Ibid.

16. Heckel, *Über die Natur und Heilart der Faulfieber*, 15. I have translated *Mittel* as "technology" here.

17. The manuscript was only published posthumously in 1817; Fichte, *J. G. Fichte Gesamtausgabe der Bayerischen Akademie der Wissenschaften*, II, 11:83. All citations to Fichte refer to part, volume, and page number.

18. Ibid..

19. Ibid., II, 11:84.

20. Ibid., II, 11:85.

21. Ibid., II, 11:84.

22. Ibid., II, 11:85. The German term Fichte used was *Bücherwesen.*

23. Ibid., II, 11:84.

24. "Editor's Preface," in Fichte, *Fichte's Early Philosophical Writings*, 138.

25. Fichte, quoted in ibid.

26. *Some Lectures Concerning the Scholar's Vocation*, in Fichte, *Fichte's Early Philosophical Writings*, 172.

27. Ibid.

28. *On the Essence of the Scholar*, in Fichte, *J. G. Fichte Gesamtausgabe der Bayerischen Akademie der Wissenschaften*, I, 8:133.

29. On proliferation of writing, see Siskin, *Work of Writing*.

30. *On the Essence of the Scholar*, in Fichte, *J. G. Fichte Gesamtausgabe der Bayerischen Akademie der Wissenschaften*, I, 8:135.

31. *Deduced Plan*, in ibid., II, 11:154.

32. *Plan zu einem periodischen schriftstellerischen Werke einer deutschen Universität*, in ibid., II, 9:350.

33. *Deduced Plan*, in ibid., II, 11:154–55.

34. *Plan zu einem periodischen schriftstellerischen Werke*, in ibid., II, 9:351.

35. Ibid.

36. For a much more thorough account of the lecture upon which I draw, see Franzel, *Fictions of Dialogue*.

37. Friesen, "Lecture as a Transmedial Pedagogical Form."

38. See Eisenstein, *Printing Revolution in Early Modern Europe*, 525.

39. Blair, "Student Manuscripts and the Textbook," 46.

40. Brockliss, "Curricula," 567, quoted in Franzel, *Fictions of Dialogue*, 39.

41. Friesen, "Lecture as a Transmedial Pedagogical Form."

42. The lecture was also a central practice beyond the university. It was, as Franzel describes it, a "public site" of knowledge.

43. Schleiermacher, "Gelegentliche Gedanken," 194.

44. For more on this "divine idea," see note 26.

45. *On the Essence of the Scholar*, in Fichte, *J. G. Fichte Gesamtausgabe der Bayerischen Akademie der Wissenschaften*, I, 8:125–26.

46. Ibid., I, 8:130. Considering reports from contemporaries like Novalis, F. Schlegel, and his numerous students (over four hundred per year in Jena), Fichte, if nothing else, did excite. As Novalis wrote after attending a lecture in Jena in 1796, "I owe Fichte my excitement. He awakened and stirred me." Even those who criticized his lecturing did so because of the strong impression that his lectures left. Reporting on an acquaintance's experience at a Fichte lecture, J. J. Wagner wrote, "*Er trug tiefe Spuren der Gewalt, die Fichtes Geist über ihn geübt hatte.*" Ehrlich, *Fichte als Redner*, 19, 12.

47. *On the Essence of the Scholar*, in Fichte, *J. G. Fichte Gesamtausgabe der Bayerischen Akademie der Wissenschaften*, I, 8:130.

48. Ibid., I, 8:131.

49. Ibid., I, 8:137.

50. Fichte, *Fichte's Early Philosophical Writings*, 140.

51. See Ehrlich, *Fichte als Redner*, 339–40.

52. Franzel, *Fictions of Dialogue*, 21, 91.

53. Ibid., 90.

54. *Deduced Plan*, in Fichte, *J. G. Fichte Gesamtausgabe der Bayerischen Akademie der Wissenschaften*, II, 11:135.

55. Ibid., II, 11:120–21.

56. Ibid., II, 11:89–90.

57. Unlike Fichte, Schleiermacher's contribution was published soon after he wrote it. Schleiermacher, "Gelegentliche Gedanken," 194.

58. Ibid., 195.

59. Ibid., 193, 178.

60. On early romanticism and foundationalism, see Frank, *Unendliche Annäherung*.

61. Schleiermacher, "Gelegentliche Gedanken," 194.

62. Ibid.

63. Wilhelm von Humboldt described the lecture in similar terms. The lecture "excites" students because it is performed before an audience that "think[s] along with the lecturer." Just as "stimulating as the lonely muse of the writer's life, or the less strictly organized life of the academy," the lecture performs "the movement of science." It "is obviously quicker and livelier at a university, where science and its problems are constantly tossed back and forth in a large group of powerful, spry, and youthful minds. Science cannot be truly presented as science without being comprehended anew every time. And thus the prospect of not happening quite regularly upon new discoveries is hard to imagine." See Humboldt, "On the Internal and External Organization," 10:257.

64. Schleiermacher quoted in Franzel, *Fictions of Dialogue*, 29.

65. Schelling, *F. W. J. Schelling, Werke*, 236, 284; Schleiermacher, "Gelegentliche Gedanken," 161–71; Humboldt, "On the Internal and External Organization," 10:257.

66. Schleiermacher, "Gelegentliche Gedanken," 194.

67. Savigny, "Review: *Gelegentliche Gedanken über Universitäten im deutschen Sinne*," in Müller, *Gelegentliche Gedanken über Universitäten*, 262.

68. Schleiermacher, "Gelegentliche Gedanken," 194.

69. Ibid.

70. Ibid.

71. Franzel, *Fictions of Dialogue*, 163.

72. Schleiermacher, "Gelegentliche Gedanken," 195.

73. Ibid., 206.

74. Derrida, *On Grammatology*; de Man, *Rhetoric of Romanticism*; Kittler, *Aufschreibesysteme 1800/1900*, 159–63. On these criticisms and the lecture, see Franzel, *Fictions of Dialogue*, 22.

75. Schleiermacher, *Occasional Thoughts*.

76. Schelling, *F. W. J. Schelling, Werke*, 526.

77. This is one reading of Charles E. McClelland in *State, Society, and University in Germany*, 152; see also Clark, *Academic Charisma*, 443–46.

78. Schleiermacher, "Gelegentliche Gedanken," 164, 165.

79. Mittelstrauß, *Wissenschaft als Lebensform*, 33.

80. My use of "practice" as I lay it out in the following paragraphs follows in large part MacIntyre's definition of a practice "as any coherent and complex form of socially established cooperative human activity through which goods internal to that activity are realized in the course of trying to achieve those standards of excellence which are appropriate to, and partly definitive of, that form of activity, with the result that human powers to achieve excellence, and conceptions of the ends and goods involved, are systematically extended." See MacIntyre, *After Virtue*, 187.

81. Schlegel, *Vorlesungen über schöne Literatur*, 523.

82. "Gedanken zu einer Antrittsrede in Erlangen," in Fichte, *J. G. Fichte Gesamtausgabe der Bayerischen Akademie der Wissenschaften*, II, 9:23.

83. Schleiermacher, "Gelegentliche Gedanken," 218, 224.

84. *Deduced Plan*, in Fichte, *J. G. Fichte Gesamtausgabe der Bayerischen Akademie der Wissenschaften*, II, 11:128.

85. Ibid., II, 11:86–87.

86. Ibid., II, 11:94.

87. Ibid.

88. *Einige Vorlesungen Über die Bestimmung des Gelehrten*, in Fichte, *J. G. Fichte Gesamtausgabe der Bayerischen Akademie der Wissenschaften*, I, 3:55.

89. Ibid.

90. *On the Essence of the Scholar*, in Fichte, *J. G. Fichte Gesamtausgabe der Bayerischen Akademie der Wissenschaften*, I, 8:65.

91. Gregory, *Unintended Reformation*, 379.

92. *On the Essence of the Scholar*, in Fichte, *J. G. Fichte Gesamtausgabe der Bayerischen Akademie der Wissenschaften*, I, 8:71–72.

93. *Deduced Plan*, in Fichte, *J. G. Fichte Gesamtausgabe der Bayerischen Akademie der Wissenschaften*, II, 11:94.

94. *On the Essence of the Scholar*, in Fichte, *J. G. Fichte Gesamtausgabe der Bayerischen Akademie der Wissenschaften*, I, 8:5.

95. Ibid., I, 8:131.

96. Ibid., I, 8:71, 125.

97. *Deduced Plan*, in Fichte, *J. G. Fichte Gesamtausgabe der Bayerischen Akademie der Wissenschaften*, II, 11:88–89.

98. Schleiermacher, "Gelegentliche Gedanken," 161.

99. Ibid.

100. Ibid., 162.

101. Higton, *Theology of Higher Education*, 60.

102. Schleiermacher, "Gelegentliche Gedanken," 167.

103. On this comparison, see Higton, *Theology of Higher Education*, 56–57.

104. See Schleiermacher, "Versuch einer Theorie des geselligen Betragens," in Meckenstock, *Friedrich Daniel Ernst Schleiermacher*, 163–84.

105. Schleiermacher, "Gelegentliche Gedanken," 178.

106. Ibid., 161.

107. Ibid., 170, 165, 202.

108. Higton, *Theology of Higher Education*, 60.

109. Schleiermacher, "Gelegentliche Gedanken," 166.

110. Weber, "Meaning of 'Value Freedom,'" 307.

111. Steffens, *Ueber die Idee der Universitäten*, 20.

112. Ibid., 38.

113. I borrow this idea from McDowell, *Mind and World*, lecture iv.

114. S. Weber, *Institution and Interpretation*, 25.

*Chapter Eight. Berlin, Humboldt, and the Research University*

1. Humboldt, "Antrag auf Errichtung der Universität Berlin," in *Wilhelm von Humboldts Gesammelte Schriften*, 10:150.

2. Tenorth, "Nur in Berlin, Die Universität in Ihrer Welt," 118.

3. For a detailed account of Humboldt's activities in this position, see Sweet, *Wilhelm von Humboldt*, 2:3–106.

4. For more on these problems, see Tenorth, "Wilhem von Humboldts Universitätskonzept."

5. The exact reasons for Humboldt's resignation after only two years in office remain unclear, but they had a great deal to do with Prussian power politics and the fact that he was never appointed interior minister. He did, however, consider his primary task in Berlin accomplished: "My primary ambition is to formulate clear principles, to act strictly in accord with them . . . and to leave the rest to nature, which needs only a push and an initial direction"; "Bericht der Sektion des Kultus und Unterrichts," in Humboldt, *Wilhelm von Humboldts Gesammelte Schriften*, 10:200.

6. For details on this event and the tensions it caused among the faculty, see Lenz, *Geschichte der Königlichen Friedrich-Wilhelms-Universität*, 407ff.

7. On a related gap in normative promises and historical facts around the contemporary university, see Lacapra, "University in Ruins?"

8. For an overview of other university reform efforts, see Cobb, "Forgotten Reforms."

9. Vom Bruch, "Die Gründung der Berliner Universität."

10. For an overview of these tensions, see Weischedel, *Idee und Wirklichkeit einer Universität.*

11. Hohendahl, "Humboldt Revisited." Hohendahl offers an excellent summary of the twentieth-century debates surrounding Humboldtian ideals and realities.

12. For an overview of the historiographical situation, see Ash, *Mythos Humboldt*; see also Spiewak, "Falsches Vorbild."

13. For a useful summary of these arguments and a cogent reply, see Tenorth, "Wilhem von Humboldts Universitätskonzept."

14. Langewiesche, "Die Humboldtsche Universität als nationaler Mythos."

15. These include Spranger, Über das Wesen der deutschen Universität; König, *Vom Wesen der deutschen Universität*; Becker, *Gedanken zur Hochschulreform.* On the invention of the Humboldt myth, see Palatschek, "Die Erfindung der Humboldtschen Universitätsidee."

16. Spranger, *Wilhelm von Humboldt und die Reform des Bildungswesen*, 134. For a similar account, see König, *Vom Wesen der deutschen Universität*, 27.

17. Spranger established this neohumanist line in his *Wilhelm von Humboldt und die Humanitätsidee*, where he juxtaposed Humboldt's purportedly humanist vision of the university to the "encyclopedism" of Enlightenment. Karl Jaspers offered a similarly humanistic interpretation of Humboldt's project in *Die Idee der Universität* (Berlin 1923) and *Die Erneuerung der Universität: Reden und Schriften 1945/46* (Heidelberg, 1986), esp. "Die Erneuerung der Universität" (Rektoratsrede, 1945).

18. Gregory, *Unintended Reformation*, 349.

19. It was not published in its entirety until 1903 in Gebhardt, *Wilhelm von Humboldts Politische Denkschriften*, 250–60.

20. Gebhardt, *Wilhelm von Humboldt als Staatsmann*, 124.

21. In his argument for the establishment of the Kaiser Wilhelm Gesellschaft as a research institution independent of the university, Adolf Harnack appealed to this text as well. See Harnack, *Geschichte der Königlich Preussischen Akademie der Wissenschaften zu Berlin*. Harnack reprinted the Humboldt text in 2:361–66 and discussed Humboldt's activities and his reorganization of the relationship between the university and the academy in 2:523–608.

22. On the experience of American academics in Germany and their appeal to German ideals of the university, see Herbst, *German Historical School in American Scholarship*, esp. chap. 1; Diehl, *Americans and German Scholarship*, esp. chap. 3.

23. Humboldt, "Über die innere und äussere Organisation," in *Wilhelm von Humboldts Gesammelte Schriften*, 10:251.

24. Ibid., 10:253.

25. Humboldt, *Wilhelm von Humboldts Gesammelte Schriften*, 10:231, 10:226.

26. This and the following paragraph draw on Wellmon, *Becoming Human*.

27. For a more contemporary expansion of this concept, see McDowell, *Mind and World*.

28. "Theorie der Bildung des Menschen," in Humboldt, *Wilhelm von Humboldts Gesammelte Schriften*, 1:283.

29. Ibid.

30. "Unmaßgebliche Gedanken über den Plan einer Errichtung eines Litthauischen Stadtschulwesens," September 27, 1809, in Humboldt, *Wilhelm von Humboldts Gesammelte Schriften*, 3:262.

31. Ibid.

32. "Antrag auf Errichtung der Universität Berlin," May 12, 1809, in Humboldt, *Wilhelm von Humboldts Gesammelte Schriften*, 10:140.

33. Ibid.

34. On this debate, see vom Bruch, "Die Gründung der Berliner Universität."

35. "Antrag auf Errichtung der Universität Berlin," July 24, 1809, in Humboldt, *Wilhelm von Humboldts Gesammelte Schriften*, 10:150, 269.

36. "Bericht der Sektion des Kultus und Unterrichts," December 1809, in Humboldt, *Wilhelm von Humboldts Gesammelte Schriften*, 10:219.

37. Nicolai, *Beschreibung der Königlichen Residenzstädte Berlin und Potsdam*, 723–28. In an amended version, Nicolai gives a list of private lectures. See his *Wegweiser für Fremde und Einheimische* (Berlin, 1793), the chapter "Von Akademien, gelehrten Gesellschaften, Vorlesungen, Gymnasien und Schulen," esp. 132–33. For a listing of these, see Köpke, *Die Gründung der Königlichen Friedrich-Wilhelms-Universität*, 141n; see also "Einleitung," *Die Vorlesungen*, xx–xxi.

38. On the regulations for the relationship between the university and these other institutions and collections, see section VII, paragraphs 1 and 2, as printed in Lenz, *Geschichte der Königlichen Friedrich-Wilhelms-Universität*, 4:256.

39. See, for example, the *Verzeichnis* for the *Wintersemester 1810/11* as reprinted in Virmond, *Die Vorlesungen der Berliner Universität*, 13–14.

40. "Antrag auf Errichtung der Universität Berlin," in Humboldt, *Wilhelm von Humboldts Gesammelte Schriften*, 10:150.

41. *Deduced Plan*, in Fichte, *J. G. Fichte Gesamtausgabe der Bayerischen Akademie der Wissenschaften*.

42. Humboldt, "Über die innere und äussere Organisation," in *Wilhelm von Humboldts Gesammelte Schriften*, 10:255.

43. Ibid., 260.

44. Ibid., 253.

45. Ibid., 251.

46. MacIntyre, *Three Rival Versions of Moral Enquiry*, 160.

47. See Marino, *Praeceptores Germaniae*.

48. Turner, "University Reformers and Professorial Scholarship in Germany," 522.

49. Quoted in Turner, "Prussian Universities and the Concept of Research," 69, 70.

50. Michaelis, *Raisonnement über die protestantischen Universitäten in Deutschland*, 1:12.

51. Clark, *Academic Charisma*.

52. In C. Bornhak, *Geschichte der preußischen Universitätsverwaltung bis 1810* (Berlin, 1900), quoted in Stichweh, *Ausdifferenzierung der Wissenschaft*, 219n.

53. Michaelis, *Raisonnement über die protestantischen Universitäten in Deutschland*, 1:12.

54. Brandes, *Über den gegenwärtigen Zustand der Universität Göttingen* (Göttingen, 1802), 160, quoted in Turner, "Prussian Universities and the Concept of Research," 82.

55. Turner, "Prussian Universities and the Concept of Research," 71.

56. See Vandermeersch, "Die Universitätslehrer," 185.

57. Turner, "Prussian Universities and the Concept of Research."

58. Ludwig Heinrich Jacob, *Ueber Universitäten in Deutschland, besonders in den königlichen preussisichen Staaten* (Berlin, 1798), 26, quoted in Turner, "University Reformers and Professional Scholarship in Germany," 516.

59. Brandes, quoted in Turner, "University Reformers and Professional Scholarship in Germany," 517.

60. Michaelis, *Raisonnement über die protestantischen Universitäten in Deutschland*, quoted in Turner, "University Reformers and Professional Scholarship in Germany," 529.

61. See Turner, "Prussian Universities and the Concept of Research," 80; Hammerstein, *Universitäten und Aufklärung*. For a detailed account of Göttingen's contributions, see Marino, *Praeceptores Germaniae*.

62. See Clark, *Academic Charisma*, 219.

63. See Tenorth, "*Was heißt Bildung in der Universität?*"

64. Turner, "University Reformers and Professional Scholarship in Germany," 506.

65. Humboldt, Über die innere und äussere Organisation," in *Wilhelm von Humboldts Gesammelte Schriften*, 10:252.

66. "Littauischen Plan," in Humboldt, *Wilhelm von Humboldts Gesammelte Schriften*, 13:261.

67. Ibid., 13:262.

68. Hadot, *Philosophy as a Way of Life*.

69. "Amtliche Arbeiten 1809–10," in Humboldt, *Wilhelm von Humboldts Gesammelte Schriften*, 13:247.

70. Humboldt, "Über die innere und äussere Organisation," in *Wilhelm von Humboldts Gesammelte Schriften*, 10:252.

71. Humboldt, *Wilhelm von Humboldts Gesammelte Schriften*, 13:278.

72. Quoted in Turner, "University Reformers and Professional Scholarship in Germany," 514.

73. Humboldt, "Über die innere und äussere Organisation," in *Wilhelm von Humboldts Gesammelte Schriften*, 10:252.

74. Schleiermacher, "Gelegentliche Gedanken," 163.

75. There are a number of thorough institutional histories of the Friedrich-Wilhelms-Universität zu Berlin. For two of the most canonical but still unparalleled in historical detail, see Köpke, *Die Gründung der Königlichen Friedrich-Wilhelms-Universität*; Lenz, *Geschichte der Königlichen Friedrich-Wilhelms-Universität*.

76. Max Lenz makes clear that the actual structures of the new university in Berlin borrowed heavily from other German universities. See Lenz, *Geschichte der Königlichen Friedrich-Wilhelms-Universität*, 1:277. See also Palatscheck, "Die Erfindung der Humboldtschen Universitätsidee," 189.

77. This is Helmut Schelsky's basic conclusion in *Einsamkeit und Freiheit*, but Schelsky doubted whether a Humboldtian university was appropriate for a modern industrial society. Humboldt's bureaucratic organization of the university, its close connection to the state, may seem hopelessly conservative today, but under the po-

litical conditions of early nineteenth-century Prussia it should not be too surprising. Following Schleiermacher, Humboldt offered a relatively pragmatic solution to the problem of the relationship of the university and the state.

78. This committee was replaced in 1816 by the *wissenschaftlichen Prüfungskommisionen*; see Spranger, *Wilhelm von Humboldt und die Reform des Bildungswesen*, 124.

79. For a detailed discussion of the committee's work on this plan, see Spranger, *Wilhelm von Humboldt und die Reform des Bildungswesen*, 240–55.

80. "Ideen zu einer Instruktion für die wissenschaftliche Deputation bei der Sektion des öffentlichen Unterrichts," in Humboldt, *Wilhelm von Humboldts Gesammelte Schriften*, 10:179.

81. See Humboldt's letter to Altenstein from October 28, 1809, in Humboldt, *Wilhelm von Humboldts Gesammelte Schriften*, 10:180, 182, 187.

82. Ibid., 10:182, 184.

83. For a detailed narrative of the university's first hires, see Lenz, *Geschichte der Königlichen Friedrich-Wilhelms-Universität*, 1:220ff.

84. Moraw, "Humboldt in Giessen."

85. "Bericht der Sektion des Kultus und Unterrichts," December 1809, in Humboldt, *Wilhelm von Humboldts Gesammelte Schriften*, 10:219.

86. Turner, "University Reformers and Professorial Scholarship in Germany," 512.

87. Humboldt, *Wilhelm von Humboldts Gesammelte Schriften*, 10:228.

88. Virmond, *Die Vorlesungen der Berliner Universität.*

89. See ibid.

90. See Veysey, *Emergence of the American University*, 121–79, 263–341.

91. This is the question that Joseph Ben-David fails to answer in his account of Humboldt's role in the early research university. See Ben-David, *Scientist's Role in Society*, esp. chap. 4.

92. For an excellent summary of the different stories that could be told about the research university and its relationship to the concept of research, see Turner, "Humboldt in North America?," 289–312.

93. The clearest argument for this can be found in McClelland, *State, Society, and University in Germany*; and, especially, McClelland, *German Experience of Professionalization.*

94. Howard, *Protestant Theology*, 273, 278. Howard, however, correctly notes that this move toward greater specialization was actually consonant with earlier romantic and idealist formulations of *Wissenschaft*. See ibid., 272.

95. For studies on the professionalization of the university, see Bledstein, *Culture of Professionalism.*

96. Howard, *Protestant Theology*, 278.

*Chapter Nine. The Disciplinary Self and the Virtues of the Philologist*

1. This chapter draws on arguments I made in Wellmon, "Crisis of Purpose."

2. Daston, "Die Akademien und die Einheit der Wissenschaften."

3. The scientist was Adolf Kußmaul. Quoted in Alexander Busch, *Die Geschichte des Privatdozent*, 26n.

4. Daston and Galison, "Image of Objectivity," 122. Daston and Galison are interested in tracing the history of various forms of "objectivity" in science, but I am more interested in the relationship between ethics and science more generally and its relationship to the university. See also Daston, "Moral Economy of Science"; Shapin, *Scientific Life*, 1–46.

5. In what follows, I draw especially on the historical work of Clark, "On the Dialectical Origins of the Research Seminar," and *Academic Charisma*, esp. 140–83; and Erben, "Die Entstehung der Universitäts-Seminare."

6. On this notion of intellectual apprenticeship, see Polanyi, *Personal Knowledge*, esp. 49–68; Matthew Crawford (personal conversation).

7. Olesko, *Physics as Calling*, 1. Olesko's book gives a broader introduction to the emergence of natural science seminars, especially, as the title notes, the Königsberger seminar for math and physics established in 1834.

8. Francke, quoted in Martens, "Hallescher Pietismus und Gelehrsamkeit."

9. For details on this training, see Paulsen, *Geschichte des Gelehrten Unterrichts*, 382.

10. Francke, *Kurtzer Bericht von der gegenwärtigen Verfassung*.

11. Carl Hinrichs, *Preußentum und Pietismus* (Göttingen, 1971), 1, quoted in Sheehan, *Enlightenment Bible*.

12. Sheehan, *Enlightenment Bible*, 60–61.

13. Quoted in Paulsen, *Geschichte des Gelehrten Unterrichts*, 435.

14. Erben, "Die Entstehung der Universitäts-Seminare," 1251.

15. Sheehan, *Enlightenment Bible*, 63.

16. From the seminar's statutes (*Schulordnung*), as quoted in Paulsen, *Geschichte des Gelehrten Unterrichts*, 431.

17. On the history of this tension, see Grafton and Jardine, *From Humanism to the Humanities*. On the particularly German critique of Latinate humanism, see Marchand, *Down from Olympus*, esp. 3–35.

18. See Legaspi, *Death of Scripture*, chap. 3.

19. Paulsen, *Geschichte des Gelehrten Unterrichts*, 434.

20. Heyne, quoted in ibid., 443.

21. Heyne, quoted in Pütter and Saalfeld, *Versuch einer academischen Gelehrten-Geschichte*, 2:249.

22. Ibid. Translation in Clark, "On the Dialectical Origins of the Research Seminar," 130.

23. Clark, "On the Dialectical Origins of the Research Seminar."

24. "Pädagogisches Seminar für höhere Schule," in K. A. Schmid, *Encyklopädie des gesammten Erziehungs- und Unterrichtswesen*, 866.

25. In what follows, I cite from Wolf, *Friedrich August Wolf's Darstellung der Altherthumswissenschaft*, 9.

26. Ibid. The "*schönen Wissenschaften*" referred to a range of sciences, such as aesthetics, history, and poetics. On the history of the concept, see Strube, "Die Geschichte des Begriffs 'schöne Wissenschaften.' "

27. Wolf, *Friedrich August Wolf's Darstellung der Altherthumswissenschaft*, 10.

28. Ibid., 20. On Wolf's Hellenism, see La Vopa, "Specialists against Specialization," 37–38.

29. Wolf, *Kleine Schriften in lateinischer und deutscher Sprache*, 878.

30. Ibid., 847.

31. Wolf, *Friedrich August Wolf's Darstellung der Altherthumswissenschaft*, 96, 50.

32. Kunoff, *Foundations of the German Academic Library*, 82.

33. Arnoldt, *Friedrich August Wolf*, 1:127. For a more detailed description of these reporting and evaluation methods, see Clark, *Academic Charisma*, 169–83.

34. Wolf, *Friedrich August Wolf*, 1:56.

35. Ibid., 1:27–28.

36. See Stichweh, *Die Ausdifferenzierung der Wissenschaft*, 117.

37. Wolf, *Friedrich August Wolf*, 1:55.

38. For more details on the forms of interaction facilitated by the seminar, see Clark, *Academic Charisma*.

39. Wolf, *Friedrich August Wolf*, 1:55.

40. "Wolfs Bericht an den Canzler von Hoffmann in Betreff des bei der halleschen Universität einzurichtenden philologischen Seminariums," September 6, 1787, printed in Arnoldt, *Friedrich August Wolf*, 249.

41. Wolf, *Prolegomena to Homer*, 55–56.

42. Ibid.

43. For an excellent overview of Wolf's argument in the *Prolegomena*, see Grafton, "Introduction," in ibid., 3–36.

44. Grafton, "Polyhistor into Philolog," 131.

45. See Marchand, *Down from Olympus*, 20.

46. Wolf, *Friedrich August Wolf's Encyclopädie der Philologie*, 16.

47. Grafton, "Introduction," in Wolf, *Prolegomena to Homer*, 29.

48. Wolf, *Friedrich August Wolf's Darstellung der Altherthumswissenschaft*, 75.

49. For an early formulation of this, see "Über das Studium des Alterthums, und des Griechischen insbesondere," in Humboldt, *Wilhelm von Humboldts Gesammelte Schriften*, 1:255–81.

50. Hoffmann, *August Böckh*, 470. For the history of the seminar's establishment, see Lenz, *Geschichte der Königlichen Friedrich-Wilhelms-Universität*, 2/2 146; Köpke, *Die Gründung der Königlichen Friedrich-Wilhelms-Universität*, 241.

51. "Seminarordnung," originally published in the *Unversitätskalender* of 1813, reprinted in Virmond, *Die Vorlesungen der Berliner Universität*, 791–92.

52. See Erben, "Die Entstehung der Universitäts-Seminare," 1258.

53. "Seminarordnung," in Virmond, *Die Vorlesungen der Berliner Universität*, 791.

54. Ibid.

55. "Pädagogisches Seminar für höhere Schule," in K. A. Schmid, *Encyklopädie des gesammten Erziehungs- und Unterrichtswesen*, 803.

56. See Herdt, *Putting on Virtue*, 26–27; and MacIntyre, *After Virtue*, 181.

57. Daston, "Die Akademien und die Einheit der Wissenschaften," 83.

58. Böckh, *Encyclopädie und Methodologie*, 172.

59. Ibid., 658.

60. McNeely and Wolverton, *Reinventing Knowledge*.

61. Olesko, *Physics as Calling*, 24. These reforms were soon followed by the edict of October 12, 1812, which required gymnasium students to pass an exam not only to gain entrance to the university but also to graduate from gymnasium.

62. See Jeismann, *Das preußische Gymnasium in Staat und Gesellschaft*; Clark, *Academic Charisma*, 125.

63. For a detailed discussion of this exam and on the history of its design, see Spranger, *Wilhelm von Humboldt und die Reform des Bildungswesen*, 235–57.

64. See McNeely and Wolverton, *Reinventing Knowledge*, 191.

65. Other seminars—Greifswald (1822), Breslau (1812), and Bonn (1819)—were modeled on the Berlin seminar. See Wiese, *Das höhere Schulwesen in Preußen*, 1:530–43.

66. K. Fricke (1903), 16, quoted in Kittler, *Aufschreibesysteme 1800/1900*, 160.

67. For more on this history, see Jeismann, *Das preußische Gymnasium in Staat und Gesellschaft*, 321–26.

68. "Pädagogisches Seminar für höhere Schule," in K. A. Schmid, *Encyklopädie des gesammten Erziehungs- und Unterrichtswesen*, 803.

69. Böckh began these lectures in Heidelberg and continued them upon arriving in Berlin in 1812. The lectures actually consist of a written series of texts that he amended and added to over the course of his career. This text was edited and published on the basis of student lecture notes.

70. See Tenorth, "Transformation der Wissensordung," 34.

71. Böckh, *Encyclopädie und Methodologie*, 1, 3, 4.

72. Ibid., 37, 20, 4.

73. Ibid., 10.

74. Ibid., 31.

75. Ibid., 33.

76. "Von der Philologie, besonders der klassischen in Beziehung zur morganländischen, zum Unterricht und zur Gegenwart," in Böckh, *August Boeckh's gesammelte kleine Schriften*, 2:192.

77. "Ueber den Sinn und Geist der Gründung der Berliner Universität," in Böckh, *Gesammelte kleine Schriften*, 2:136.

78. On this helping role, see Horstmann, "Die Forschung in der klassischen Philologie des 19. Jahrhunderts."

79. Thus, he consistently uses the term "researcher" [*Forscher*] and not "erudite" [*Gelehrte*] to describe the philologist. See, for example, Böckh, *Encyclopädie und Methodologie*, 48, 177, 248, 249, 253, 306, 533, 560, and 745.

80. Ibid., 16.

81. "Ueber die Pflichten der Männer der Wissenschaft," in Böckh, *Gesammelte kleine Schriften*, 2:128.

82. "Von der Philologie, besonders der klassischen in Beziehung zur morganländischen zum Unterricht und zur Gegenwart," in ibid., 2:190.

83. "Ueber den Sinn und Geist der Gründung der Berliner Universität," in Böckh, *Gesammelte kleine Schriften*, 2:139.

84. Ibid., 140.

85. "Ueber Pflichten der Männer der Wissenschaft," in Böckh, *Gesammelte kleine Schriften*, 2:128; and "Ueber den Sinn und Geist der Gründung der Berliner Universität," in ibid., 2:139.

86. Ibid., 145.

87. MacIntyre, *After Virtue*, 194.

88. "Ueber den Sinn und Geist der Gründung der Berliner Universität," in Böckh, *Gesammelte kleine Schriften*, 2:139.

89. J. H. Voss quoted in Grafton, "Polyhistor into Philolog," 173.

90. See Turner, "Prussian Universities and the Concept of Research," 84. As Grafton puts it, there was less a "revolution in teaching than a gradual increase in the attention given to controverted problems and technical methods for their solution and a gradual decrease in the attention given to the traditional objects of humanist study" ("Polyhistor into Philolog," 168).

91. Quoted in Howard, *Protestant Theology*, 273.

92. Diesterweg, Über das Verderben auf den deutschen Universitäten, 10.

93. Ibid., 35.

94. See Paulsen, *Geschichte des Gelehrten Unterrichts*, 674–77; O'Boyle, "Klassische Bildung und soziale Struktur." For a broader account of this tension upon which I draw, see La Vopa, "Specialists against Specialization."

95. Böckh, *Gesammelte kleine Schriften*, 5:248.

96. “Rede zur Eröffnung der eilften Versammlung Deutscher Philologen, Schülmänner und Orientalisten,” in Böckh, *Gesammelte kleine Schriften*, 2:190.

97. Ibid., 191.

98. Ibid., 190.

99. “Ueber den Sinn und Geist der Gründung der Berliner Universität,” in Böckh, *Gesammelte kleine Schriften*, 2:139.

100. Daston, “Die Akademien und die Einheit der Wissenschaften,” 64.

101. Ibid., 73.

102. On this continued tension, see Most, “On the Use and Abuse of Ancient Greece for Life”; see also Marchand, *Down from Olympus*, 116–51.

103. Mommsen, *Reden und Aufsätze*, 37, 38. On Mommsen’s account of philology’s task, see Landfester, “Ulrich von Wilamowitz-Moellendorff und die hermeneutische Tradition,” 158.

104. Böckh, *Encyclopädie und Methodologie*, 9.

105. “Von der Philologie, besonders der klassischen in Beziehung zur morganländischen, zum Unterricht und zur Gegenwart,” in Böckh, *Gesammelte kleine Schriften*, 2:190.

106. Grafton, “Polyhistor into Philolog,” 174.

107. Howard, *Protestant Theology*, 273.

108. Böckh, quoted in Most, “On the Use and Abuse of Ancient Greece for Life,” 14.

109. Böckh, Über das Verhältniss der Wissenschaft zum Leben, in Böckh, *August Boeckh’s Gesammelte kleine Schriften*, 2:22.

110. Böckh, *Encyclopädie und Methodologie*, 9. I translate Böckh’s *Humanitätsstudium* loosely as “humanities.”

111. For a broader and more contemporary discussion of this, see Shapin, *Scientific Life*, 1–46.

112. M. Weber, *From Max Weber*, 45. I have slightly altered this translation to highlight the virtue language that Weber uses in the original.

113. On the intimate relationship between the German research university and masculinity, see Mazon, *Gender and the Modern Research University*.

*Afterword. Too Many Links*

1. Laura Pappano, “The Year of the MOOC,” *New York Times*, November 2, 2012.

2. The first MOOC seems to have been produced at the University of Manitoba in 2008, although various forms of such courses and collaborations precede even that date. On some of these early efforts as the first developed in Canada, see McAuley et al., “MOOC Model for Digital Practice.”

3. Bowen, *Higher Education in the Digital Age*, 46.

4. For Thrun's original comment, see Steven Leckart, "The Stanford Education Experiment Could Change Higher Learning Forever," www.wired.com/wiredscience/2012/03/ff_aiclass/all/; for his second thoughts, see Max Chafkin, "Udacity's Sebastian Thrun, Godfather of Free Online Education, Changes Course," www.fastcompany.com/3021473/udacity-sebastian-thrun-uphill-climb.

5. Aaron Bady, "The MOOC Moment and the End of Reform," http://thenewinquiry.com/blogs/zunguzungu/the-mooc-moment-and-the-end-of-reform/.

6. See http://blog.coursera.org/post/51696469860/10-us-state-university-systems-and-public-institutions.

7. For an overview of higher education cost scholarship, see Bowen, *Higher Education in the Digital Age*, 1–41.

8. Abbott, "Disciplines and Authority," 210.

9. Summers, foreword to *An Avalanche Is Coming*, 1.

10. Gilman, *University Problems in the United States*, 49.

11. Kerr, *Uses of the University*, 18.

12. All three of the international rankings of universities—Times Higher Education, QS Top Universities, and Academic Ranking of World Universities—weigh research-related activity as at least 50 percent of its valuation. See Summer, foreword to *An Avalanche Is Coming*, 21.

13. Menand, *Marketplace of Ideas*, 52.

14. W. Wilson, *Princeton for the Nation's Service*, 32.

15. Ibid., 12.

16. Menand, "Limits of Academic Freedom," 19. See also Taylor, *Crisis on Campus*.

17. C. N. Davidson, "MOOCs and the Future of the Humanities," *Los Angeles Review of Books*, June 15, 2013.

18. Ibid.

19. C. N. Davidson, *Now You See It*, 10.

20. Daston and Galison, *Objectivity*, 204.

21. C. N. Davidson, *Now You See It*, 7.

22. Davidson and Goldberg, *Future of Thinking*, 17.

23. Abbott, "Disciplines and Authority," 222.

24. See Andrew Piper, "The Internet Killed Books Again," http://iasc-culture.org/THR/channels/Infernal_Machine/2014/03/the-internet-killed-books-again/.

25. For the department's letter, see http://chronicle.com/article/The-Document-an-Open-Letter/138937/; for a similar point, see David Theo Goldberg, "The Afterlife of the Humanities," http://humafterlife.uchri.org.

26. Bethany Nowviskie, "Toward a New Deal," http://nowviskie.org/2013/new-deal/; McGann, *New Republic of Letters*, 1–3.

27. Clark, *Academic Charisma*, 446.

28. Grafton, "Humanities and Inhumanities."

29. Bowen, *Higher Education in the Digital Age*.

# BIBLIOGRAPHY

Abbott, Andrew. "The Disciplines and Authority." In *The Future of the City of Intellect: The Changing American University*. Edited by Steven Brint, 205–30. Palo Alto, CA: Stanford University Press, 2002.

Adelung, Johann Christoph. *Grammatisch-kritisches Wörterbuch der hochdeutschen Mundart*. Vienna: Bauer, 1811.

Albrecht, Peter, and Ernst Hirsch, eds. *Das niedere Schulwesen im Übergang vom 18. zum 19. Jahrhundert*. Tübingen: Wolfenbütteler Studien zur Aufklärung, 1995.

Anderson, Wilda. "Encyclopedic Topographies." *MLN* 101, no. 4 (1986): 912–29.

Anon. "Bemerkungen über den Werth der Akademien." In *Jahrbuch der Universitäten, Gymnasien, Lyceen und andern gelehrten Bildungsanstalten in und ausser Deutschland*, vol. 1:1. Erfurt: Henningssche Buchhandlung, 1798.

Anon. "Vorschlag und Aufforderung an die deutsche Nation über die Notwendigkeit eines allgemeinen Repertoriums der Literatur und Bücherkunde, und über die Möglichkeit diesen Gedanken zu realisieren." *Intelligenzblatt der Allgemeinen Literatur-Zeitung*, June 16, 1790.

Aquinas, Saint Thomas. *Summa Theologiae, Questions on God*. Edited by Brian Davies and Brian Leftow. Cambridge: Cambridge University Press, 2006.

Aristotle. *The Art of Rhetoric*. Translated by J. H. Freese. Cambridge, MA: Harvard University Press, 1926.

———. *Metaphysics*. In *The Basic Works of Aristotle*, edited by Richard McKeon, translated by W. D. Ross. New York: Random House, 1941.

———. *Nicomachean Ethics*. Translated by H. Rackham. Cambridge, MA: Harvard University Press, 1968.

———. "On Interpretation." In *The Organon*, translated by Harold P. Cooke, 112–81. Cambridge, MA: Harvard University Press, 1938.

Arnoldt, J. F. J. *Friedrich August Wolf in seinem Verhältnisse zum Schulwesen und zur Paedagogik*. Braunschweig: C. A. Schwetschke, 1861.

Arum, Richard, and Josipa Roksa, *Academically Adrift: Limited Learning on College Campuses*. Chicago: University of Chicago Press, 2011.

Ash, Mitchell G., ed. *Mythos Humboldt*. Vienna: Bohlau, 1999.

Auletta, Ken. "Get Rich U." *New Yorker*, April 30, 2012.

Bacon, Francis. *De Augmentis*. In *The Works of Francis Bacon*, vol. 4, edited by James Spedding. London: Spottiswoode, 1885.

———. "A Description of the Intellectual Globe." In *Philosophical Studies VI: c. 1611–c. 1619, The Oxford Francis Bacon*, edited by Graham Rees and Lisa Jardine, 96–169. Oxford: Clarendon Press, 1996.

———. *The New Organon*. Edited by Lisa Jardine and Michael Silverthorne. Cambridge: Cambridge University Press, 2000.

Bates, David. "Cartographic Aberrations: Epistemology and Order in the Encyclopedic Map." *Studies on Voltaire and the Eighteenth Century* 5 (2002): 1–20.

Beaujean, Marion. *Der Trivialroman in der zweiten Hälfte des 18. Jahrhunderts: Die Ursprünge des modernen Unterhaltungsroman*. Bonn: H. Bouvier, 1964.

Becker, Carl Heinrich. *Gedanken zur Hochschulreform*. Leipzig: Quelle & Meyer, 1919.

Behler, Ernst. "Friedrich Schlegels Enzyklopädie der literarischen Wissenschaften im Unterschied zu Hegels Enzyklopädie der philosophischen Wissenschaften." In *Studien zur Romantik und idealistischen Philosophie*, 1:236–63. Paderborn: Schöningh, 1988.

Beiser, Frederich. *The Fate of Reason: German Philosophy from Kant to Fichte*. Cambridge, MA: Harvard University Press, 1996.

Ben-David, Joseph. *The Scientist's Role in Society*. Chicago: University of Chicago Press, 1971.

Bergk, Johann Adam. *Die Kunst Bücher zu lesen: Nebst Bemerkungen über Schriften und Schriftsteller*. Jena: Hempelsche Buchhandlung, 1799.

Bernhard, Johann Adam. *Kurtzgefasste Curieuse Historie derer Gelehrten*. Frankfurt am Main: Johann Maximilian von Sand, 1718.

Beutler, Johann Heinrich Christoph, and Johann Christoph Friedrich Gutsmuth, eds. *Allgemeines Sachregister ueber die wichstigsten deutschen Zeit- und Wochenschriften*. Leipzig, 1790.

Birtsch, Günther. "Die Berliner Mittwochsgesellschaft." In *Über den Prozeß der Aufklärung in Deutschland im 18. Jahrhundert. Personen, Institutionen und Medien*, edited by Hans Erich Bödeker and Ulrich Herrmann. Göttingen: Vandenhoeck & Ruprecht, 1987.

Blair, Ann. "Student Manuscripts and the Textbook." In *Scholarly Knowledge: Textbooks in Early Modern Europe*, edited by Emidio Campi, Simone de Angelis, Anja-Silvia Goeing, and Anthony Grafton, 39–74. Geneva: Libraire Droz, 2008.

———. *Too Much to Know: Managing Scholarly Information before the Modern Age*. New Haven, CT: Yale University Press, 2010.

Blair, Ann, and Anthony Grafton. "The Practices of Erudition according to Morhof." In *Mapping the World of Learning: The Polyhistor of Daniel Georg Morhof*, edited by Françoise Waquet, 53–74. Wiesbaden: Harrassowitz, 2000.

———. "Reassessing Humanism and Science," *Journal of the History of Ideas* 53, no. 4 (1992).
Blair, Ann, and Peter Stallybras. "Mediating Information, 1450–1800." In *This Is Enlightenment*, edited by Clifford Siskin and William Warner, 139–63. Chicago: University of Chicago Press, 2010.
Bledstein, Burton. *The Culture of Professionalism: The Middle Class and the Development of Higher Education in America*. New York: W. W. Norton, 1976.
Blum, Rudolf. "Bibliographia: eine Wort- und begriffsgeschichtliche Untersuchung." In *Archiv für Geschichte des Buchwesens*, 10:1010–246. Frankfurt: Buchhändler-Vereinnigung, 1969.
Böckh, August. *August Boeckh's Gesammelte kleine Schriften: Boeckh's Reden Gehalten auf der Universität und in der Akademie der Wissenschaften zu Berlin*. Vol. 2. Edited by Ernst Bratuscheck. Leipzig: Teubner, 1859.
———. *Encyclopädie und Methodologie der philologischen Wissenschaften von August Boeckh*. Edited by Ernst Bratuscheck. Leipzig: Teubner, 1877.
———. *Gesammelte kleine Schriften: Akademische Abhandlungen*. Vol. 5. Edited by Paul Eichholtz and Ernst Brautscheck. Leipzig: Teubner, 1871.
Bodmer, Johann Jacob. *Anklagung des verderbten Geschmack, Oder Critische Anmerkungen Ueber Hamburgischen* PATROTEN *und die Hallischen* TADLERINNEN. Frankfurt and Leipzig, 1728.
Böll, Friedrich Phillip Karl. *Das Wesen der Universität in Briefen*. 1782.
Bowen, William G. *Higher Education in the Digital Age*. Princeton, NJ: Princeton University Press, 2013.
Bowen, William G., Matthew M. Chingos, and Michael S. McPherson. *Crossing the Finish Line: Completing College at America's Public Universities*. Princeton, NJ: Princeton University Press, 2011.
Bowie, Andrew. *Schelling and Modern European Philosophy: An Introduction*. London: Routledge, 1993.
Breidbach, Olaf, and Paul Ziche, eds. *Naturwissenschaften um 1800: Wissenschaftskultur in Weimar-Jena*. Weimar: Böhlaus Nachfolger, 2001.
Brentano, Clemons. *Universitati Litterariae: Kantate auf den 15ten October 1810*. Berlin: Julius Eduard Hitzig, 1910.
Brewer, Talbot. "The Coup That Failed: How the Near-Sacking of a University President Exposed the Fault Lines of American Higher Education." *Hedgehog Review* 16, no. 2 (2014): 65–83.
Brockliss, Laurence. "Curricula." In *A History of the University in Europe*, vol. 2, *Universities in Early Modern Europe*, edited by Hilde de Ridder-Symoens, 565–620. Cambridge: Cambridge University Press, 1996.
Brucker, Jakob. *Ehrentempel der deutschen Gelehrsamkeit*. Augspurg: Haid, 1747.
Bumann, Waltraud. "Der Begriff der Wissenschaft im deutschen Sprach- und Den-

kraum." In *Der Wissenschaftsbegriff: Historische und systematische Untersuchungen*, edited by Alwin Diemer, 56–106. Meisenheim: Hain, 1970.
Burdach, Karl Friedrich. *Der Organismus menschlicher Wissenschaft und Kunst*. Leipzig: Mitzky, 1809.
Bürger, Christa. "Literarischer Markt und Öffentlichkeit am Ausgang des 18. Jahrhunderts in Deutschland." In *Aufklärung und literarische Öffentlichkeit*, edited by Christa Bürger, Peter Bürger, and Jochen Schulte-Sassen, 162–218. Frankfurt am Main: Surhkamp, 1980.
Buringh, Eltjo, and Jan Luiten van Zanden. "Charting the 'Rise of the West': Manuscripts and Printed Books in Europe, a Long-Term Perspective from the Sixth through the Eighteenth Centuries." *Journal of Economic History* 69, no. 2 (June 2009): 409–45.
Busch, Alexander. *Die Geschichte des Privatdozent*. New York: Arno Press, 1977.
Calhoun, Craig, ed. *Habermas and the Public Sphere*. Cambridge, MA: MIT Press, 1993.
Campe, Johann H., ed. *Allgemeine Revision des gesammten Schul- und Erziehungswesen von einer Gesellschaft practischer Erzieher*. Vienna: Gräfer, 1785–1792.
Carels, Peter, and Dan Flory. "Johann H. Zedler's Universal Lexicon." In *Notable Encyclopedias of the Seventeenth and Eighteenth Centuries: Nine Predecessors of the Encyclopédie*, edited by Frank A. Kafkar, 165–96. Oxford: Voltaire Foundation, 1981.
Carr, Nicholas. "Is Google Making Us Stupid? What the Internet Is Doing to Our Brains." *Atlantic*, July–August 2008. www.theatlantic.com/magazine/archive/2008/07/is-google-making-us-stupid/6868/.
———. *The Shallows: What the Internet Is Doing to Our Brains*. New York: W. W. Norton, 2010.
Casper, Gerhard. "Teaching and Research." *Stanford Today*, March/April 1998.
Cassirer, Ernst. *Das Erkenntnisproblem in der Philosophie und Wissenschaft der neueren Zeit*. 3 vols. Berlin: Verlag Bruno, 1922.
———. *Kants Leben und Lehre*. Darmstadt: Wissenschaftliche Buchgesellschaft, 1977.
———. *The Philosophy of the Enlightenment*. Translated by Fritz C. A. Koelln. Boston: Beacon Press, 1955.
Chartier, Roger. *The Order of Books: Readers, Authors and Libraries in Europe between the Fourteenth and the Eighteenth Centuries*. Translated by Lydia G. Cochrane. Cambridge: Polity, 1994.
Christensen, Clayton M., and Henry J. Eyring. *The Innovative University: Changing the DNA of Higher Education from the Inside Out*. San Francisco: Jossey-Bass, 2011.
Clark, William. *Academic Charisma and the Origins of the Research University*. Chicago: University of Chicago Press, 2006.

———. "On the Dialectical Origins of the Research Seminar." *History of Science* 27 (1989): 111–54.

Cobb, James Denis. "The Forgotten Reforms: Non-Prussian Universities, 1797–1817." PhD diss., University of Wisconsin–Madison, 1980.

Collins, Randall. *The Sociology of Philosophies: A Global Theory of Intellectual Change*. Cambridge, MA: Harvard University Press, 2000.

Compayré, Gabriel. *Abelard and the Origin and Early History of Universities*. New York: Scribner, 1910.

d'Alembert, Jean Le Rond. *Preliminary Discourse to the Encyclopedia of Diderot*. Translated by Richard N. Schwab. Indianapolis: Bobbs-Merrill Educational Publishing, 1983.

Dann, Otto. "Vom Journal des Şcavans zur wissenschaftlichen Zeitung." In *Gelehrte Bücher vom Humanismus bis zur Gegenwart*, edited by Bernhard Fabian and Paul Raabe, 63–80. Wiesbaden: Harrassowitz, 1983.

———, ed. *Lesegesellschaften und bürgerliche Emanzipation. Ein Europäischer Vergleich*. München: Beck, 1981.

Darnton, Robert. *The Business Of Enlightenment: A Publishing History of the Encyclopédie, 1775–1800*. Cambridge, MA: Belknap, 1979.

———. "'What Is the History of Books?' Revisited." *Modern Intellectual History* 4, no. 3 (2007): 495–508.

Daston, Lorraine. "Die Akademien und die Einheit der Wissenschaften. Die Disziplinierung der Disziplinen." In *Die Königlich Preußische Akademie der Wissenschaften zu Berlin im Kaiserreich*, edited by Jürgen Kocka, Rainer Hohlfeld, and Peter Walther, 61–84. Berlin: Akademie Verlag, 1999.

———. "The Moral Economy of Science." *Osiris* 10 (1995): 2–24.

Daston, Lorraine, and Peter Galison. "The Image of Objectivity." *Representations* 40 (1990): 81–128.

———. *Objectivity*. New York: Zone, 2010.

Davidson, Arnold. "Ethics as Ascetics: Foucault, the History of Ethics, and Ancient Thought." In *Foucault and the Writing of History*, edited by Jay Goldstein, 63–80. Oxford: Blackwell, 1994.

Davidson, Cathy N. *Now You See It: How Technology and Brain Science Will Transform Schools and Business in the 21st Century*. New York: Viking, 2011.

Davidson, Cathy N., and David Theo Goldberg. *The Future of Thinking: Learning Institutions in a Digital Age*. Cambridge, MA: MIT Press, 2010.

Delblanco, Andrew. *College: What It Was, Is, and Should Be*. Princeton, NJ: Princeton University Press, 2012.

de Man, Paul. *The Rhetoric of Romanticism*. New York: Columbia University Press, 1984.

de Ridder-Symoens, Hilde, and Walter Rüegg, eds. *A History of the University in Europe*. 4 vols. Cambridge: Cambridge University Press, 1992–2011.

Derrida, Jacques. *On Grammatology*. Translated by Gayatri Chakravorty Spivak. Baltimore: Johns Hopkins University Press, 1998.
Diderot, Denis. "Encyclopedia." In *Rameau's Nephew and Other Works*, edited and translated by Jacques Barzun, 291–323. New York: Doubleday, 1956.
Diehl, Carl. *Americans and German Scholarship, 1770–1870*. New Haven, CT: Yale University Press, 1978.
Dierse, Ulrich. *Encyklopädie. Zur Geschichte eines philosophischen und wissenschaftlichen Begriffs*. Bonn: Bouvier Verlag, 1977.
Diesterweg, Adolph. *Über das Verderben auf den deutschen Universitäten*. Essen: Bädeker, 1836.
Donoghue, Frank. *The Last Professors: The Corporate University and the Fate of the Humanities*. New York: Fordham University Press, 2008.
Eberhard, Johann August. "Prüfung der Frage, über die Dunkelheit der Kantischen Philosophie." *Neues philosophisches Magazin* 1 (1789).
———. *Synonymisches Handwörterbuch der deutschen Sprache*. Reutlingen: Job. Macken, 1805.
Eco, Umberto. *Semiotics and the Philosophy of Language*. Bloomington: University of Indiana Press, 1984.
Ehrlich, Adelhied. *Fichte als Redner*. Munich: tuduv Studie, 1977.
Eisenstein, Elizabeth L. *The Printing Revolution in Early Modern Europe*. Cambridge: Cambridge University Press, 2005.
Engelsing, Rolf. *Analphabetentum und Lektüre: Zur Sozialgeschichte des Lesens in Deutschland zwischen feudaler und industrieller Gesellschaft*. Stuttgart: J. B. Metzlersche Verlagsbuchhandlung, 1973.
———. *Der Bürger als Leser: Lesergeschichte in Deutschland 1500–1800*. Stuttgart: Metzler, 1974.
Enos, Richard Leo. "Ancient Greek Writing Instruction and Its Oral Antecedents." In *A Short History of Writing Instruction: From Ancient Greece to Contemporary America*, edited by James J. Murphy, 1–35. New York: Routledge, 2012.
Erben, Wilhelm. "Die Entstehung der Universitäts-Seminare." *Internationale Monatsschrift für Wissenschaft, Kunst, und Technik* 7 (1913): 1247–64.
Ersch, Johann Samuel, ed. *Allgemeines Repertorium der Literatur für die Jahre 1785 bis 1790*. Jena, 1793.
———, ed. *Allgemeines Repertorium der Literatur für die Jahre 1791–1795*. Vol. 3. Weimar: Industrie-Comptoirs, 1800.
Ersch, Johann Samuel, and Johann Gottfried Gruber, eds. *Allgemeine Encyclopädie der Wissenschaften und Künste in alphabetischer Folge*. Leipzig: Gleditsch, 1818.
Eschenburg, Joachim. *Lehrbuch der Wissenschaftskunde: Ein Grundriß encyklopädischer Vorlesungen*. Berlin: Friedrich Nicolai, 1800.
Fabian, Bernhard. "Im Mittelpunkt der Bücherwelt. Über Gelehrsamkeit und ge-

lehrtes Schriftum um 1750." In *Wissenschaften im Zeitalter der Aufklärung*, edited by Rudolf Vierhaus, 249–74. Göttingen: Vandenhoeck & Ruprecht, 1985.

Fabricius, Johann Andreas. *Abriß einer allgemeinen Historie der Gelehrsamkeit*. 3 vols. Leipzig: Weidmann, 1752–1754.

Febvre, Lucien, and Henri-Jean Martin. *The Coming of the Book: The Impact of Printing 1450–1800*. New York: Verso, 1997.

Feder, Johann Georg Heinrich. *Grundriß der Philosophischen Wissenschaften nebst der nöthigen Geschichte zum Gebrauch seiner Zuhörer*. Coburg: Johann Carl Findeisen, 1767.

Fichte, J. G. *Fichte's Early Philosophical Writings*. Edited by Daniel Breazeale. Ithaca, NY: Cornell University Press, 1988.

———. *J. G. Fichte Gesamtausgabe der Bayerischen Akademie der Wissenschaften*. Edited by Reinhard Lauth, Hans Jacob, and Hans Gliwitsky. Stuttgart-Bad Constatt: Friedrich Frommann Verlag, 1964–.

Foucault, Michel. *Technologies of the Self: A Seminar with Michel Foucault*. Amherst: University of Massachusetts Press, 1988.

Francke, August Hermann. *Kurtzer Bericht von der gegenwärtigen Verfassung des Paedagogii Regii zu Glaucha vor Halle*. Halle: Waysenhaus, 1710.

Frank, Manfred. *Unendliche Annäherung: Die Anfänge der Frühromantik*. Frankfurt am Main: Suhrkamp, 1977.

Franzel, Sean. *Fictions of Dialogue: The Media, Pedagogy and Politics of the Romantic Literature*. Evanston, IL: Northwestern University Press, 2013.

Frasca-Spada, Marina, and Nick Jardine, eds. *Books and the Sciences in History*. Cambridge: Cambridge University Press, 2003.

Friesen, Norm. "The Lecture as a Transmedial Pedagogical Form: A Historical Analysis." *Educational Researcher* 40, no. 3 (2011): 95–102.

Garve, Christian. *Gesammelte Werke*. Edited by Kurt Wölfel. Hildesheim: Georg Olms Verlag, 1985.

Gebhardt, Bruno. *Wilhelm von Humboldt als Staatsmann*. Vol. 1. Stuttgart: J. G. Cotta, 1896.

———, ed. *Wilhelm von Humboldts Politische Denkschriften*. 2 vols. Berlin: B. Behr, 1903.

Gierl, Martin. "Bestandsaufnahme im gelehrten Bereich: Zur Entwicklung der 'Historia literaria' im 18. Jahrhundert." In *Denkhorizonte und Handlungsspielräume. Historische Studien für Rudolf Vierhaus zum 70. Geburtstag*, 53–80. Göttingen: Wallstein, 1992.

———. "Historia literaria. Wissenschaft, Wissensordnung und Polemik." In *Historia Literaria: Neuordnungen des Wissens im 17. und 18. Jahrhundert*, edited by Frank Grunert and Friedrich Vollhardt, 114–227. Berlin: Akademie Verlag, 2007.

———. "Kompilation und die Produktion von Wissen im 18. Jahrhundert." In *Die*

*Praktiken der Gelehrsamkeit in der Frühen Neuzeit*, edited by Helmut Zedlemeier, 63–94. Tübingen: Niemeyer, 2001.
Giesecke, Michael. *Der Buchdruck in der frühen Neuzeit*. Frankfurt am Main: Suhrkamp, 1991.
Gilman, Daniel Coit. *University Problems in the United States*. New York: Arno Press, 1969.
Gitelman, Lisa. *Always Already New: Media, History, and the Data of Culture*. Cambridge, MA: MIT Press, 2006.
Glick, Thomas, Steven J. Livesey, and Faith Willis, eds. *Medieval Science, Technology, and Medicine: An Encyclopedia*. New York: Francis & Taylor, 2005.
Goetschel, Willi, Catriona Macleod, and Emery Snyder. "The *Deutsche Encyclopädie*." In *Notable Encyclopedias of the Late Eighteenth Century: Eleven Successors of the Encyclopédie*, edited by Frank A. Kafker, 257–333. Oxford: Voltaire Foundation, 1994.
Goldgar, Anne. *Impolite Learning: Conduct and Community in the Republic of Letters 1680–1750*. New Haven, CT: Yale University Press, 1995.
Gottsched, Johann Christoph. *Die vernünftigen Tadlerinnen*, 3rd ed. Hamburg, 1748.
Graff, Gerald. *Professing Literature: An Institutional History*. Chicago: University of Chicago Press, 1987.
Grafton, Anthony. "Humanities and Inhumanities." *New Republic*, February 17, 2010.
———. "Polyhistor into Philolog: Notes on the Transformation of German Classical Scholarship." *History of Universities* 3 (1983): 159–92.
———. "A Sketch of a Lost Continent: The Republic of Letters." In *Worlds Made by Words*, 9–34. Cambridge, MA: Harvard University Press, 2009.
———. "The World of the Polyhistors: Humanism and Encyclopedism." *Central European History* 18 (1985): 31–47.
Grafton, Anthony, and Lisa Jardine. *From Humanism to the Humanities*. Cambridge, MA: Harvard University Press, 1986.
Gregory, Brad S. *The Unintended Reformation: How a Religious Revolution Secularized Society*. Cambridge, MA: Harvard University Press, 2012.
Grimm, Gunter E. Grimm. *Letternkultur: Wissenschaftskritik und antigelehrtes Dichten in Deutschland von der Renaissance bis zum Sturm und Drang*. Tübingen: Max Niemeyer, 1998.
Grunert, Frank, and Friedrich Vollhardt, eds. *Historia Literaria: Neuordnungen des Wissens im 17. und 18. Jahrhundert*. Berlin: Akademie Verlag, 2007.
Gundling, Nicolaus Hieronymous. *Vollständige Historie der Gelahrtheit*. Vol. 1. Frankfurt/Leipzig: Wolfgang L. Spring, 1734.
Guyer, Paul. *Kant and the Claims of Taste*. Cambridge: Cambridge University Press, 1997.
Habel, Thomas. *Gelehrte Journale und Zeitungen der Aufklärung: zur Entstehung,*

*Entwicklung und Erschließung deutschsprachiger Rezensionszeitschriften des 18. Jahrhunderts*. Bremen: Edition Lumiere, 2007.

Habermas, Jürgen. "Die Idee der Universität—Lernprozesse." In *Eine Art Schadensabwicklung: Kleine Politische Schriften IV*, 71–100. Frankfurt: Suhrkamp, 1987.

———. *The Structural Transformation of the Public Sphere: An Inquiry into a Category of Bourgeois Society*. Translated by Thomas Burger. Cambridge, MA: MIT Press, 1999.

Habermas, Jürgen, and Niklas Luhmann. *Theorie der Gesellschaft oder Sozialtechnologie*. Frankfurt am Main: Suhrkamp, 1990.

Hadot, Pierre. *Philosophy as a Way of Life*. Translated by Arnold Davidson. Oxford: Blackwell, 1995.

Hahn, Roger. "Scientific Research as an Occupation in Eighteenth-Century Paris." *Minerva* 13 (1975): 501–13.

Hammerstein, Notker. "Die Universitätsgründungen im Zeichen der Aufklärung." In *Beiträge zu Problemen deutscher Universitätsgründungen der frühen Neuzeit*, edited by Peter Baumgart, 263–98. Nendeln: KTO Press, 1978.

———. "Epilogue: The Enlightenment." In *A History of the University in Europe*, vol. 2, *Universities in Early Modern Europe 1500–1800*, edited by Hilde de Ridder-Symoens, 621–40. Cambridge: Cambridge University Press, 1996.

———, ed. *Universitäten und Aufklärung*. Göttingen: Wallstein Verlag, 1995.

Hansen, Mark. "New Media." In *Critical Terms for Media Studies*, edited by M. Hansen and W. J. T. Mitchell, 172–85. Chicago: University of Chicago Press, 2010.

Harnack, Adolf. *Geschichte der Königlich Preussischen Akademie der Wissenschaften zu Berlin*. 2 vols. Berlin: Reichsdruckerei, 1900.

Haskell, Thomas L. *The Emergence of Professional Social Science: The American Social Science Association and the Nineteenth Century Crisis of Authority*. Baltimore: Johns Hopkins University Press, 2000.

Heckel, August. *Über die Natur und Heilart der Faulfieber*. Berlin: Friedrich Maurer, 1809.

Heilbron, John. "Das Regime der Disziplinen." In *Interdisciplinarität als Lernprozess*, edited by Hans Joas, 23–46. Göttingen: Wallstein Verlag, 2005.

Heinzmann, Johann Georg. *Appel an meine Nation: Über die Pest der deutschen Literatur*. Bern, 1795.

Henningsen, Jürgen. "Enzyklopädie: Zur Sprach- und Bedeutungsgeschichte eines pädagogischen Begriffs." In *Archiv für Begriffsgeschichte*, vol. 10, edited by Erich Rothacker. Bonn, 1966.

Herbst, Jürgen. *The German Historical School in American Scholarship: A Study in the Transfer of Culture*. Ithaca, NY: Cornell University Press, 1965.

Herder, Johann Gottfried. *Briefe zur Beförderung der Humanität*. Berlin: Aufbau Verlag, 1971.

Herdt, Jennifer A. *Putting on Virtue: The Legacy of the Splendid Vices*. Chicago: University of Chicago Press, 2010.

Heubaum, A. "Die Reformbestrebungen unter dem preussichen Minister J. v. Massow auf dem Gebiet des höheren Bildungswesen." *Mitteilungen der Gesellschaft für deutsche Erziehungs- und Schulgeschichte* 14 (1904): 186–225.

Heumann, Christoph August. *Conspectus reipublicae literariae*. Hannover, 1718.

Heusinger, Johann Heinrich Gottlieb. *Versuch einer Encyklopädie der Philosophie, verbunden mit einer praktischen Anleitung zu dem Studium der kritischen Philosophie vorzüglich auf den Universitäten*. Weimar: Industrie-Comptoir, 1796.

Higton, Mike. *A Theology of Higher Education*. Oxford: Oxford University Press, 2012.

Hoche, J. G. *Vertraute Briefe über die jetzige Lesesucht und über den Einfluß derselben auf die Verminderung des häuslichen und öffentlichen Glücks*. Hannover: Ritscher, 1794.

Hoffmann, Maximillian. *August Böckh. Lebensbeschreibungen und Auswahl aus seinem wissenschaftlichen Briefwechsel*. Leipzig: B. G. Teubner, 1901.

Hofstetter, Michael J. *The Romantic Idea of a University: England and Germany, 1770–1850*. New York: Palgrave, 2001.

Hohendahl, Peter Uwe. "Humboldt Revisited: Liberal Education, University Reform, and the Opposition to the Neoliberal University." *New German Critique* 38, no. 2 113 (2011): 159–96.

Horstmann, Axel. "Die Forschung in der klassischen Philologie des 19. Jahrhunderts." In *Konzeption und Begriff der Wissenschaften des 19. Jahrhunderts*, edited by Alwin Diemer, 27–57. Maisenhain am Glan: Anton Hain, 1978.

Howard, Thomas Albert. *Protestant Theology and the Making of the Modern German University*. Oxford: Oxford University Press, 2009.

Humboldt, Wilhelm von. *Wilhelm von Humboldts Gesammelte Schriften*. Edited by Bruno Gebhardt. Berlin: B. Behr's Verlag, 1968.

Hunter, Ian. *Rival Enlightenments: Civial and Metaphysical Philosophy in Early Modern Germany*. Cambridge: Cambridge University Press, 2001.

Iselin, Isaak. *Ueber die Geschichte der Menschheit*. 2 vols. Frankfurt: Harscher, 1764.

Israel, Jonathan I. *Enlightenment Contested: Philosophy, Modernity, and the Emancipation of Man 1670–1752*. Oxford: Oxford University Press, 2006.

Jeismann, Karl-Ernst. *Das preußische Gymnasium in Staat und Gesellschaft*. Vol. 2. Stuttgart: Klett-Cotta, 1996.

Jentzsch, Rudolf. *Der deutsch-lateinische Büchermarkt nach den Leipziger Ostermeßkatalogen von 1740, 1770 und 1800 in seiner Gliederung und Wandlung*. Leipzig: Voigtländer, 1912.

Jöcher, Christian Gottlieb. *Allgemeines Gelehrten-Lexicon*. 2 vols. Leipzig: Geditsch, 1750.

———. *Compendiöses Gelehrten-Lexicon*. 2 vols. Leipzig: Johann Friedrich Gleditschens, 1726/1733.

Johns, Adrian. *The Nature of the Book: Print and Knowledge in the Making*. Chicago: University of Chicago Press, 1998.

Jordheim, Helge. "The Present of the Enlightenment." In *This Is Enlightenment*, edited by Clifford Siskin and William Warner, 189–208. Chicago: University of Chicago Press, 2010.

Justi, Johann Heinrich Gottlobs von. *Grundsätze der Policeywissenschaft*. 3rd ed. Göttingen: Vandenhoeck, 1782.

Kant, Immanuel. *The Conflict of the Faculties*. In *Immanuel Kant: Religion and Rational Theology*, edited and translated by Allen W. Wood and George di Giovanni. Cambridge: Cambridge University Press, 2001.

———. *Critique of Pure Reason*. Translated and edited by Paul Guyer and Allen W. Wood. Cambridge: Cambridge University Press, 1998.

———. *Critique of the Power of Judgment*. Edited by Paul Guyer. Translated by Paul Guyer and Eric Matthews. Cambridge: Cambridge University Press, 2000.

———. *Gesammelte Schriften*. Edited by the Königlich Preussischen Akademie der Wissenschaften. 29 vols. to date. Berlin: Walter de Gruyter, 1902–.

———. *Practical Philosophy*. Translated and edited by Mary J. Gregor. Cambridge: Cambridge University Press, 1996.

———. *Prolegomena zu einer jeden künftigen Metaphysik, die als Wissenschaft wird auftreten können*. Edited by Rudolf Malter. Reclam: Stuttgart, 1989.

Kappens, Johann Erhard. "Vorrede." In *Versuch einer Geschichte der schönen und anderen Wissenschaften, wie auch der freyen und einiger mechanischen Künste*, vol. 1, by Juvenel de Carlencas, translated by Johann Erhard Kappens. Leipzig: Gleditschische Buchhandlung, 1749.

Karabel, Jerome. *The Chosen: The Hidden History of Admission and Exclusion at Harvard, Yale, and Princeton*. New York: Mariner, 2005.

Kelly, Donald R., ed. *History and the Disciplines: The Reclassification of Knowledge in Early Modern Europe*. Rochester, NY: University of Rochester Press, 1997.

Kerr, Clark. *The Uses of the University*. Cambridge, MA: Harvard University Press, 1963.

Kim, You Sin. *Als die Lumpen Flügel bekamen: Frühromantik im Zeitalter des Buchdrucks*. Würzburg: Königshausen & Neumann, 2004.

Kirchner, Joachim. *Bibliographie der Zeitschriften des deutschen Sprachgebietes bis 1900*. Vol. 1, part 4. Stuttgart: Hiersemann, 1966.

———. *Das deutsche Zeitschriftenwesen, seine Geschichte und Seine Probleme. Teil I: Von den Anfängen bis zum Zeitlater der Romantik*. Wiesbaden: Otto Harrassowitz, 1958.

———. *Die Grundlagen des deutschen Zeitschriftenwesens. Mit einer Gesamtbib-*

*liographie der deutschen Zeitschriften bis zum Jahre 1790*. 2 vols. Leipzig: K. W. Hiersemann, 1928–1931.

Kittler, Friedrich. *Aufschreibesysteme 1800/1900*. Munich: Fink, 1995.

Kluge, Alexander. *Die Universitäts-Selbstverwaltung*. New York: Arno Press, 1977.

Knodt, Eva. *Negative Philosophie und dialogische Kritik: Zur Struktur poetischer Theorie bei Lessing und Herder*. Tübingen: M. Niemeyer, 1988.

Kocka, Jürgen. "Bildungsbürgertum—Gesellschaftliche Formation oder Historikerkonstruct?" In *Bildungsbürgertum im 19. Jahrhundert, Teil IV*, edited by Jürgen Kocka. Stuttgart: Klett-Cotta, 1989.

———, ed. *Das Bildungsbürgertum im 19. Jahrhundert*. 4 vols. Stuttgart: Klett-Cotta, 1985–1992.

König, Rene. *Vom Wesen der deutschen Universität*. Edited by Peter Thurn. Opladen: Leske & Budrich, 2000.

Köpke, Rudolf. *Die Gründung der Königlichen Friedrich-Wilhelms-Universität zu Berlin*. Berlin: Gustav Schade, 1860.

Koschorke, Albrect. *Körperströme und Schriftverkehr. Mediologie des 18. Jahrhunderts*. Munich: Wilhelm Fink Verlag, 2003.

Koselleck, Reinhart. *Preußen zwischen Reform und Revolution: Allgemeines Landrecht, Verwaltung und soziale Bewegung von 1791 bis 1848*. Stuttgart: Klett, 1975.

Köster, Heinrich, Martin Gottfried, and Johann Friedrich Roos, eds. *Deutsche Enzyklopädie, oder Allgemeines Real-Wörterbuch aller Künste und Wissenschaften*. 23 vols. Frankfurt am Main: Varrentrap Sohn & Wenner, 1778–1807.

Kronman, Anthony T. *Education's End: Why Our Colleges and Universities Have Given Up on the Meaning of Life*. New Haven, CT: Yale University Press, 2007.

Krug, Wilhelm Traugott. *Über den Zusammenhang der Wissenschaften unter sich und mit den höchsten Zwecken der Vernunft*. Jena: Akademisches Leseinstitut, 1795.

———. *Versuch einer neuen Eintheilung der Wissenschaften zur Begründung einer besseren Organization höheren gelehrten Bildungsanstalten*. Züllichau: Darnmann, 1805.

———. *Versuch einer systematischen Enzyklopädie der Wissenschaften*. Leipzig: Winckelmann & Barth, 1796.

Kühlmann, Wilhelm. *Gelehrtenrepublik und Fürstenstaat: Entwicklung und Kritik des deutschen Späthumanismus in der Literatur des Barockzeitalters*. Tübingen: Nieymeyer, 1982.

Kuklick, Bruce. *A History of Philosophy in America: 1720–2000*. Oxford: Oxford University Press, 2003.

Kunoff, Hugo. *The Foundations of the German Academic Library*. Chicago: American Library Association, 1982.

Lacapra, Dominick. "The University in Ruins?" *Critical Inquiry* 25, no. 1 (Autumn 1998): 32–55.

Landau, Albert, ed. *Rezensionen zur Kantischen Philosophie 1781–87*. Bebra: Albert Landau Verlag, 1991.

Landfester, Manfred. "Ulrich von Wilamowitz-Moellendorff und die hermeneutische Tradition des 19. Jahrhunderts." In *Philologie und Hermeneutik im 19. Jahrhundert: Zur Geschichte und Methodologie der Geisteswissenschaften*, edited by Hellmut Flashar. Göttingen: Vandenhoeck & Ruprecht, 1979.

Langewiesche, Dieter. "Die Humboldtsche Universität als nationaler Mythos." *Historische Zeitschrift* 290 (2010): 53–88.

La Vopa, Anthony J. *Grace, Talent, and Merit: Poor Students, Clerical Careers, and Professional Ideology in Eighteenth-Century Germany*. Cambridge: Cambridge University Press, 1988.

———. "Specialists against Specialization: Hellenism as Professional Ideology." In *German Professions 1800–1950*, edited by Geoffrey Cocks and Konrad H. Jarausch, 27–45. Oxford: Oxford University Press, 1990.

Lear, Jonathan. *Aristotle: The Desire to Understand*. Cambridge, MA: Harvard University Press, 1988.

Legaspi, Michael. *The Death of Scripture and the Rise of Biblical Studies*. Oxford: Oxford University Press, 2010.

Lenz, Max. *Geschichte der Königlichen Friedrich-Wilhelms-Universität zu Berlin*. 2 vols. Halle: Buchhandlung des Waisenhauses, 1910.

Lessing, Gotthold Ephraim. *Der junge Gelehrte*. In *Gotthold Ephraim Lessing Werke 1743–1750*, edited by Jürgen Stenzel. Frankfurt am Main: Deutscher Klassiker Verlag, 1999.

Lichtenberg, Georg. *Gedanken, Satiren, Fragmenten*. Vol. 1. Edited by Wilhelm Herzog. Jena: Diederichs, 1907.

Lindemann, Margot. *Deutsche Presse bis 1815. Geschichte der deutschen Presse Teil 1*. Berlin: Colloquium Verlag, 1969.

Luhmann, Niklas. *Soziologische Aufklärung: Aufsätze zur Theorie sozialer Systeme*. Köln: Westdeutscher Verlag, 1970.

———. *Universität als Milieu*. Edited by Andre Kieserling. Bielefeld: Haux, 1992.

Lupton, Christina. *Knowing Books: The Consciousness of Mediation in Eighteenth-Century Britain*. Philadelphia: University of Pennsylvania Press, 2011.

Maatsch, Jonas. "Jenaer Vorlesungen zur Enzyklopädie und Wissenschaftskunde." In *Gelehrte Wissenschaft: Das Vorlesungsprogramm der Universität Jena um 1800*, edited by Thomas Bach, Jonas Maatsch, and Ulrich Rasche, 125–40. Franz Steiner Verlag: Stuttgart, 2008.

———. *Naturgeschichte der Philosopheme: Frühromantische Wissensordnungen im Kontext*. Heidelberg: Universitätsverlag Winter, 2008.

MacIntyre, Alasdair, *After Virtue: A Study in Moral Theory*. 3rd ed. Notre Dame, IN: Notre Dame University Press, 2008.

———. *Three Rival Versions of Moral Enquiry: Encyclopedia, Genealogy and Tradition*. Notre Dame, IN: University of Notre Dame Press, 1988.

Marchand, Susan. *Down from Olympus: Archaeology and Philhellinism in Germany, 1750–1970*. Princeton, NJ: Princeton University Press, 1996.

Marino, Luigi. *Praeceptores Germaniae: Göttingen 1770–1820*. Göttingen: Vandenhoeck & Ruprecht, 1995.

Markus, Gygory. "Changing Images of Science." *Thesis Eleven* 33, no. 1 (August 1992): 1–56.

Martens, Wolfgang. *Die Botschaft der Tugend: Die Aufklärung im Spiegel der deutschen moralischen Wochenschriften*. Stuttgart: Metzlersche Verlagsbuchhandlung, 1968.

———. "Hallescher Pietismus und Gelehrsamkeit, oder vom 'allzu großen Mißtrauen in den Wissenschaften." In *Res Publica Litteraria*, 2:497–523. Wiesbaden: Otto Harrassowitz, 1987.

———. "Lasterhaftes Lesen." In *Einladung ins 18. Jahrhundert: Ein Almanach aus dem Verlag C. H. Beck*, edited by Ernst-Peter Wickenberg, 97–103. Munich: C. H. Beck, 1987.

———. *Print, Manuscript and the Search for Order, 1450–1830*. Cambridge: Cambridge University Press, 2003.

———. "Vom Thomasius bis Lichtenberg: Zur Gelehrtensatire der Aufklärung." *Lessing Yearbook* 10 (1978): 7–34.

Martini, Friedrich Heinrich Wilhelm. *Allgemeine Geschichte der Natur in alphabetischer Ordnung*. Berlin: Joachim Pauli, 1774.

Mauch, Otto. *Der lateinische Begriff Disciplina: Eine Wortuntersuchung*. Freiburg: Paulusdruckerei, 1941.

Mazon, Patricia M. *Gender and the Modern Research University: The Admission of Women to German Higher Education, 1865–1914*. Palo Alto, CA: Stanford University Press, 2003.

McAuley, Alexander, et al. "The MOOC Model for Digital Practice." Canadian Social Science and Research Council, 2010.

McClellan, James E. *Science Reorganzied: Scientific Societies in the Eighteenth Century*. New York: Columbia University Press, 1985.

McClelland, Charles E. *The German Experience of Professionalization: Modern Learned Professions and Their Organization from the Early Nineteenth Century to the Hitler Era*. Cambridge: Cambridge University Press, 1991.

———. *State, Society, and University in Germany: 1700–1914*. Cambridge: Cambridge University Press, 1980.

McDowell, John. *Mind and World*. Cambridge, MA: Harvard University Press, 1994.

McGann, Jerome. *A New Republic of Letters*. Cambridge, MA: Harvard University Press, 2014.

McKitterick, David. "Bibliography, Bibliophily and Organization of Knowledge." In *The Foundations of Scholarship: Libraries and Collecting, 1650–1750*. Los Angeles: Clark Memorial Library, 1992.

McLuhan, Marshall. *Understanding Media: The Extensions of Man*. Cambridge, MA: MIT Press, 1994.

McNeely, Ian F., and Lisa Wolverton. *Reinventing Knowledge: From Alexandria to the Internet*. New York: W. W. Norton, 2008.

Meckenstock, Günter, ed. *Friedrich Daniel Ernst Schleiermacher: Schriften aus der Berliner Zeit, 1796–1799*. New York: Walter de Gruyter, 1984.

Meiners, Christoph. *Über die Verfassung und Verwaltung deutscher Universitäten*. 2 vols. Göttingen: Röwer, 1801–1802.

Meld-Shell, Susan. *Kant and the Limits of Autonomy*. Cambridge, MA: Harvard University Press, 2009.

Menand, Louis. "The Limits of Academic Freedom." In *The Future of Academic Freedom*, edited by Louis Menand, 3–20. Chicago: University of Chicago Press, 1996.

———. *The Marketplace of Ideas: Reform and Resistance in the American University*. New York: W. W. Norton, 2010.

Mencke, Johann Burkhard. *The Charlatanry of the Learned*. Translated by H. L. Mencken. New York: Knopf, 1937.

Michaelis, Johann David. *Raisonnement über die protestantischen Universitäten in Deutschland*. 4 vols. Frankfurt am Main, 1768–1776.

Minnis, A. J. *Medieval Theory of Authorship: Scholastic Attitudes in the Later Middle Ages*. London: Scholar Press, 1984.

Mittelstrauß, Jürgen. *Wissenschaft als Lebensform*. Frankfurt am Main: Suhrkamp, 1982.

Mommsen, Theodor. *Reden und Aufsätze*. Edited by Otto Hirschfeld. Berlin: Weidmann, 1905.

Moraw, Peter. "Humboldt in Giessen: Zur Professorenberufung an einer deutschen Universität des 19. Jahrhunderts." *Geschichte und Gesellschaft* 10 (1984): 47–71.

Most, Glenn W. "On the Use and Abuse of Ancient Greece for Life." *Cultura tedesca* 20 (200): 31–53.

Müller, Ernst, ed. *Gelegentliche Gedanken über Universitäten*. Leipzig: Reclam, 1990.

Müller, Gerhard, Klaus Ries, and Paul Ziche, eds. *Die Universität Jena. Tradition und Innovation um 1800*. Stuttgart: Franz Steiner, 2001.

Nancy, Jean-Luc. *The Discourse of the Syncope: Logodaedalus*. Translated by Saul Anton. Palo Alto, CA: Stanford University Press, 2008.

Nardi, Paolo. "Relations with Authority." In *History of the University in Europe*, vol. 2, *Universities in the Middle Ages*, edited by Hilde de Ridder-Symoens, 77–107. Cambridge: Cambridge University Press, 1996.

Neiman, Susan. *The Unity of Reason: Rereading Kant*. Oxford: Oxford University Press, 1994.

Nelles, Paul. “Historia litteraria and Morhof: Private Teaching and Professorial Libraries at the University of Kiel.” In *Mapping the World of Learning: The Polyhistor of Daniel Georg Morhof*, edited by Francoise Waquet, 31–56. Wiesbaden: Harrassowitz, 2000.

Nicolai, Friedrich. *Beschreibung der Königlichen Residenzstädte Berlin und Potsdam*. Vol. 2. Berlin: Friedrich Nicolai, 1786.

———. *Das Leben und die Meinungen des Herrn Magister Sebaldus Nothanker*. Vol. 1. Berlin: Friedrich Nicolia, 1776.

———. *Wegweiser für Fremde und Einheimische*. Berlin: Friedrich Nicolia, 1793.

Nipperdey, Thomas. “Verein als soziale Struktur in Deutschland im späten 18. und frühen 19. Jahrhundert. Eine Fallstudie zur Modernisierung.” In *Gesellschaft, Kultur, Theorie*, edited Thomas Nipperdey, 174–205. Göttingen: Vandenhoeck & Ruprecht, 1976.

Novalis [Friedrich von Hardenberg]. *Novalis Schriften. Die Werke Friedrich von Hardenbergs*. 6 vols. Edited by Paul Kluckhohn and Richard Samuel. Darmstadt: Wissenschaftliche Buchhandlung, 1960–1988.

Nussbaum, Martha. *Not for Profit: Why Democracy Needs the Humanities*. Princeton, NJ: Princeton University Press, 2012.

O’Boyle, Lenore. “Klassische Bildung und soziale Struktur in Deutschland zwischen 1800 und 1848.” *Historische Zeitschrift* 207 (1969): 584–609.

Olesko, Kathryn. *Physics as Calling: Discipline and Practice in the Königsberg Seminar for Physics*. Ithaca, NY: Cornell University Press, 1991.

Palatschek, Sylvia. “Die Erfindung der Humboldtschen Universitätsidee in der ersten Hälfte des 20. Jahrhunderts.” *Historische Anthropologie* 10 (2002): 183–205.

Papst, Stephan. “Vollständigkeit und Totalität: Die Allgemeine Literatur-Zeitung und die Ordnung des Wissens um 1800.” In *Organisation der Kritik: Die Allgemeine Literatur-Zeitung in Jena 1785–1803*, edited by Stefan Matuschek, 55–76. Heidelberg: Universitätsverlag Winter, 2004.

Parkes, M. B. “Reading, Copying and Interpreting a Text in the Early Middle Ages.” In *A History of Reading in the West*, edited by Guglielmo Cavallo and Roger Chartier, 90–102. Amherst: University of Massachusetts Press, 1999.

Parsons, Talcott. *The American University*. Cambridge, MA: Harvard University Press, 1973.

Paulsen, Friedrich. *Geschichte des Gelehrten Unterrichts auf den deutschen Schulen und Universitäten vom Ausgang des Mittelalters bis zur Gegenwart*. 2 vols. 3rd ed. Leipzig: Veit, 1919–1921.

Peters, John Durham. *Speaking into the Air: A History of the Idea of Communication*. Chicago: University of Chicago Press, 1999.

Petrus, Klaus. “Beschrieene Dunkelheit und Seichtigkeit: Historisch-systematische Voraussetzungen zwischen Kant und Garve im Umfeld der Göttinger Rezension.” *Kant-Studien* 85, no. 3 (1994): 280–302.

Piper, Andrew. *Dreaming in Books*. Chicago: University of Chicago Press, 2009.

Plato. *Phaedrus*. Translated by Christopher Rowe. New York: Penguin Classics, 2005.

Polanyi, Michael. *Personal Knowledge: Towards a Post-Critical Philosophy*. Chicago: University of Chicago Press, 1968.

Polheim, Konrad. "Studien zu Friedrich Schlegels poetischen Begriffen." *DVjS* 35 (1961): 363–98.

Postman, Neil. *Technopoloy: The Surrender of Culture to Technology*. New York: Knopf, 1992.

Price, Leah. *How to Do Things with Books in Victorian Britain*. Princeton, NJ: Princeton University Press, 2012.

Puschner, Uwe. "Mobil gemachte Feldbibliotheken. Deutsche Enzklopädien und Konversationslexika im 18. Und 19. Jahrhundert." *Internationales Archiv für Sozialgeschichte der deutschen Literatur* 8 (1997): 62–77.

Pütter, Johann Stephan. *Neuer Versuch einer Juristischen Encyclopädie und Methodologie*. Göttingen, 1767.

Pütter, Johann Stephan, and Friedrich Saalfeld. *Versuch einer academischen Gelehrten-Geschichte der Georg-Augustus Universität zu Göttingen*. 4 vols. Göttingen: Vandenhoeck & Ruprecht, 1765–1838.

Raabe, Paul. "Die Zeitschrift als Medium der Aufklärung." *Wolfenbütteler Studien zur Aufklärung* 1 (1974): 99–136.

Reimmann, Jakob Friedrich. *Versuch einer Einleitung in die Historiam Literariam so wohl insgemein als auch die Historiam Literariam derer Teutschen insonderheit*. Halle, 1708.

Rietzschel, Evi, ed. *Gelehrsamkeit ein Handwerk? Bücherschreiben ein Gewerbe? Dokumente zum Verhältnis von Schriftsteller und Verleger im 18. Jahrhundert in Deutschland*. Leipzig: Reclam, 1982.

Rosenberg, Daniel. "An 18th-Century Time Machine: The *Encyclopédie* of Denis Diderot." In *Postmodernism and the Enlightenment*, edited by Daniel Gordon, 227–50. London: Routledge, 2001.

Rüegg, Walter. "Foreword." In *A History of the University in Europe*, vol. 1, *Universities in the Middle Ages*, edited by Hilde de Ridder-Symoens, xix–xxviii. Cambridge: Cambridge University Press, 1992.

———. "Themes." In *A History of the University in Europe*, vol. 1, *Universities in the Middle Ages*, edited by Hilde de Ridder-Symoens, 3–33. Cambridge: Cambridge University Press, 1992.

Salzmann, Christian Gotthilf. *Carl von Carlsberg, oder über das menschliche Elend*. Vol. 1. Leipzig: Christian Gottlieb Schmieder, 1784.

Schaper, Eva. "Taste, Sublimity, and Genius: The Aesthetics and Nature of Art." In *The Cambridge Companion to Kant*, edited by Paul Guyer, 367–93. Cambridge: Cambridge University Press, 1992.

Scheel, Otto. "*Die deutsche Universität in ihren Anfängen bis zur Gegenwart.*" In Michael Doeberl, *Das akademische Deutschland*, vol. 1. Berlin: C. A. Weller, 1930.

Schelling, Friedrich. *F. W. J. Schelling, Werke: Auswahl in drei Bänden*. Leipzig: Fritz Eckardt Verlag, 1907.

———. "System of Philosophy in General and of the Philosophy in Nature in Particular." In *Idealism and the Endgame of Theory*, edited and translated by Thomas Pfau, 139–85. Albany: State University of New York Press, 1994.

Schelsky, Helmut. *Einsamkeit und Freiheit: Zur sozialen Idee der deutschen Universität*. Düsseldorf: Bertelsmann-Universitäts-Verlag, 1960.

Schenda, Rudolf. *Volk ohne Buch: Studien zur Sozialgeschichte der populären Lesestoffe 1770–1910*. Frankfurt am Main: Vittorio Klostermann, 1970.

Schlegel, August Wilhelm. *Vorlesungen über das akademische Studium: Bonner Vorlesungen I*. Heidelberg: Lothar Stiehm Verlag, 1971.

———. *Vorlesungen über Encyklopädie*. In *Vorlesungen Über Ästhetik II, 1803–1827*, edited by Ernst Behler. Vol. 2/1 of *August Wilhelm Schlegel Kritische Ausgabe der Vorlesungen*, edited by Georg Braungart. Paderborn: Ferdinand Schöningh, 2007.

———. *Vorlesungen über schöne Literatur*. In *Vorlesungen Über Ästhetik I: 1798–1803*, edited by Ernst Behler. Vol. 1 of *August Wilhelm Schlegel Kritische Ausgabe der Vorlesungen*, edited by Georg Braungart. Paderborn: Ferdinand Schöningh, 1989.

Schleiermacher, Friedrich. "Gelegentliche Gedanken über Universitäten in deutschem Sinn, Nebst einem Anhang über eine neue zu errichtende." In *Gelegentliche Gedanken über Universitäten*, edited by Ernst Müller, 159–258. Leipzig: Reclam, 1990.

Schmid, Carl Christian Erhard. *Allgemeine Encyklopädie und Methodologie der Wissenschaften*. Jena: Akademische Buchhandlung, 1810.

Schmid, Carl Heinrich. "Ueber die Klassifikation und Rangordnung der Wissenschaften." *Gothaisches Magazin der Künste und Wissenschaften* 2 (1777).

Schmid, Karl Adolf. *Encyklopädie des gesammten Erziehungs- und Unterrichtswesen, bearbeitet von einer Anzahl Schulmänner und Gelehrten*. Vol. 5. Gotha: Rudolf Besser, 1866.

Schmidt-Biggemann, Wilhelm. "New Structures of Knowledge." In *A History of the University in Europe*, vol. 2, *Universities in Early Modern Europe*, edited by Hilde de Ridder-Symoens, 489–530. Cambridge: Cambridge University Press, 1996.

———. *Topica universalis: Eine Modellgeschichte humanistischer und barocker Wissenschaft*. Hamburg: Felix Meiner, 1983.

Schneider, Notker. "Experientia-ars-scientia-sapientia: Zu Weisen und Arten des Wissens im Anschluß an Aristoteles und Thomas von Aquin." In *Scientia und ars in Hoch- und Spätmittelalter*, edited by Ingrid Craemer-Ruegenberg and Andreas Speer, 171–88. Berlin: Walter de Gruyter, 1994.

Schneider, Ulrich Johannes. "Die Konstruktion des allgemeinen Wissens in Zedlers Universal-Lexicon." In *Wissenssicherung, Wissensordnung und Wissensverbreitung: Das Europäische Modell der Enzyklopädien*, edited by Theo Stammen, 81–101. Berlin: Akademie Verlag, 2004.

Schneiders, Werner. "Akademische Weltweisheit: Die deutsche Philosophie im Zeitalter der Aufklärung." In *Frankreich und Deutschland im 18. Jahrhundert*, edited by Gerhard Sauder and Joachim Schlobach, 25–44. Heidelberg: Winter, 1986.

———. "Der Philosophiebegriff des philosophischen Zeitalters: Wandlungen im Selbstverständnis der Philosophie von Leibniz bis Kant." In *Wissenschaften im Zeitalter der Aufklärung*, edited by Rudolf Vierhaus, 58–92. Göttingen: Vandenhoeck & Ruprecht, 1985.

Schrimpf, G. "Disciplina." In *Historisches Wörterbuch der Philosophie*, vol. 2, edited by Joachim Ritter and Karl Gründer, 256–61. Basel: Schwabe, 1972.

Schulte-Sassen, Jochen. "Das Konzept bürgerlich-literarischer Öffentlichkeit und die historischen Gründe seines Zerfalls." In *Aufklärung und literarische Öffentlichkeit*, edited by Christa Bürger, Peter Bürger, and Jochen Schulte-Sassen, 83–115. Frankfurt am Main: Surhkamp, 1980.

Schwartz, Paul. *Die Gelehrtenschulen Preussens unter dem Oberschulkollegium (1787–1806) und das Abiturientenexamen*. 3 vols. Berlin: Weidmann, 1910–1912.

Shapin, Steven. *The Scientific Life: A Moral History of a Late Modern Vocation*. Chicago: University of Chicago Press, 2008.

Sheehan, Jonathan. *The Enlightenment Bible: Translation, Scholarship, Culture*. Princeton, NJ: Princeton University Press, 2005.

Siskin, Clifford. *The Work of Writing: Literature and Social Change In Britain, 1700–1830*. Baltimore: Johns Hopkins University Press, 1998.

Siskin, Clifford, and William Warner, eds. *This Is Enlightenment*. Chicago: University of Chicago Press, 2010.

Smith, Adam. *The Wealth of Nations*. New York: Bantam Classics, 2003.

Spiewak, Martin. "Falsches Vorbild." *HU 200 Jubiläumsmagazin*, 2009, 72–73.

Spranger, Eduard. *Fichte, Schleiermacher, Steffens über das Wesen der deutschen Universität*. Leipzig: Dürr, 1910.

———. *Wilhelm von Humboldt und die Humanitätsidee*. Berlin: Reuther & Reichard, 1909.

———. *Wilhelm von Humboldt und die Reform des Bildungswesen*. 3rd ed. Tübingen: Max Niemeyer Verlag, 1965.

Steffens, Henrik. *Ueber die Idee der Universitäten*. Berlin: Realschulbuchhandlung, 1809.

Steig, Reinhold. *Heinrich von Kleists Berliner Kämpfe*. Berlin: W. Spemann, 1901.

Stern, Albert. *Über die Beziehungen Christian Garve's zu Kant nebst mehreren ungedruckten Briefen Kant's, Feder's, Garve's*. Leipzig: Denicke, 1884.

Stichweh, Rudolf. *Der frühmoderne Staat und die europäische Universität: zur In-*

*teraktion von Politik und Erziehungssystem im Prozess ihrer Ausdifferenzierung (16.–18. Jahrdundert)*. Frankfurt am Main: Suhrkamp, 1991.

———. *Die Ausdifferenzierung der Wissenschaft: Eine Analyse am deutschen Beispiel*. Grünwald: Kleine Verlag, 1981.

———. *Zur Entstehung des modernen Systems wissenschaftlicher Disziplinen in Deutschlands 1740–1890*. Frankfurt am Main: Suhrkamp, 1984.

Stolle, Gottlieb. *Anleitung zur Historie der Gelahrtheit*. Jena: Meyer, 1724.

Stölzel, Adolf. "Die Berliner Mittwochsgesellschaft über Aufhebung oder Reform der Universitäten (1795)." *Forschungen zur brandenburgischen und preußischen Geschichte* 2 (1889): 201–22.

Strube, Werner. "Die Geschichte des Begriffs 'schöne Wissenschaften.'" *Archiv für Begriffsgeschichte* 33 (1990): 136–216.

Sulzer, Johann Georg. *Kurzer Begriff aller Wissenschaften und anderer Theilen der Gelehrsamkeit*. 2nd ed. Leipzig: Johann Christian Lagenheim, 1759.

Summers, Lawrence. Foreword to *An Avalanche Is Coming: Higher Education and the Revolution Ahead*, by Michael Barber, Katelyn Donnelly, and Saad Rizni. Institute for Public Policy Research, March 2013.

Sweet, Paul. *Wilhelm von Humboldt: A Biography*. 2 vols. Columbus: Ohio State University Press, 1978.

Taylor, Mark C. *Crisis on Campus: A Bold Plan for Reforming Our Colleges and Universities*. New York: Knopf, 2010.

Tenorth, Heinz-Elmar, ed. *Geschichte der Universität unter den Linden: Genese der Disziplinen*. Vol. 4. Berlin: Akademie Verlag, 2010.

———. "Nur in Berlin, Die Universität in Ihrer Welt." In *Welt Wissen*, edited by Jochen Hennig and Udo Andraschke, 118–32. München: Hirmer Verlag, 2011.

———. "Transformation der Wissensordung. Die Berliner Universität vom ausgehenden 19. Jahrhundert bis 1945. Zur Einleitung." In *Geschichte der Universität Unter den Linden: 1810–2010: Transformation der Wissensordnung*, edited by Heinz-Elmar Tenorth, 9–52. Berlin: Akademie Verlag, 2010.

———. "*Was heißt Bildung in der Universität?* Oder: Transzendierung der Fachlichkeit als Aufgabe universitärer Studien." *Die Hochschule* 1 (2010): 119–34.

———. "Wilhem von Humboldts Universitätskonzept und die Reform in Berlin—eine Tradition jenseits des Mythos." *Zeitschrift für Germanistik* 20 (2010): 15–28.

Terlinder, Reinhard F. *Versuch einer Vorbereitung zu der heutigen positiven in Teutschland üblichen gemeinen Rechtsgelahrtheit für angehende Rechtsgelehrte*. Münster: Perrenon, 1787.

Thomasius, Christian. *Einleitung zu der Vernunft-Lehre*. Halle: C. Salfeld, 1719.

———. *Freimütige, lustige und ernsthafte, jedoch vernunftmässige Gedanken oder Monatsgespräche über allerhand, fürnehmlich aber neue Bücher*. Vol. 3 (January–June 1689). Frankfurt am Main, 1689. Reprint, Frankfurt am Main: Athenäum, 1972.

Trunz, Erich. "Der deutsche Späthumanismus um 1600 als Standeskultur." In *Deutsche Barockforschung. Dokumentation einer Epoche*, edited by R. Alewyn, 148–59. Köln: Kiepenheuer & Witsch, 1966.

Turner, R. Steven. "The *Bildungsbürgertum* and the Learned Professions in Prussia, 1770–1830: The Origins of a Class." *Histoire Sociale/Social History* 13 (May 1980): 108–35.

———. "Historicism, Kritik, and the Prussian Professoriate, 1790–1840." In *Philologie und Hermeneutik im 19. Jahrhundert II*, edited by Mayotte Bollack, Heinz Wismann, and Theodor Lindken, 450–78. Göttingen: Vandenhoeck & Ruprecht, 1983.

———. "Humboldt in North America? Reflections on the Research University and Its Historians." In *Humboldt International: Der Export des deutschen Universitätsmodells im 19. und 20. Jahrhundert*, edited by Rainer Christoph Schwings, 289–312. Basel: Schwabe, 2001.

———. "Prussian Universities and the Concept of Research." *Internationales Archiv für Sozialgeschichte der deutschen Literatur* 5 (1980): 68–93.

———. "University Reformers and Professional Scholarship in Germany, 1760–1806." In *Europe, Scotland, and the United States from the 16th to the 20th Century*, vol. 2 of *University in Society*, edited by Lawrence Stone, 495–531. Princeton, NJ: Princeton University Press, 1974.

Valenza, Robin. *Literature, Language, and the Rise of the Intellectual Disciplines in Britain, 1680–1820*. Oxford: Oxford University Press, 2009.

Vandermeersch, Peter A. "Die Universitätslehrer." In *Geschichte der Universität in Europa: Von der Reformation zur Französisichen Revolution (1500–1800)*, edited by Walter Rüegg, 181–212. Munich: C. H. Beck, 1996.

van der Zande, Johann. "In the Image of Cicero: German Philosophy between Wolff and Kant." *Journal of the History of Ideas* 56 (1995): 419–42.

Velkley, Richard. *Freedom and the End of Reason: On the Moral Foundations of Kant's Critical Philosophy*. Chicago: University of Chicago Press, 1989.

Verger, Jacques. "Patterns." In *A History of the University in Europe*, vol. 1, *Universities in the Middle Ages*, edited by Hilde de Ridder-Symoens, 35–67. Cambridge: Cambridge University Press, 1992.

Veysey, Laurence R. *The Emergence of the American University*. Chicago: University of Chicago Press, 1965.

Vidal, Fernando. *Sciences of the Soul: The Early Modern Origins of Psychology*. Chicago: University of Chicago Press, 2011.

Vierhaus, Rudolf. "Einleitung." In *Wissenschaften im Zeitalter der Aufklärung*, edited by Rudolf Vierhaus, 7–17. Göttingen: Vandenhoeck & Ruprecht, 1985.

———, ed. *Wissenschaften im Zeitalter der Aufklärung*. Göttingen: Vandenhoeck & Ruprecht, 1985.

Virmond, Wolfgang, ed. *Die Vorlesungen der Berliner Universität 1810–1834 nach dem deutschen und lateinischen Lektionskatalog sowie den Ministerialakten*. Berlin: Akademie Verlag, 2011.

Voltaire, François-Marie Arouet de. "Men of Letters." In *The Encyclopedia of Diderot & d'Alembert Collaborative Translation Project*, translated by Dena Goodman. Ann Arbor: MPublishing, University of Michigan Library, 2003. http://hdl.handle.net/2027/spo.did2222.0000.052 (accessed July 24, 2012). Originally published as "Gens de lettres," *Encyclopédie ou Dictionnaire raisonné des sciences, des arts et des métiers*, 7:599–600 (Paris, 1757).

vom Bruch, Rüdiger. "Die Gründung der Berliner Universität im Kontext der deutschen Universitätslandschaft um 1800." In *Die Universität Jena: Tradition und Innovation um 1800*, edited by Gerhard Müller, Klaus Ties, and Paul Ziche, 63–73. Stuttgart: Franz Steiner Verlag, 2001.

von Arnim, Achim. "Der Studenten erstes Lebehoch bei der Ankunft in Berlin am 15. Oktober 1810." In *Achim von Arnims Werke*, vol. 3, edited by Reinhold Steig, 418–20. Leipzig: Insel, 1903.

Wachler, Ludwig. *Aphorismen über die Universitäten und über ihr Verhältniß zum Staate*. Marburg: Neue Akademische Buchhandlung, 1802.

Wakefield, Andrew. *The Disordered Police State: German Cameralism as Science and Practice*. Chicago: University of Chicago Press, 2009.

Walch, Johann Georg. *Philosophisches Lexikon*. Leipzig: Gleditschens Buchhandlung, 1733.

———. *Philosophisches Lexikon*. 4th ed. Edited by Justus Christian Hennings. Leipzig: Gleditschens Buchhandlung, 1775.

Walker, Mack. *German Home Towns: Community, State, and General Estate 1648–1871*. Ithaca, NY: Cornell University Press, 1971.

Waquet, Francoise. *Mapping the World of Learning: The Polyhistor of Daniel Georg Morhof*. Wiesbaden: Harrassowitz, 2000.

Weber, Max. *From Max Weber: Essays in Sociology*. Translated by H. H. Gerth and C. Wright Mills. Oxford: Oxford University Press, 1946.

———. "The Meaning of 'Value Freedom' in the Sociological and Economic Sciences." In *Max Weber: Collected Methodological Writings*, translated by Hans Heinrik Bruun. New York: Routledge, 2012.

Weber, Samuel. *Institution and Interpretation*. Minneapolis: University of Minnesota Press, 1987.

Weischedel, Wilhelm. *Idee und Wirklichkeit einer Universität: Dokumente zur Geschichte der Friedrich-Wilhelms-Universität zu Berlin*. Berlin: De Gruyter, 1960.

Wellmon, Chad. *Becoming Human: Romantic Anthropology and the Embodiment of Freedom*. University Park: Penn State University Press, 2010.

———. "A Crisis of Purpose: Knowledge, Virtue and the University of the Future." *Hedgehog Review* 15, no. 2 (Summer 2013): 79–91.

———. "Touching Books: Diderot, Novalis and the Encyclopedia of the Future." *Representations* 114:1 (2011): 65-102.

———. "Why Google Isn't Making Us Stupid . . . or Smart." *Hedgehog Review* 14, no. 1 (Spring 2012).

Werle, Dirk. "Die Bücherflut in der frühen Neuzeit." In *Frühneuzeitliche Stereotype: Zur Produktivität und Restriktivität sozialer Vorstellungsmuster*, edited by Miroslawa Czarnecka, 469–86. Berlin: Peter Lang, 2010.

Wiese, Ludwig Adolf. *Das höhere Schulwesen in Preußen: Historisch-statistische Darstellung*. 4 vols. Berlin: Wiegandt & Grieben, 1864/1869/1874/1902.

Wilson, Holly L. *Kant's Pragmatic Anthropology: Its Origin, Meaning, and Critical Significance*. Albany: State University of New York Press, 2006.

Wilson, Woodrow. *Princeton for the Nation's Service*. Princeton, NJ: Princeton University Press, 1896.

Wittmann, Reinhard. *Geschichte des deutschen Buchhandels: Ein Überblick*. Munich: C. H. Beck, 1991.

Wolf, Friedrich August. *Friedrich August Wolf: Ein Leben in Briefen*. 3 vols. Stuttgart: J. B. Metzlersche Verlagsbuchhandlung, 1935.

———. *Friedrich August Wolf's Darstellung der Altherthumswissenschaft nebst einer Auswahl seiner kleinen Schriften*. Edited by S. W. F. Hoffmann. Leipzig: A. Lehnhold, 1833.

———. *Friedrich August Wolf's Encyclopädie der Philologie: Vorlesungen im Winterjahre 1798–99*. Edited S. M. Stockmann. Leipzig: Serig, 1845.

———. *Kleine Schriften in lateinischer und deutscher Sprache: Deutsche Aufsätze*. Vol. 2. Edited by Gottfried Bernhardy. Halle: Buchhandlung des Waisenhauses, 1869.

———. *Prolegomena to Homer* (1795). Translated by Anthony Grafton, Glenn W. Most, and James E. G. Zetzel. Princeton, NJ: Princeton University Press, 1985.

Wolff, Christian. *Philosophia Rationalis*. Frankfurt, 1740.

———. *Vernünftige Gedanken* in *Gesammelte Werke*. Edited by J. Ecole et al. Hildesheim: Georg Olms Verlag, 1978.

Yeo, Richard. *Encyclopedic Visions: Scientific Dictionaries and Enlightenment Culture*. Cambridge: Cambridge University Press, 2001.

Zammito, John. *Kant, Herder, and the Birth of Anthropology*. Chicago: University of Chicago, 2002.

———. "Rousseau-Kant-Herder and the Problem of *Aufklärung* in the 1760s." In *New Essays on the Pre-Critical Kant*, edited by Tom Rockmore. Amherst, NY: Humanity Press, 2001.

Zedelmeier, Helmut. "Schelling's Vorlesung *Ueber das Studium der Historie und Jurisprudenz*—Eine historische Lektüre." In *Die bessere Richtung der Wissenschaften: Schellings "Vorlesungen über die Methode des akademischen Studiums" als*

*Wissenschafts- und Universitätsprogramm*, edited by Paul Ziche and Gianfranco Frigo, 185–206. Stuttgart: Frommann-Holzboog, 2011.

Zedler, Johann Heinrich [then Carl Günther Ludovici], ed. *Grosses vollständiges Universal-Lexicon aller Wissenschaften und Künste*. Leipzig/Halle: Zedler, 1732–1750.

Zimietzki, Friedrich W. *Das akademische Leben im Geiste der Wissenschaft: Eine freie Gabe an die Brüder und Genossen deutscher Universitäten*. Berlin, 1812.

Ziolkowski, Theodor. *German Romanticism and Its Institutions*. Princeton, NJ: Princeton University Press, 1991.

Zöller, Günter. "*Die 'Bestimmung alles Wissens'*—Absolutes, Wissenschaft und Handeln in Schelling's *Vorlesungen über die Methode des akademischen Studium*." In *Die bessere Richtung der Wissenschaften: Schellings "Vorlesungen über die Methode des akademischen Studiums" als Wissenschafts- und Universitätsprogramm*, edited by Paul Ziche and Gianfranco Frigo, 65–88. Stuttgart: Frommann-Holzboog, 2011.

# INDEX